T0345329

The Science of Global Warming Remediation

The Science of Global Warming Remediation examines the workings of a complex chemical system using concepts such as chemical kinetics, thermodynamics, and oxidation/reduction. It focuses on preventing environmental deterioration as well as using environmental chemistry for environmental cleanup or remediation. Further, it describes how to utilize mechanical, chemical, and biological methods to detoxify contaminated land or water. This book also considers how environmental legislation aims to modify human behavior so as to reduce or eliminate the environmental threats identified through science.

Features:

- Presents multiple methods for water treatment
- Explains the physiological dangers of exposure to various toxic materials
- Illustrates the mechanisms of major partitioning systems and sinks for carbon dioxide
- Examines the mechanics of global warming and the potential long-term effects
- Provides step-by-step solutions to empower individuals to act locally

The Science of Global Warming Remediation

Mark Harris

CRC Press
Taylor & Francis Group
Boca Raton London New York

CRC Press is an imprint of the
Taylor & Francis Group, an **informa** business

Designed cover image: Shutterstock

First edition published 2024
by CRC Press
6000 Broken Sound Parkway NW, Suite 300, Boca Raton, FL 33487-2742

and by CRC Press
4 Park Square, Milton Park, Abingdon, Oxon, OX14 4RN

CRC Press is an imprint of Taylor & Francis Group, LLC

ISBN: 978-1-032-37767-4 (hbk)
ISBN: 978-1-032-37769-8 (pbk)
ISBN: 978-1-003-34182-6 (ebk)

DOI: 10.1201/9781003341826

Typeset in Times
by KnowledgeWorks Global Ltd.

Contents

SECTION I   Sequestration of Carbon Dioxide

SECTION II   Other Potent Greenhouse Gases

SECTION III   Atmospheric Emissions

SECTION IV   Water

SECTION V   Remediation of Polluted Soils

SECTION VI Appendices

About the Author

Mark Anglin Harris is a Professor of Applied Climatology and Environmental Chemistry at the Northern Caribbean University (NCU), Jamaica, where he has been on the faculty since 2002. His baccalaureate degree, conferred in 1977, majored in geology and physical geography at the University of Windsor, Canada. He received his PhD in environmental geoscience from the University of Adelaide in 2001 after tutoring there from 1992 to 1995 earth science and physical geography. He researches mainly on remediation of polluted land, water, and air. His previous books, titled *Geobiotechnological Solutions to Anthropogenic Disturbances* and *Confronting Global Climate Change: Experiments & Applications in the Tropics*, were published, respectively, in 2016 by Springer-Nature and in 2019 by CRC-Taylor & Francis.

Professor Harris has been the senior author of approximately 30 research articles, having become, so far, the only multiple recipients of the NCU Distinguished Faculty Award for research or scholarship, having won it four times, in 2003, 2007, 2012, and 2016. He received the 2020–2021 Musgrave silver medal for science.

Acknowledgments

In 2019–2020, two individuals, who had neither met each other, nor whom the author has ever met, reviewed the 2019 book titled *Confronting Global Climate Change: Experiments and Application in the Tropics*. Yet, each of these reviewers, unknown to each other, lauded the problem-and-solution section of that book. Because similar conclusions from independent, reputable, unsolicited sources unknown to each other are usually correct, the problem-and-solution approach became the backbone of the current work. Therefore, the author thanks those two reviewers: Professor Larry Erickson, PhD, Emeritus Professor at Kansas State University and Christopher Serju, Senior Gleaner Writer, Jamaica Gleaner.

Based on his review of the author's above-mentioned 2019 book, Mr. Ronald Clahar, who rose from rural Manchioneal, Jamaica, become a successful educator who founded a school in New York, reminded the author of the purpose of any book of "science": clear explanations to a wide audience. The author bore in mind this comment when explaining concepts in the current work.

Several problems in this book were solved by students in the author's class of 2022 (Biology Seminar 1). Their names are as follows:

Saadatu Abubakar, Sonia Bailey, Alicia Baldwin, Rondel Brown, Caslyn Campbell, Dana Cousins, Kadian Elliott, Brittania Fagon, Abigail Grant, Toni-Ann Grant, Evernisha Lewis Roen Lobban, Sudie-Ann Myers, Emily Palmer, Melloie Popo, Abbygayle Powell, Lazonia Powell, Daniel Richards, Jontae Robinson, Breana Stewart, Jessica White, and Dayna Wilson.

Introduction
The Global Situation

THREAT OF RISING SEA LEVELS

Millions of people live at a level near, or in some cases below, sea level. Consequently, any substantial temporary or permanent rise in sea level creates risks to lives and property. Such an event occurred on February 1, 1953, when high tides and strong winds combined to breach the system of dikes protecting much of the Netherlands from seawater, flooding about one-sixth of the Netherlands as far inland as 64 km, killing about 2000 people, and leaving approximately 100,000 homeless (Manahan 2005). The tidal surge, causing the North Sea to rise up to 5 m above its average level, caused widespread flooding along the east coast of Britain.

The following factors could raise ocean levels to destructive highs due to greenhouse warming (Manahan 2005; NOAA 2022):

- Simple expansion of warmed oceanic water could raise sea levels by about 1/3 m over the next century.
- The melting of glaciers, such as those in the Alps, has probably raised ocean levels by about 5 cm during the last century and the process is continuing.
- Of greatest concern is that global warming could cause the great West Antarctic ice sheet to melt, which would raise sea levels by as much as 6 m.

This book presents four subheadings, Sections 1–4, titled chronologically:

1. Greenhouse gases and sequestration
2. Air quality
3. Water: quality and quantity
4. Land: remediation

SECTION 1 GREENHOUSE GASES AND SEQUESTRATION

Chapter 1 discusses the evidence of global warming, including rising sea levels and acidification of oceans. Chapter 2 identifies several ways of measuring some of these changes highlighted in Chapter 1. Sequestration (pulling back or locking up of CO_2 from the atmosphere) remains the most effective remediation process, and Chapters 3–5 introduce, respectively, three such major conduits: forest, water, and soil. Sequestration of atmospheric CO_2 by soils and vegetation remains the major focus. The carbon footprint of an entity (individual, group, product, or process) depicts its impact on atmospheric carbon. Knowing how to measure a carbon footprint is the starting point for remediating it. Hence, Chapter 6 identifies methods of calculating the carbon footprints of various aspects of production, usage, and the eventual destruction of consumer products. Chapter 6 identifies alternative ways to accomplish a particular task, for example, that of daily getting oneself to work. Worked examples are sectionalized and problems are stated for the reader to investigate. For instance, investigating the subsection "pets," an individual who owns a large dog and three cats fed by pre-packaged chemically preserved commercial feeds, incorporates, for the dog, a carbon footprint equivalent to that of a large SUV, while having one cat is more comparable to the carbon footprint of a sub-compact car. The remediation tactic in such a case could include (a) feeding pets free restaurant food scraps (which the owners would otherwise have dumped as garbage) or (b) reducing the number of pets.

If fossil fuel combustion rates were reduced by a factor of 25 and deforestation were reduced, carbon dioxide levels could be kept permanently below 500 parts per million (ppm). Hence, Chapter 8 identifies some of the processes and procedures required for operating carbon capture systems (CCS). Yet even the maximal amount of carbon which could be captured and stored over the course of the century would be exceeded by the enormity of (projected) fossil fuel emissions. Not inconsequential in this global overheating condition, methane and nitrous oxide, each much more intrinsically effective as a global blanket than CO_2, appear, respectively, in Chapters 9 and 10, with approaches for their reduction thereof. Solving the technical difficulties entailed in utilizing clathrates (frozen aggregates of methane, which occupy a volume on the sea floor so large as to be one reason for such currently high sea levels) as fuel provides an opportunity to reduce sea levels.

SECTION 2 AIR QUALITY

When oxides of sulfur from burnt fossil fuels pollute damp air, droplets of water containing sulfurous and sulfuric acids also form thus:

$$SO_2 + H_2O = H_2SO_3 \qquad (0.2.1)$$

$$SO_3 + H_2O \rightarrow H_2SO_4 \qquad (0.2.2)$$

Chapter 11 explores the effects of sulfur dioxide, formed by the combustion of sulfur, sulfides, and those of various organic sulfur compounds.

Chapter 12 presents the disproportionately massive tonnage of CO_2 which cement manufacturing places into the atmosphere. Pozzolana – naturally occurring pozzolans of volcanic origin – are a broad class of siliceous and/or aluminous materials which, in themselves, possess little or no cementitious value but which will, in finely divided form and in the presence of water, react chemically with calcium hydroxide [$Ca(OH)_2$] at ordinary temperatures to form compounds possessing cementitious properties. The capacity of a pozzolan to react with calcium hydroxide and water is determined by measuring its pozzolanic activity, and vast quantities of fly ash (from the burning of fossil fuels), being pozzolanic, can reduce the calcium oxide and/or hydroxide required in cement manufacture. This replacement, by extension, would substantially decrease the tonnage of carbon dioxide being placed in the skies annually.

Chapter 13 studies the sources of automobile and factory emissions and presents strategies to avoid inhaling concentrated forms of those emissions. Combustion sources generally include (1) power plants and domestic heating equipment, whether coal, oil, or gas-fired; these produce sulfur oxides, nitrogen oxides, and particulates; (2) motor vehicles, which produce both the photochemical oxidant type of pollution (oxidants are corrosive) and the pollution by carbon monoxide and carbon dioxide, as found in many cities; and (3) in addition to common pollutants such as sulfur dioxide and particulate matter, specific industrial activities producing pollutants related to the processes and products of the industry concerned (Manahan 2005).

Motor vehicles operating at sub-ideal internal engine temperature can adversely affect climate change and human health. Removing a thermostat from an internal combustion engine allows water to circulate and cool the engine block at a time when cooling is not required, thereby delaying the operating temperature of the engine, thus unnecessarily releasing more unburnt emissions into the atmosphere. Such a practice frequently occurs in some parts of the tropics under the misguided notions of engine thermostats being for "cold countries" and that those who remove thermostats are "climatizing the car." Chapter 14 documents the root causes for these erroneous ideas and suggests correctives. If these pollutants from incorrectly operated automobiles move inside buildings, they become restricted in movement, thereby exacerbating human morbidities. Chapter 15 identifies sources of indoor air pollution and appropriate remediation.

Chapter 16 examines the role of excessive atmospheric air movements (and the lack of movements, such as during temperature inversions), chimney-stack parameters, and atmospheric stability categories on the dispersion of pollutants. Particulate matter (PM) is one of the major air pollutants in urban areas (Zigler et al. 2018). The effects of PM pollution on human and environmental systems have been discussed by many scientists. The relationship between aerosol concentration and meteorological variables should be investigated for better control and monitoring applications. Aerosol concentrations are controlled by atmospheric mixing, chemical transformation, emissions, etc. (Zigler et al. 2018). Hence, Chapter 16 includes ways to understand the transport and delivery of meteorological hazards brought on by burning fossil fuels.

Heat, its environmental ill-effects, and its quantitative measurement characterizes Chapter 17. Dry or humid heat is the climatic stress most widely encountered in the world since it affects not only many kinds of industrial and agricultural work but also, more particularly, those in developing countries outside temperate zones. Acclimatization of the animal body and its measures to combat heat serve to maintain thermoregulation. As similar measures to those encountered in animal physiology operate to maintain the heat balance in the environment, safeguarding the integrity of natural processes is necessary. Correctives to overheating (Chapter 17) include using the wood from fire-resistant, easily worked invasive tropical and subtropical tree species such as *S. campanulata*, as interior building-partition walls in corresponding latitudes.

Section 3 Water: Quality and Quantity

Section 4 (Chapters 18–25) discusses water pollution from agricultural practices due to animal wastes, material eroded from the land, plant nutrients, inorganic salts and minerals resulting from irrigation, herbicides, and pesticides; to these may be added various infectious agents contained in wastes which pollute water. With potentially increasing applications of pesticides to enlarge crop yields if global warming restricts rainfall, surface- and groundwater sources of potable water may exhibit lowered yields (Deshmukh et al. 2021). However, air contamination causes rainwater to become acidic and cloudy and adds heavy metals such as Pb into rainwater. Water treatment and protection are also presented.

SECTION 4 LAND: REMEDIATION

Contamination by organic chemicals such as gasoline, heavy metals (coming from the burning of fossil fuels), PAHs (polycyclic aromatic hydrocarbons) as a class of organic chemicals, shale oil extraction, contamination from the dumping of petroleum-based industrial waste and pesticides into soils (Chapter 26), and mine wastes (Chapters 27–29) are discussed. The requirements and quantification for bioremediation of soils include water, oxygen, and nutrients for soil microbes. As quantification generally requires statistical evaluations, Chapter 30 recaps some basic procedures of statistical measurements using worked examples. With water availability being potentially a major challenge of global warming, Chapter 31 calculates several hydrological problems of water resources, while Chapter 32 presents renewable energy sources.

SECTION 5 BIOCHAR REMEDIATION OF CARBON DIOXIDE

Having presented some negative outcomes of large-scale fossil fuel burning, several correctives can be discussed. Thus, in addition to a program of global reforestation, biochar not only removes CO_2 but also prevents the removed CO_2 from re-entering the atmosphere for a much longer time (Chapters 6, 8, 26) than it otherwise would, while restricting N_2O (Chapter 10). Biochar is a soil amendment (Chapter 26) that locks carbon in the soil by slowing down the rate of decay, allowing organic matter to last longer in the soil. Based on pyrolyzing biomass, biochar may represent the single most important initiative for humanity's environmental future, allowing potential solutions to food security, the fuel crisis, and the climate problem (Flannery 2007). Half of the emission reductions and the majority of CO_2 removal result from the one to two orders of magnitude longer persistence of biochar than the biomass it is made from Lehman et al. (2021).

PROBLEM 1.5.1
The biochar process

What is biochar and how is "biochar" made?

SOLUTION 1.5.1

Biochar is produced by the prolonged thermal treatment (>350°C) (Smebye et al. 2017) of biomass under low-oxygen conditions (pyrolysis) which drives off gases and leaves a residue of mainly carbon behind which immobilizes contaminants in water, soils, or sediments, as well as improving crop productivity in weathered and eroded soils.

Feedstock biomass can include forestry and agricultural waste products, municipal green waste, biosolids, animal manures, and some industrial wastes such as papermill wastes (DPIWA 2022). Being a stable, carbon-rich form of charcoal that can be added to soil to sequester carbon and reduce net greenhouse gas emissions, biomass is heated at temperatures greater than 250°C (DPIWA 2022), such that thermochemical conversion of biomass to a solid carbonaceous product occurs.

PROBLEM 1.5.2
Carbon richness of biochar

How is carbon richness in biochar achieved?

SOLUTION 1.5.2

When biomass is left to smolder in layers on the forest floor following a forest fire, when plant and animal matter bake slowly in a nearly oxygen-free environment, carbon-rich biochar remains. Similarly, there are no roaring flames in a biochar kiln. Instead, biomass of different kinds is slowly baked until it becomes a carbon-rich char. This process is pyrolysis, which refers to the chemical decomposition of organic material when exposed to elevated temperatures in an atmosphere with restricted levels of oxygen (Chan et al. 2007)

By driving off the volatiles while retaining the carbon, pyrolysis produces about 50% carbon whereas ash, produced in oxygen-rich conditions and hence composed mainly of metal oxides, is about 5% carbon. Non-volatiles include lignocellulosic recalcitrant organic molecule frameworks. The higher the proportion of these in the biomass, the higher the eventual carbon content of the biochar. Thus, decreasing the oxygen-related reactions spares the carbon. Thus, a high proportion of non-volatiles plus slow pyrolysis produces carbon richness.

PROBLEM 1.5.3
Stability of biochar

How stable is biochar, i.e., how long does it remain before being fully decomposed?

SOLUTION 1.5.3

Studies of charcoal from natural fire and ancient anthropogenic activity indicate millennial-scale stability (DPIWA 2022). However, the stability of modern biochar products is uncertain: it is difficult to establish the half-life of newly produced biochar through short-term experiments, and aging processes are expected to affect turnover in the longer term. The limited data available suggests that the turnover time of newly produced biochar ranges from decades to centuries (DPIWA 2022), depending on feedstock and process conditions. Presently there is no established method to artificially age biochar and assess likely long-term stability.

PROBLEM 1.5.4
Biochar production in the tropics

Compared with those of extra-tropical regions, tropical soils have a larger loading of biochar. What attributes of plant species enhance their suitability for biochar in the tropics?

SOLUTION 1.5.4

Suitability varies directly with the rapidity of biomass acquisition and level of species invasiveness. In this regard are two invasive species: *Bambusa* spp. (bamboo), the fastest growing plant on Earth, and *Spathodea campanulata*, a very fast-growing tree.

PROBLEM 1.5.5
Effectiveness of bamboo as biochar

Why is bamboo highly favored for making biochar?

SOLUTION 1.5.5

Biochar is expensive (production and transport), and economic returns are unlikely for commodity agriculture (DPIWA 2022). Obtaining enough biochar is difficult – there is no large-scale production (DPIWA 2022) but

- The invasiveness of bamboo, being the fastest-growing plant, provides for an almost inexhaustible source of biochar in the tropics.
- Bamboo forests are growing while tropical and subtropical forest coverage is declining (Chaturvedi et al. 2023).
- Bamboo biochar is a stable form of soil carbon with a naturally porous structure that improves aeration, water-holding capacity, and nutrient retention of soils and acts as a refuge for beneficial soil microbiology. It is the building block of resilient soils.
- Bamboo can tolerate a wide range of temperatures, is efficient in its water and soil usage, and needs no fertilizers or pesticides.
- Bamboo is an excellent absorber of carbon dioxide and releases a large amount of oxygen (approximately 30%) back into the atmosphere, thus reducing global warming.
- High strength bamboo grows quickly and can be collected for usage for charcoal in just 3–6 years, unlike trees, which can be harvested every 20–50 years, making it far more desirable than wood, which takes a lot longer to grow.
- Easy, no-maintenance management, and high biomass production
- Bamboo may be harvested yearly and regenerated without needing to be replanted.
- Bamboo produces 30% more oxygen than trees and aids in lowering carbon dioxide levels, contributing to global warming. Some bamboo replenishes fresh air by capturing up to 12 tons of carbon dioxide per hectare.

It is therefore advisable to grow more bamboo to assist the environment.

PROBLEM 1.5.6
Tree age and carbon sequestration

Based on the lack of apparent growth of old trees, it is widely held that they are of little value in removing carbon from the atmosphere. Hence, such trees are most useful for biochar, thereby facilitating the planting of new trees. What is the evidence?

SOLUTION 1.5.6

Retrospective tropical dendrochronology (tree ring analyses) of 61 species of trees growing in unmanaged tropical wet forests of Suriname and reaching ages from 84 to 255 years shows positive trends of diameter growth and carbon accumulation over time (Kohl et al. 2017). In the last quarter of their lifetime, some trees accumulate on average between 39% (*C. odorata*) and 50% (*G. glabra*) of their final carbon stock, thereby suggesting that trees maintain high rates of carbon accumulation at later stages of their lifetime (Kohl et al. 2017). They found that such attributes vary among tree species and are even non-existent in others.

Thus, Kohl et al. (2017) also found that the 61 mature trees observed showed the following four general growth patterns:

- Category 1: Those which maintain an increase of C-accumulation with increasing age.
- Category 2: As above, but with a clear depression between the age period from 110 to 140 years.
- Category 3: Those showing a C-accumulation pattern that is characterized by relatively uniform rates after an initial phase and show no trend.
- Category 4: Trees exhibiting an increase in the first half of life followed by a continuous decrease that is maintained until the end of life.

Therefore, to avoid the inadvertent conversion to biochar of actively carbon sequestering "old" trees into biochar, retrospective dendrochronology (Kohl et al. 2017) can determine the appropriate category listed above of each tree, thereby avoiding haphazard and counterproductive selections of mature trees for conversion to biochar.

The relatively low correlation coefficients between age and biomass addition indicates that for the trees in their sample, age alone is not decisive for the amount of C accumulated at the end of their lifetime. Hence, old-growth trees in tropical forests not only contribute to carbon stocks by long carbon resistance times but also maintain high rates of carbon accumulation at later stages of their lifetime. Luyssaert et al. (2008) found that forests between 15 and 800 years of age usually had a positive net ecosystem productivity (the net carbon balance of the forest including soils), such that their results demonstrated that old-growth forests can continue to accumulate carbon, contrary to the long-standing view that they are carbon neutral. One explanation could be that the density of added wood increased with increasing age, thereby concealing the full extent of added biomass: Luyssaert et al. (2008) discovered that harvest age itself is a determinant for obtaining wood of higher density. Hence, the selection of trees for biochar based on directly linking lower sequestration ability with increasing age is unfounded. For biochar, studying the sequestration attributes of individual species is more beneficial.

PROBLEM 1.5.7
Types of biochar kilns: negative and positive effects

What are the benefits and disadvantages of the various types of biochar kilns?

SOLUTION 1.5.7

Earth mound or earth covered pit kilns

These have been used most frequently. They are free of investment cost, merely requiring some poles and sand to cover the pyrolyzing biomass. However, they are slow, requiring several days according to Duku et al. (2011), releasing gases while partially burnt, and generating significant gas/aerosol emissions with negative potential environmental impacts (Pennise et al. 2001; Sparrevik et al. 2015), while volatilizing soil organic matter. Moreover, the accompanying excessive heat of several hundred degrees destroys soil organisms within a certain radius of the center.

Pyrolytic cook-stoves and gasifiers

These show the most positive potential environmental due to avoided firewood consumption and emissions from electricity generation. Advantages include that they burn cleanly, thus reducing indoor air emissions, can use various biomass residues as feedstock, and are fuel-efficient, reducing emissions of CO, CH_4, and aerosols by around 75% (Jetter et al. 2012) compared to open-fire or three-stone cooking.

In the whole life cycle perspective, the generation of biochar in cook-stoves or gasifiers was observed to provide the most beneficial alternative due to avoided impacts. Even though they do not require any material or investment, Smebye et al. (2017), in a life-cycle assessment (LCA) using end point methods, caution that the use of earth-mound kilns should not be advocated because they are slow, laborious, and negatively impact the environment.

Flame curtain and retort kilns

On the other hand, Smebye et al. (2017) report that biochar generation per se does not result in significantly positive life cycle impacts for flame curtain and retort kilns, and thus additional environmental benefits are needed to warrant their generation in a life cycle perspective.

Gasifier units and pyrolytic stoves

Gasifier units and pyrolytic cook-stoves favorably impacted the environment to a significant degree because of the avoided wood consumption and emissions from electricity generation, respectively (Smebye et al. 2017).

PROBLEM 1.5.8
Biochar in developing countries

The production of biochar in modern industrial devices can be a highly controlled process with low gas emissions However, achieving the same results under rural tropical conditions, i.e., with poorly maintained technologies in very low-income settings, is more challenging (Smebye et al. 2017). What are the available techniques for efficient biochar production in developing countries?

SOLUTION 1.5.8

Traditionally, earth-bound kilns (earth-covered pit kilns) have constituted the most rudimentary and popular modes of biochar production in developing countries. But despite not requiring any ex-situ material or investment, Smeybe et al. (2017) found them to be slow (several days) and laborious with negative potential impacts on the environment. Moreover, the added soil loses physical and chemical integrity, as excessive heat converts it to clinker.

Smeybe et al. (2017) also observed that it is easier to make a garden waste stove from a 200-L metal barrel, particularly if the walls are thick, to increase longevity by reducing rusting.

PROBLEM 1.5.9
Risks of biochar

What are the drawbacks of biochar applications?

SOLUTION 1.5.9

Biochar poses several environmental risks (DPIWA 2022):

- All biochars are not the same – their characteristics and value in the soil depend on the base material and the pyrolysis technique.
- Not all soil and biochar combinations will provide a positive result.
- Some material produced as by-products of the industry may contain impurities and toxins, with an unknown impact on the food web, microbial processes, and nitrification.
- Biochar absorbs and concentrates herbicides and pesticides in the root zone.

PROBLEM 1.5.10
Biochar and soils

What are the effects of bamboo biochar on soils?

SOLUTION 1.5.10

Global warming-based rising sea levels indirectly bring into production thinner, less mature soils, often with lower cation exchange capacities and hence lacking the ability to store macronutrient bases such as Ca^{2+}, Mg^{2+}, and K^+. Because of its negative charge, bamboo biochar aids in boosting soils' base exchange capacity, improving nutrient availability for plants, as facilitated by its excellent adsorption properties, huge surface area, and highly porous structure. For these reasons, a bamboo is a superior option to other wood planks regarding soil carbon, fertility, and tilth.

In an unpublished paper titled "Benefits of papermill biochar" (Agrichar TM), Dr Lukas Van Zwieten, who tested the impact of biochar with and without a complete fertilizer in factorial combination, observed the following:

- Benefits to soil properties included increased pH in the ferrosol of up to two units.

- There was significantly increased total soil carbon (between 0.5% and 1%) in both soil types.
- Increased CEC in the ferrosol and reduced Al availability (from 2 cmol (+) · kg^{-1} to <0.1 centi-mol (+) · kg^{-1}).
- The biochars significantly increased crop growth (measured as height and weight of plants) in the ferrosol: wheat biomass was up to 2.5 times higher when biochar plus fertilizer was compared to fertilizer treatment alone.
- Results suggest improved fertilizer use efficiency with biochar application, especially in the ferrosol.
- Earthworms showed a preference for biochar-amended soil over control soils.
- This was particularly evident in the ferrosol where up to 92% of the worms migrated to the biochar-amended soil. The results from this work demonstrate that biochars derived from papermill wastes are valuable soil amendments.

PROBLEM 1.5.11
Protection of organic matter

How can biochar help to protect the organic matter of soil?

SOLUTION 1.5.11

Excellent fire resistance is one benefit of the high-temperature biochar's robust C–C covalent bonds. High pyrolysis temperatures (>500°C) yield biochar with high fire resistance properties due to the development of strong C–C covalent bonds and the absence of volatile matter. This helps to protect soil organic matter during fires.

PROBLEM 1.5.12
Benefits of bamboo

What are the main medical, health, and environmental advantages of bamboo biochar?

SOLUTION 1.5.12

Activated carbon contains micropores, mesopores, and macropores within its structure. These structures largely determine the properties of activated carbon such as being an adsorbent. The combination of the large surface area, charged groups, and functional groups magnifies the potential to absorb heavy metallic elements and other contaminants (mainly organic ones) (Spokas 2010; Buss et al. 2012).

Bamboo charcoal

- It has four times the absorption rate of conventional charcoal.
- It has ten times more surface area.
- Bamboo charcoal's porous structure offers microscopic holes that efficiently absorb odors, moisture, and airborne pollutants like formaldehyde, ammonia, and benzene and act as a carbon dioxide absorbent (Gupta & Ghosh 2015; Jiang et al. 2019).

- These benefits were the driving force behind several uses for this material, including blood purification, electromagnetic wave absorbers, and water purifiers. These benefits are influenced by the activation and carbonization procedures used to make bamboo charcoal.
- Being hollow, split bamboo dries very rapidly under passive solar energy (direct sunlight), thereby avoiding extra energy inputs prior to biocharring.

PROBLEM 1.5.13
Charcoal yields of different bamboos

Which are the most efficient biochar-yielding bamboo varieties?

SOLUTION 1.5.13

Chaturvedi et al. (2022) observed biochar yields occurring in the order bamboo chopsticks waste > fiber bundle bamboo > *Bambusa vulgaris* > *Phyllostachys mazel* > moso bamboo > thorny bamboo and as %, respectively, 74, 69, 36, 34, 32, 24.

PROBLEM 1.5.14
Drawbacks of agricultural wastes: an argument for biochar

What harmful effects do accumulating agricultural wastes inflict on the environment? What are the alternative options?

SOLUTION 1.5.14

They cause harm as they accumulate in water sources, resulting in pollution such as unpleasant odors (Gholz 1987), eutrophication, and high levels of biological oxygen demand and chemical oxygen demand (Kanu & Achi 2011). Therefore, sustainably converting such wastes can alleviate the problem. One research trial turns the biomass waste from industry and agricultural by-products into a precursor to produce activated carbons, energy pellets and biochar production, and soil amendments (Prahas et al. 2008; Stella et al. 2016).

PROBLEM 1.5.15
Required attributes of biochar

What are the ideal attributes of biochar stock materials?

SOLUTION 1.5.15

In general, the selection of raw material for biochar depends on seven important criteria (Menendez-Diaz & Martin-Gullon 2006):

- High carbon content
- Low inorganic matter content for low ash result
- High density and volatile matter content
- Abundant availability so that the raw material is always at a very low cost.

- Potentially low temperature of activation
- Low degradation rate upon storage
- Possibility of producing an activated carbon with a high percent yield

PROBLEM 1.5.16
Carbonization

What is carbonization?

SOLUTION 1.5.16

Carbonization is the thermal decomposition and removal of non-carbon species, particularly nitrogen, oxygen, and hydrogen (Yahya et al. 2015), from raw material by reducing volatile matter content. This produces charcoal with high fixed carbon content by breaking down the cross-linkage in the raw material, the decomposition of lignocellulosic content leading to the elimination of non-carbon elements, followed by deposition of tars (Menendez-Diaz & Martin-Gullon 2006). Basically, volatile matter content with low molecular weight will be diffused first followed by light aromatics and hydrogen gas. At the same time, pore structures start to develop, and the tars produced will fill the pore structures (Stella et al. 2016).

REFERENCES

Buss W, Kammann C, Koyro HW (2012) Biochar reduces copper toxicity in Chenopodium quinoa wild: In a sandy soil. J Environ Qual 41, 1157–1165. https://doi.org/10.2134/jeq2011.0022

Chan KY, Van Zwieten L, Meszaros I, Downie A, Joseph S (2007) Agronomic values of greenwaste biochar as a soil amendment. Aust J Soil Res 45(8), 629–634.

Chaturvedi K, Singhwane A, Dhangar M. et al (2023) Bamboo for producing charcoal and biochar for versatile applications. Biomass Conv. Bioref. https://doi.org/10.1007/s13399-022-03715-3

Deshmukh SP, Mevada KD, Poonia TC, Saras P (2021) A Textbook of Rainfed Agriculture and Watershed Management. SK Kataria Publishers, New Delhi

DPIWA (2022) Carbon farming: applying biochar to increase soil carbon. Department of primary industries, Western Australia Government. https://www.agric.wa.gov.au/soil-carbon/carbon-farming-applying-biochar-increase-soil-carbon

Duku MH, Gu S, Hagan EB (2011) Biochar production potential in Ghana a review. Renew Sustain Energy Rev 15(8), 3539–3551.

Flannery T (2007) Tim Flannery awarded Australian of the Year – 26/01/2007 – ABC Jan 26, 2007 Friday, 26 January 2007. https://au.search.yahoo.com/yhs/search?p=Prof.%20Tim%20Flannery%20biochar%www.abc.net.ausciencenews

Gholz HL (1987) Agroforestry: Realities, Possibilities and Potentials. Springer Dordrecht, Dordrecht, Holland.

Gupta T, Ghosh R (2015) Rotating bed adsorber system for carbon dioxide capture from fue gas. Int J Greenhouse Gas Control 32, 172–188.

Jiang L, Gonzalez-Diaz A, Ling-Chin J, Roskilly AP, Smallbone AJ (2019) Post-combustion CO_2 capture from a natural gas combined cycle power plant using activated carbon adsorption. Appl Energy 245, 1–15.

Kanu I, Achi OK (2011) Industrial effluents and their impact on water quality of receiving rivers in Nigeria. J. Appl. Technol. Environ. Sanit 1, 5–86.

Kohl M, Neupane PR, Lotfiomran, N (2017) The impact of tree age on biomass growth and carbon accumulation capacity: A retrospective analysis using tree ring data of three tropical tree species grown in natural forests of Suriname. PLOS ONE, 12(8). https://doi.org/10.1371/journal.pone.0181187

Luyssaert S, Schulze ED, Börner A, Knohl A, Hessenmöller D (2008) Old-growth forests as global carbon sinks. Nature 455(7210), 213–215. doi: 10.1038/nature07276.

Manahan SE (2005) CRC Press Environmental Chemistry, 8th ed. CRC Press.

Menendez-Diaz JA, Martin-Gullon I (2006) Types of carbon adsorbents and their production. In Activated Carbon Surfaces in Environmental Remediation, edited by T. J. Bandosz. Academic Press, New York, pp. 1–47.

NOAA (2022) Climate Change: Global Sea Level. NOAA Climate.gov. www.climate.gov

Pennise DM, Smith KR, Kithinji JP, Rezende ME, Raad TJ et al. (2001) Emissions of greenhouse gases and other airborne pollutants from charcoal making in Kenya and Brazil. J Geophys Res Atmos 106(D20), 24143–24155.

Prahas D, Kartika Y, Indraswati N, Ismadji S (2008) Activated carbon from jackfruit peel waste by H_3PO_4 chemical activation: Pore structure and surface chemistry characterization. Chem Eng J 140, 32–42. doi: 10.1016/j.cej.2007.08.032.

Smebye AB, Sparrevik M, Schmidt HP, Cornelissen G (2017) Life-cycle assessment of biochar production systems in tropical rural areas: Comparing flame curtain kilns to other production methods. Biomass and Bioenergy, Vol 101, Pages 35–43

Sparrevik M, Adam C, Martinsen V, Jubaedah J, Cornelissen G (2015) Emissions of gases and particles from charcoal/biochar production in rural areas using medium-sized traditional and improved "retort" kilns. Biomass Bioenerg 72, 65–73.

Spokas KA (2010) Review of the stability of biochar in soils: Predictability of O:C molar ratios. Carbon Manag 1, 289–303.

Stella MG, Sugumaran P, Niveditha S (2016) Production, characterization and evaluation of biochar from pod (Pisum sativum), leaf (Brassica oleracea) and peel (Citrus sinensis) wastes. Int J Recycl Org Waste Agric 5, 43–53 (2016). https://doi.org/10.1007/s40093-016-0116-8

Yahya Z, Al-Qodah CW, Ngah Z (2015) Agricultural bio-waste materials as potential sustainable precursors used for activated carbon production: A review. Renew Sust Energ Rev 46, 218–235. https://doi.org/10.1016/j.rser.2015.02.051

Zigler CM, Choira C, Dominici F (2018) Impact of national ambient air quality standards nonattainment designations on particulate pollution and health. Epidemiology 29(2), 165–174. https://doi.org/10.1097/EDE.0000000000000777

Section I

Sequestration of Carbon Dioxide

1 The Evidence

CARBON DIOXIDE: MAIN POINTS

- Amount in earth's atmosphere: 416 ppm in 2021
- Atmospheric levels have been rising for 200 years
- Earth's ideal CO_2: <350 ppm
- Scientific Consensus (18 Scientific Associations): earth's climate is warming
- Amount of CO_2 in earth's atmosphere: 418 ppm in May 2022

1.1 EXPECTED TEMPERATURE RISE

If you add to the blanket, expect to feel warmer. How much warmer?

Historically, we have a 7°C effect from CO_2. Put another way, CO_2 up to the Industrial Revolution accounted for 7°C of the atmosphere's heat. But we have gone from 280 to 400 ppm (10/7 times as much or 3/7 increase). This should translate into $7 \times 3/7 = 3$°C change, but it takes some time because oceans are slow to respond, having an enormous heat capacity.

Consequently, the oceans are predicted to rise approximately half a meter by 2100, maybe as much as 1 m, drowning much of Bangladesh, several atolls, the Nile valley, and Louisiana. But it doesn't stop there: it won't stabilize until maybe 2300, by which time the rise could be several meters. This is even if humans stop CO_2 production today (Figure 1.1).

1.2 GLOBAL WARMING AND FEEDBACK SYSTEMS

As is the case with the human body (Figure 1.2), feedback systems preserve a "set point" temperature on earth. There are two positive feedback loops: the water vapor feedback loop and the snow/ice albedo feedback loop. Water vapor locks in infrared rays, and the extra heat, in turn, vaporizes more water, which locks in even more heat, resulting in positive feedback. The albedo effect (heat loss by reflectance) from snow cools the earth, thereby producing more snow (another positive feedback loop). Earth's climate is stable because of the infrared flux, where negative feedback loops dampen the chances of runaway positive loops. Negative feedback in response to excess earth heat is the formation of more low clouds in response to increased heat, as low clouds reflect incoming radiation to increase the albedo effect.

PROBLEM 1.2.1
Homeostatic temperature of planet Earth

Where and why are the signs of excess global warming most intense?

DOI: 10.1201/9781003341826-3

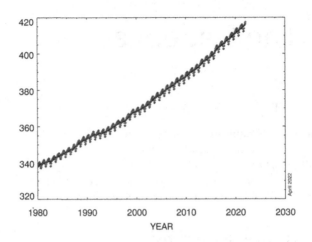

FIGURE 1.1   Global mean CO_2 since 1980.

Source: Allan (2022).

SOLUTION 1.2.1

According to Water Encyclopedia (2022a), precipitation amounts have changed in different ways in various regions during the last 80 years, but they generally have increased in the middle and high latitudes, often more than 10%. They say that in the United States, annual rainfall has increased by about 10% during the 20th century, on an average, the largest increases in precipitation being expected to occur near polar regions for two reasons. First, the laws of diffusion (energy moving from locations of high concentration to areas of low concentration), observations, and climate models indicate that the warming rate has been, and will continue to be, the highest there (in polar regions), and warmer air can hold more water vapor. Second, the warming will reduce the extent of sea ice, and because the vapor pressure over liquid water exceeds that over ice, thereby allowing more evaporation from open water, greater precipitation is expected in the polar regions (as vapor removed from water gathers over the ice).

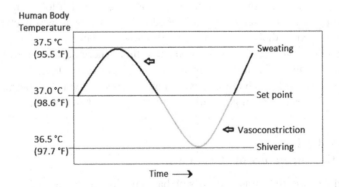

FIGURE 1.2   Homeostatic body mechanisms controlling the temperature at around 37°C.

Contrastingly, decreases in precipitation have been occurring in some locations, including in the Northern Hemisphere tropics, especially in Africa, where a significant decrease in rainfall has occurred since 1950 (Water Encyclopedia 2022b). Intense drought plagues the African Sahel region more than ever before. Yet, in the tropical Pacific, evaporation rates and rainfall amounts all have 47 increased since 1950 despite the lack of ice-derived vapor (Water Encyclopedia 2022b).

1.3 QUANTIFYING CO_2 RISE

Radiation consists of oscillating electrical and magnetic fields. Hence, a molecule that generates an oscillating electric field can interact with incoming electromagnetic energy. Only when a vibrating diatomic molecule is heteronuclear will it have a dipole and, hence, an oscillating dipole and an oscillating electric field as it vibrates. As it interacts with the radiation, the molecule may take energy from the radiation field. The oxygen molecule, on the other hand, is homonuclear, is not a dipole, and cannot interact with such radiation. The reactive molecule gains one of three energy states: rotational, vibrational, or excited. Hence, oxygen is not a greenhouse gas.

As a change of dipole moment in the lower and upper vibrational states always accompanies infrared (IR) radiation, the major constituent gases of the atmosphere, N_2, O_2, and Ar, do not absorb IR. It is the trace gases, such as CO_2, CH_4, H_2O, O_3, N_2O, and CFCs (chlorofluorocarbons), which absorb IR in the troposphere. Without IR absorption, the average surface temperature of earth would be approximately $-30°C$ rather than approximately $+15°C$. By having increased the concentrations of CO_2, CH_4, N_2O, and CFCs, anthropogenic activities increased IR trapping and created what is commonly referred to as the "Greenhouse effect."

PROBLEM 1.3.1
The greenhouse effect process

The greenhouse effect = trapping of outgoing infrared radiation by "radiatively active gases."

a. How does it occur?
b. What is the energy of a photon of light of wavelength 530 nm?
c. How much energy is carried by a mole of photons of wavelength 530 nm?

SOLUTION 1.3.1

a. The energy of a photon is inversely proportional to the wavelength of a photon. Light causes or accelerates photochemical reactions. Hence, sunlight can supply energy (as J = joules):

$$E_{photon} = hc/\lambda \text{ (E is inversely proportional to } \lambda) \tag{1.3.1}$$

where
E_{photon} = energy of a photon
  h = Planck's constant ($6.62607015 \times 10^{-34}$)
  c = speed of light (3×10^{-8} m $\cdot$ s^{-1})
  λ = wavelength of light (nanometers or nm)

$$\Delta E, \text{kJ mole}^{-1} = \left(1.19 \times 10^5\right) / \lambda, \text{nm} = \left(1.19 \times 10^5\right) / 530 \text{ nm}$$

b. ΔE, kJ mole^{-1} = (hc) / λ (nm) = (hc) / 530 $\times$ 10^{-9}
   ΔE, kJ mole^{-1} = (6.626 $\times$ 10^{-34} $\times$ 3 $\times$ 10^8) / 530 $\times$ 10^{-9}
   ΔE, kJ mole^{-1} = (6.626 $\times$ 3 $\times$ 10^{-17}) / 530

   Ans. = 3.7 $\times$ 10^{-19} J

c. A mole of anything is Avogadro's number of particles, which is 6.022 $\times$ 10^{23}. We multiply by Avogadro's constant to get the energy per mole:

$$= (hc / \lambda = 3.7 \times 10^{-19} \text{ J}) \times 6.022 \times 10^{23} = 3.7 \times 6.022 \times 10^4$$

   Ans. = 2.2 $\times$ 10^5 J

1.4 SOURCES OF CO$_2$

Calcining, a thermal treatment process bringing about a thermal decomposition below the melting point of the product, releases CO_2 from limestone to produce calcium oxide, which reacts with pozzolans during the manufacture of Portland cement (Chapter 12), a decomposition process that occurs at 900–1050°C. The chemical reaction is:

$$CaCO_3 \text{(s)} \rightarrow CaO \text{(s)} + CO_2 \text{(g)} \qquad (1.4.1)$$

Coal-based power plants produce greenhouse emissions and fine ash (fly ash). As cement manufacture is the largest single industrial source of atmospheric CO_2, the use of fly ash in concrete which dates to the late 20th century, has been widely researched, with up to 35% by mass replacing cement (Hemalathaa and Ramaswamy 2017). In 2015, fly ash utilization rates were 70% for China, 43% for India, and 53% for the United States (Yao et al. 2015). This leaves a large potential for increased utilization, thereby decreasing disposal volumes and costs, as well as replacing non-renewable and climate-change-exacerbating processes.

1.5 DEEP OCEAN SINK

According to Manahan (2005), it appears that at the present time, the injection of carbon dioxide from combustion processes into deep ocean regions is the only viable alternative for sequestering carbon dioxide, and this approach remains an unproven technology on a large scale. One potential drawback rests on the slight increase in ocean water pH that would result. Though only of the order of a tenth of a pH unit, it could adversely affect many of the organisms that live in the ocean (Manahan 2005).

PROBLEM 1.5.1
CO$_2$: atmosphere and ocean

How fast will the oceans absorb injections of CO_2 from the atmosphere if $t_{1/2}$ (i.e., the half-life) for uptake into surface water = 1.3 years and for exchange between surface and deep-water $t_{1/2}$ = 35 years?

SOLUTION 1.5.1

$$CO_2\,(g) \leftrightarrow CO_2\,(aq) \equiv H_2CO_3\,(aq) \leftrightarrow HCO_3^-\,(aq) \leftrightarrow CO_3^{2-}\,(aq) \leftrightarrow CaCO_3\,(s) \quad (1.5.1)$$

Find the equation of the line if it is a straight line. The graph of a linear straight-line equation is depicted by (time is always depicted on the horizontal axis):

$$y = mx + b\,(i.e., \text{``b'' is how far up on the ``y'' line ``x'' = 0 exists}) \quad (1.5.2)$$

where
m = the slope of the line (Figure 1.3)
b = the y-intercept of a line (how far up or down the line cuts the y axis)

By physical measurement on the line depicting the amount of CO_2 taken into the deep ocean, e.g., assuming the rise/run ($y_2 - y_1/x_2 - x$) = 5/4, the equation of this line is:

$$y = 5/4\,(x) \quad (1.5.3)$$

Extrapolating from $t_{1/2} = 35$,

$$t_{total} = 70 = x$$

When x = 70, y = 5/4 (70) = 87.5 (i.e., on the y axis).
Amount of CO_2 taken into the deep ocean after 70 years = 87.5%.
The deep ocean absorbs the maximum amount of CO_2 sequestered by surface water from the atmosphere after 35 years or 87.5% of the CO_2.

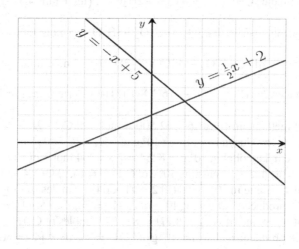

FIGURE 1.3 Linear equation showing positive slope (going upward) and negative slope (pointing downward).

Source: Modified from: by Jim.belk – own work, public domain, https://commons.wikimedia.org/w/index.php?curid=12077234

PROBLEM 1.5.2
CO_2 solubility

What would be the effect of an increase in water temperature on the rate of CO_2 sequestration?

SOLUTION 1.5.2

Henry's law equilibrium constant is a gas law stating that the amount of dissolved gas is proportional to its partial pressure in the gas phase and that the partial pressure of CO_2 in the gas phase is currently 0.04%. Hence, CO_2 dissolves according to CO_2 (gas)/CO_2 (liquid).

At 20°C, 1 L water dissolves about 1.7 g CO_2 at normal pressure (1 atm). If the pressure is twice as large, the amount of dissolved CO_2 is twice as much, that is, 3.4 g.

The solubility of gases in liquids decreases with increasing temperature. Ice-cold water can hold much more dissolved oxygen as warm water (LIMCOM 2023). CO_2 sequestration in water, therefore, decreases as the water gets warmer.

PROBLEM 1.5.3
CO_2 waste: proportions of burning fuels

What are the comparative amounts of CO_2 formed by burning a ton each of methane (the main component of natural gas), ethyl alcohol, kerosene, and isooctane (a component in gasoline)?

SOLUTION 1.5.3

$$\text{Methane}: CH_4 + 2O_2 \rightarrow CO_2 + 2H_2O \quad (\text{mole ratio} = 1:1) \qquad (1.5.4)$$

$$\text{Ethyl alcohol}: C_2H_6O + 3O_2 \rightarrow 2CO_2 + 3H_2O \quad (\text{mole ratio} = 1:2) \quad (1.5.5)$$

$$\text{Kerosene}: C_{11}H_{24} + 17O_2 \rightarrow 11CO_2 + 12H_2O \quad (\text{mole ratio} = 1:11) \quad (1.5.6)$$

$$\text{Isooctane}: 2C_8H_{18} + 25O_2 \rightarrow 16CO_2 + 18H_2O \quad (\text{mole ratio} = 1:8) \quad (1.5.7)$$

1. Moles of methane in 10^6 g/16 = 62,500 = moles of CO_2
   Mass of CO_2 produced by burning 10^6 g of CH_4 = 62,500 × 44 g = 2700 kg
2. Moles of ethyl alcohol in 10^6 g = 8264 = 2 × 8264 moles of CO_2
   Mass CO_2 produced by burning 10^6 g of C_2H_6O = 16,528.9 × 44 g = 727.3 kg
3. Moles of kerosene in 10^6 g = 6410.2 = 11 × 6410.2 moles of CO_2
   Mass CO_2 produced by burning 10^6 g of $C_{11}H_{24}$ = 70,512.8 × 44 g = 3102.5 kg
4. Moles of isooctane in 10^6 g = 8,771 = 8 × 8771 moles of CO_2
   Mass CO_2 produced by burning 10^6 g of C_8H_{18} = 70,168 × 44 g = 3087.392 kg

1.6   FLUOROCARBONS

The role of chlorofluorocarbons in destroying the ozone layer has been known since the 1980s. Fluorocarbons are strong greenhouse gases, some of which create hazardous compounds with high longevity in the environment. Fluorocarbons, or

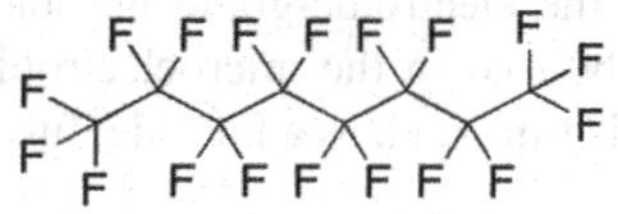

FIGURE 1.4 Perfluorooctane, a linear perfluoroalkane.

perfluorocarbons (PFCs), are organofluoride compounds having the formula $C_{(x)}F_{(y)}$. However, many fluorine-containing organic compounds are called fluorocarbons. As carbon carriers, they, and compounds that contain many C–F bonds often have distinctive properties, e.g., enhanced stability (Figure 1.4), which has negative impacts on stratospheric ozone.

PROBLEM 1.6.1
Identifying fluorocarbons

A 91.12-g sample of a gaseous fluorocarbon contains 14.80 g of carbon and 75.32 g of fluorine and occupies 15.82 L at STP (P = 1.00 atm and T = 273.15 K). What is the approximate molar mass of the fluorocarbon? State its molecular formula.

SOLUTION 1.6.1

The original ideal gas law applies the formula:

$$PV = nRT$$

The density version of the ideal gas law is:

$$PM = dRT$$

where
P = pressure measured in atmospheres (atm)
T = temperature measured in kelvin (K)
R = the ideal gas law constant (0.0821 L $\cdot$ atm $\cdot$ mole^{-1} K^{-1} × 273/1 atm)
M = now the molar mass (g $\cdot$ mole^{-1}) d is the density (g $\cdot$ L^{-1})

So,

$M = d \ (RT / P)$

$$= 91.12 \text{ g} / 14.80 \left(0.082 \text{ L} \cdot \text{atm} \cdot \text{mole}^{-1}\text{K}^{-1} \times 273 / 1 \text{ atm}\right) = 136.55 \text{ g} \cdot \text{mole}^{-1}$$

n = number of moles of gas

$n_{carbon} = 15.82$ g carbon $\times (1$ mole carbon $/ 12$ g carbon$) = 1.318$ mole

$n_{fluorine} = 75.32$ g fluorine $\times (1$ mole fluorine $/ 19$ g fluorine$) = 3.96$ mole

Ratio of molecules $= 1.318 / 3.96 = 1 : 3$

Molar mass (of fluorine + carbon) $= 12 + (3 \times 19) = 69$ g

Carbon trifluoride CF_3: the electronic-grade fluorine analog is one of several plasma etching gases widely used in the microelectronics industry, especially for the etching of silicon dioxide film, and, as a fluoride, functions as a low temperature mixed refrigerant.

1.7 THE EARLY INDUSTRIAL CARBON CYCLE

During the natural, early industrial carbon cycle, about one-twelfth of the carbon dioxide in the atmosphere is cycled annually by forest gross photosynthesis, which accounts for 50% of terrestrial photosynthesis (Met Office 2021).

PROBLEM 1.7.1
Pre-industrial and early industrial CO_2 concentrations

During the 1750s, the CO_2 concentration was around 280 ppm (Lindsey 2022). What is the rate at which it increased if in the 1990s the CO_2 levels were 335 ppm?

SOLUTION 1.7.1

1. Work out the difference between the two values (final minus the initial value = increase).
2. Divide the increase by the initial number.
3. Multiply by 100.

$$\% \text{ increase} = (\text{increase} / \text{initial value}) \times 100$$

$$\% \text{ increase} = ((335 - 280) / 280) \times 100 = (55 / 280) \times 100$$

$$\% \text{ increase} = 0.196\% = 0.2\%$$

PROBLEM 1.7.2
Changed rate of change

Problem 1.7.1, what is the average rate of change from the 1990s if in the 2000s the CO_2 concentration was about 384 ppm?

SOLUTION 1.7.2

1. Work out the difference between the two values (final minus the initial value)
2. Divide the change in y-values by the change in x-values

$$\text{Average rate of change} = (y_2 - y_1 / x_2 - x_1)$$
$$= (384 - 335 \text{ ppm}) / (2000 - 1990)$$
$$= 49 / 10 \tag{1.7.1}$$
$$= 4.9 \text{ ppm} / \text{year of } CO_2 \text{ (ppm = parts per million)}$$

1.8   SEA LEVEL RISE DUE TO THERMAL EXPANSION

The warming of earth is primarily due to the accumulation of heat-trapping green-house gases, and more than 90% of this trapped heat is absorbed by the oceans. As this heat is absorbed, ocean temperatures rise, and water expands. A greater volume of ocean water due to thermal expansion will lead to a rise in sea level. Thermal expansion is calculated using the following equation:

$$\Delta V = \beta V_o \Delta T$$

where
   ΔV = change in volume
   β = 1/K (kelvin) is the coefficient of volumetric thermal expansion
   V_o = initial volume
   ΔT = change in temperature

PROBLEM 1.8.1
Temperature and sea levels

Calculate the sea level rise of the world's ocean if the $\Delta T = 0.3°C$.

SOLUTION 1.8.1

Formula for volume expansion:

$$\Delta V = \beta V_o \Delta T \qquad (1.8.1)$$

$$\Delta V = Area \times Height$$

$$Height = \Delta V / Area$$

$$\text{Volume of the ocean} = 1.4 \times 10^{18} \ m^3$$

$$\text{Area of the ocean} = 3.6 \times 10^{14} \ m^2$$

$$\beta_{avg} \text{ deep ocean} = 0.00021 \text{ per } °C$$

Calculate the change in volume.

$$\Delta V = \beta \cdot V_o \cdot \Delta T$$

where
   $\beta = 0.00021/°C$
   $V_o = 1.4 \times 10^{18} \ m^3$
   $\Delta T = 0.2°C$
   $\Delta V = (0.00021/°C) \times (1.4 \times 10^{18} \ m^3) \times (0.3°C)$
   $\Delta V = 8.88 \times 10^{13} \ m^3$

Therefore, change in volume = 88,200,000,000,000 m³ (88.2 trillion cubic meters). Calculate the sea level rise.

$$\Delta V = \text{Area} \times \text{Height}$$

$$H = \Delta V \, / \, A$$

$$H = \left(8.88 \times 10^{13}\, m^3\right) / \left(3.6 \times 10^{14}\, m^2\right) = 2.46 \times 10^{-1}$$

$$H = 0.246 \text{ m}$$

$$H = 24.6 \text{ cm}$$

Therefore, the sea level of the world's ocean with a ΔT of 0.3°C would rise by 24.6 cm (in inches, 2.5 times x = 1, so, x = 1/2.5 = 0.4. Ans. = 9.84 inches).

PROBLEM 1.8.2
Heat distribution in the oceans

Assess the movement and implications of heat vertically in the oceans.

SOLUTION 1.8.2

Increasing ocean heat is closely linked to increases in atmospheric greenhouse gas concentrations, making the ocean an excellent indicator of how much earth is warming. Since 1971, the ocean has absorbed 90% of the excess energy added to earth's climate by burning fossil fuels and other human activities (Borunda 2021). The uppermost part of the ocean has absorbed the bulk of the extra heat because the expanded, lighter water tends to remain on top. Buoyancy and density variations are determined by variations of temperature in a fluid. Moreover, water is a poor conductor of heat. Only radiation is left, and radiative heat transfer does not occur greatly in liquids. Therefore, the top 250 ft, which is warming up the fastest, has been heating up by an average of about 0.11°C each decade since the 1970s (Borunda 2021), and only 33% (approximately) of that heat reaches below the ocean's skin.

PROBLEM 1.8.3
Heat capacity and its impact on marine life

If the temperature increase is 0.11°C, the mass of the ocean is 1.4×10^{21} kg, and the specific heat capacity of the sea water is 3900 J/kg/°C, what is the heat capacity that will impact marine organisms over a period of more than 50 years?

SOLUTION 1.8.3

The specific heat capacity of sea water is about 3900 J/kg/°C and the total mass of the oceans is 1.4×10^{21} kg (Woods Hole Oceanographic Institution 2021).

$$Q = mc\ \Delta t \tag{1.8.2}$$

where
  Q = heat
  m = mass
  c = specific heat capacity
  t = temperature

$$Q = mc\ \Delta t$$

$$Q = \left(1.4 \times 10^{21}\,\text{kg}\right)\left(3900\,\text{J}\,/\,\text{kg}\,/\,^\circ\text{C}\right)\left(0.11^\circ\text{C}\right) = 6.006 \times 10^{23}\,\text{J}$$

1.9 OCEAN ACIDIFICATION

Acidification effects on marine animals (particularly corals and calcareous plankton), ecosystems, and biogeochemical cycles may feedback into climate change (Equation 1.9.1). CO_2 acidification of the ocean has destroyed coral reefs which support marine life and provide livelihoods for a quarter-billion people. Bicarbonate and carbonic acid are formed whenever carbon dioxide dissolves in water. Therefore, there is a balance between these compounds. When carbon dioxide mixes with water, it partially converts into carbonic acid, hydrogen ions (H^+), bicarbonate $\left(HCO_3^-\right)$, and carbonate ions $\left(CO_3^{2-}\right)$. Upon diffusing into the surface, carbon dioxide reacts with water to form carbonic acid (H_2CO_3):

$$CO_2 + H_2O \leftrightarrow H_2CO_3 \tag{1.9.1}$$

And carbonic acid further dissociates into a hydrogen ion and a bicarbonate ion:

$$H_2CO_3 \leftrightarrow HCO_3^- + H^+ \tag{1.9.2}$$

Bicarbonate may further dissociate into another hydrogen ion and a carbonate ion:

$$HCO_3^-B \leftrightarrow H^+ + CO_3^{2-} \tag{1.9.3}$$

In CO_2, the mass of carbon is 12 whereas that of oxygen is 32. Hence, the ratio of their masses is 12:32 = 3:8.

PROBLEM 1.9.1
Acidification of upper levels of the oceans

Calculate the acidification pH at upper levels of ocean waters of the open seas.

SOLUTION 1.9.1

Once carbon dioxide has dissolved into the ocean, carbon dioxide molecules can react with water molecules to create carbonic acid, according to the equation of ocean acidification.

As a diprotic (containing two protons) acid, carbonic acid can then react "twice" with water to form carbonate.

$$H_2CO_3(aq) + H_2O(l) \leftrightarrow HCO_3^-(aq) + H_3O^+(aq) \qquad (1.9.4)$$

$$HCO_3^-(aq) + H_2O(l) \leftrightarrow CO_3^{2-}(aq) + H_3O^+(aq) \qquad (1.9.5)$$

Square brackets depict concentration in moles per liter, i.e., molar proportions, or the molarity (number of moles, i.e., the concentration) of a substance. For example, the expression $[K^+]$ refers to the number of moles of the potassium ion (dissociated) in the solution. Also, K_{a1}, K_{a2}, etc. signify acid dissociation constants (the proportion of hydrogen ions released) referring to the equilibrium constant for loss of the first, second, third proton (H^+) and so on. Sulfuric acid has a large K_{a2}, which makes it strong. Overall, acids that dissociate multiple times are weak (as is H_2CO_3), and K_{a1} can be used as an approximation because K_{a2} is a considerably smaller number. H_2SO_4 is an exception because both its K_{a1} and K_{a2} are large. Therefore, it is a strong acid.

Using those equations above, respectively, regarding the dissociation of carbonic acid, we find equilibrium constants:

$$K_{a1} = [HCO_3^-][H_3O^+] / [H_2CO_3] \qquad (1.9.6)$$

$$K_{a2} = [CO_3^{2-}][H_3O^+] / [HCO_3^-] \qquad (1.9.7)$$

Water is not placed in the dissociation constant expression because it has such a large concentration that no matter how much acid or base is added to the system, its concentration remains relatively unchanged. By applying Le Chatelier's principle (which states that when factors like concentration, pressure, temperature, and inert gases that affect equilibrium are changed, the equilibrium shifts in the opposite direction to offset the change), increasing the atmospheric concentration of carbon dioxide will affect the dissolution of carbon dioxide into the ocean. This will cause the equilibria of the ocean to shift, changing the pH of the ocean, as the hydronium ion concentration changes.

From the pH of the ocean, the concentrations of the various carbonate species ($H_2CO_3^*$, HCO_3^-, and CO_3^{2-}) can be determined. The concentration of hydronium ion is most easily measured as a pH value. The pH of a solution is the negative logarithm of the hydronium ion concentration. So:

$$pH = -\log[H_3O^+] \qquad (1.9.8)$$

$$[H_3O^+] = 10^{-pH} \qquad (1.9.9)$$

PROBLEM 1.9.2
pH and molarity

What is the pH of a solution that has a concentration of 5.0×10^{-2} M H_2CO_3 only, if the dissociation constant $K_{a1} = 4.3 \times 10^{-7}$ and the second dissociation constant

$$K_{a2} = 4.8 \times 10^{-11}?$$

SOLUTION 1.9.2

$$CO_2(aq) + H_2O(l) \leftrightarrow H_2CO_3(aq) \tag{1.9.10}$$

Both equations show that carbonic acid can then react twice (K_{a1} and K_{a2}) to form carbonate:

$$H_2CO_3(aq) + H_2O(l) \leftrightarrow HCO_3^-(aq) + H_3O^+(aq) \tag{1.9.11}$$

$$HCO_3^-(aq) + H_2O(l) \leftrightarrow CO_3^{2-}(aq) + H_3O^+(aq) \tag{1.9.12}$$

$$K_{a1} = [H_3O] \times [HCO_3] / [H_2CO_3]$$

$$H_2CO_3(aq) + H_2O(l) \leftrightarrow HCO_3(aq) + H_3O(aq)$$

$$4.3 \times 10^{-7} = \left[H_3O^+\right] \times \left[HCO_3^-\right] / 5.0 \times 10^{-2}$$

$$\left[4.3 \times 10^{-7}\right] \times \left[5.0 \times 10^{-2}\right] = \left[H_3O^+\right] \times \left[HCO_3^-\right]$$

$$2.15 \times 10^{-8} = \left[H_3O^+\right] \times \left[HCO_3^-\right] \ldots \text{Or, 1 mole quantity times a similar mole quantity.}$$

$$\text{Square root of } 2.15 \times 10^{-8} = 1.47 \times 10^{-4} = \left[H_3O^+\right] = \left[HCO_3^-\right]$$

$$K_{a2} = \left[H_3O^+\right] \times \left[CO_3^{2-}\right] / \left[HCO_3^-\right]$$

$$HCO_3^-(aq) + H_2O(l) \leftrightarrow CO_3^{2-}(aq) + H_3O(aq)$$

$$4.8 \times 10^{-11} = \left[H_3O^+\right]\left[CO_3^{2-}\right] / \left[1.47 \times 10^{-4}\right]$$

$$4.8 \times 10^{-11} \times 1.47 \times 10^{-4} = \left[H_3O^+\right]\left[CO_3^{2-}\right]$$

$$= 7.04 \times 10^{-16} = \left[H_3O^+\right] \times \left[CO_3^{2-}\right]$$

$$\text{Square root of } 7.04 \times 10^{-16} = 8.39 \times 10^{-8} = \left[H_3O^+\right] = \left[CO_3^{2-}\right]$$

Therefore, the concentration of oxonium ions [H_3O^+], which determine the pH, is approximately $(1.47 \times 10^{-4}) + (8.39 \times 10^{-8})$.

As 8.39×10^{-8} is such a small quantity (being from k_{a2}), we can ignore it.

In other words, $0.000147 + 0.0000000839 = 0.0001470839$

So, the answer is 1.47×10^{-4}.

Using the equation to find pH:

$$pH = -\log\left[H_3O^+\right]$$

$$pH = -\log\left[1.47 \times 10^{-4}\right]$$

$$\log \text{ of } \left[1.47 \times 10^{-4}\right] = -3.8324$$

$$pH = 3.83$$

PROBLEM 1.9.3
Ocean acidification and pH strength

Marine animals, ecosystems, and biogeochemical cycles are affected by acidification. If the H_2CO_3* concentration is 1.8×10^{-5} M, the HCO_3^- concentration is 2.5×10^{-4} M, and the density of the P_{co2} is 2.18 g $\cdot$ L^{-1}, what is the acidification pH and acidity strength (K_{a1}) of the ocean (excluding other factors)?

SOLUTION 1.9.3

$$K_{CO2} = [H_2CO_3 *] / P_{CO2} \qquad (1.9.13)$$

$$K_{CO2} = 1.8 \times 10^{-5} M / 2.18 \text{ g} \cdot L^{-1}$$

$$K_{CO2} = 1.8 \times 10^{-5} M / 0.0495 M$$

$$K_{CO2} = 8.3 \times 10^{-6} M / g \cdot L^{-1}$$

$$= 3.6 \times 10^{-4} M$$

Or "8.3×10^{-6} M/g $\cdot$ L^{-1} CO_2 moles always come from 1.8×10^{-5}M of H_2CO_3."

$$K_{a1} = [HCO_3^-][H_3O^+] / [H_2CO_3 *]$$

Using $K_{a1} = 10^{-6}$ to find H_3O^+ [being very close to 0.0000083],

$$10^{-6} \times [1.8 \times 10^{-5} M] = [2.5 \times 10^{-4} M][H_3O^+] / [\cancel{1.8 \times 10^{-5} M}] \times [\cancel{1.8 \times 10^{-5} M}]$$

$$10^{-6} \times [1.8 \times 10^{-5} M] = [2.5 \times 10^{-4} M][H_3O^+]$$

$$1.8 \times 10^{-11} M / 2.5 \times 10^{-4} M = [2.5 \times 10^{-4} M][H_3O^+] / 2.5 \times 10^{-4} M$$

$$7.2 \times 10^{-8} M = [H_3O^+]$$

The pH of K_{a1} is pH $= -\log [H_3O^+]$

$$pH = -\log[7.2 \times 10^{-8}]$$

$$pH = 7.14$$

Using $[H_3O^+] = 10^{-pH}$

$$K_{a1} = 10^{-pH}[HCO_3^-] / [H_2CO_3 *]$$

$$K_{a1} = 10^{-7.14}[2.5 \times 10^{-4} M] / [1.8 \times 10^{-5} M]$$

$$K_{a1} = 1.0 \times 10^{-16} M = \text{acidity strength of the ocean}$$

1.10 WARMER AND MOISTER?

Due to climate change, earth will become warmer, and moisture events (or the lack thereof) may be more intense. A warmer atmosphere holds more moisture – about 7% more per 1.8°F (1°C) of warming. CO_2 produced by human activities is the largest.

PRACTICE PROBLEM

The two sets of data below are for average monthly wave incursions (in meters) on a particular beach for eight months in the years 1995 and 2015 ("incursions" are the distance that the water reaches up on the beach during a wave swash):

2015: 1.17, 0.70, 0.82, 0.94, 0.86, 0.91, 0.79, 1.03
1995: 0.63, 0.49, 0.53, 0.51, 0.57, 0.60, 0.48, 0.64

   a. With the aid of a selected statistical test, (b) find out if if there is a sig-
      nificant difference between the two sets of data and, by extension, make a
      statement on the evidence for sea level rise.

SOLUTION

   a. For test selection, the t-test is appropriate (see Chapter 30: Appendix A for
      the explanation and for the procedure for part b).

PRACTICE PROBLEM

"I have 100 g of ice. It takes 406 J to raise the temperature of ice 2°C. What is the specific heat of ice?"

SOLUTION

Heat capacity for 100 g ice = 406 J/2C
Heat capacity for 100 g ice = 203 J/C
Heat capacity for 1 g ice = 2.03 J/C per gram

   It takes 2.03 J to raise a gram of ice 1°C. So, if we have 100 g of ice, we need 100 times as many joules to heat it all.
   As the main heat reservoir for incoming solar energy, there has been an accelerating change in sea temperatures over the last few decades, which has contributed to rising sea levels.

REFERENCES

Allan R (2022) *The role of the ocean in tempering global warming.* NOAA Climate. gov. Retrieved October 30, 2022, from https://www.climate.gov/news-features/blogs/enso/role-ocean-tempering-global-warming#:~:text=Making%20a%20rough%20approximation%2C%20assuming,Celsius%20(1.8%20degrees%20F)

Borunda A (2021, May 3) *Ocean warming facts and information.* Environment. Retrieved October 30, 2022, from https://www.nationalgeographic.com/environment/article/critical-issues-sea-temperature-rise

Hemalatha T, Ramaswamy A (2017) A review on fly ash characteristics – Towards promoting high volume utilization in developing sustainable concrete, *Journal of Cleaner Production*,Volume 147, 2017, Pages 546–559, ISSN 0959-6526, https://doi.org/10.1016/j.jclepro.2017.01.114

LIMCOM (2023) Dissolved oxygen https://limpopocommission.org/the-basin/the-river-basin/water-quality/principles-of-water-quality/physical-characteristics/dissolved-oxygen/

Lindsey R (2022, June 23) *Climate change: Atmospheric carbon dioxide.* Retrieved October 2022, from https://www.climate.gov/news-features/understanding-climate/climate-change-atmospheric-carbon-dioxide

Manahan S (2005) *Environmental chemistry.* Taylor and Francis, Boca Raton, Fla. Pp 355–363.

Met Office (2021, May 22) Atmospheric CO2 now hitting 50% higher than pre-industrial levels. World Economic Forum. Retrieved October 2022, from https://www.weforum.org/agenda/2021/03/met-office-atmospheric-co2-industrial-levels-environment-climate-change/

Water Encyclopedia (2022a) http://www.waterencyclopedia.com/Ge-Hy/Global-Warming-and-the-Hydrologic-Cycle.html#ixzz7ZDPlsHrl

Water Encyclopedia (2022b) http://www.waterencyclopedia.com/Ge-Hy/Global-Warming-and-the-Hydrologic-Cycle.html#ixzz7ZDO9l1U5

Woods Hole Oceanographic Institution (2021). *Ocean warming.* Retrieved October 30, 2022, from https://www.whoi.edu/know-your-ocean/ocean-topics/climate-weather/ocean-warming/

Yao ZT, Ji XS, Sarker PK, Tang JH, Ge LQ, et al. (2015) A comprehensive review on the applications of coal fly ash, *Earth-Science Reviews*, 141, 105–121, ISSN 0012-8252, https://doi.org/10.1016/j.earscirev.2014.11.016

2 Measurements

2.1 WATER: SCALE CONVERSIONS

Sodium chloride (NaCl) is the dominant salt usually found in stream sampling; however, other salts will also occur with EC readings (e.g., carbonate and bicarbonate salts, magnesium and calcium sulfates, potassium).

Some commonly used units of measurement are as follows:

ppm = parts per million*
mg/L = milligrams per liter*
μS/cm = microsiemens per centimeter (recognized EC units)
mS/cm = millisiemens per centimeter
dS/m = decisiemens per meter
1 dS/m = 100 mS/m
1 dS/m = 1 mS/cm = 1000 EC μS/cm = approx. 550 ppm

*Depending on the types of salts present

PROBLEM 2.1.1
Reporting spills and disseminating toxicity information often require conversion among different modalities. For aqueous solutions, (a) convert $0.03\ \mu g \cdot L^{-1}$ to the ppb and ppm scales and (b) convert 5 ppm to the $\mu g \cdot L^{-1}$ scale.

SOLUTION 2.1.1

a. One liter of water contains $10^9\ \mu g$.
   $0.03\ \mu g \cdot L^{-1} = 0.03 \times 10^{-9}\ \mu g = 0.03/10^9$ ("of" = "×")
   $= 0.03$ parts per $1000,000,000 = 0.03$ ppb.
   $0.03/1000$ parts per $1000,000,000/1000 = 0.00003$ ppm.
b. Converting 5 ppm to $\mu g\ L^{-1} = 5/10^6$ of $10^9 = 5 \times 10^3\ \mu g \cdot L^{-1}$

PROBLEM 2.1.2
Ethanol as fuel

Ethanol, an important industrial chemical and a component of gasohol, the mixture containing 10% ethyl alcohol ("E10"), raises the octane rating of lead-free automobile fuel and significantly decreases the carbon monoxide and carbon dioxide released from exhaust pipes (Encyclopedia.com 2018). Ethanol has an LD_{50} of 7060 mg/kg. What would be the lethal dosage for a child who weighs 35 kg?

SOLUTION 2.1.2

$$7060\ mg \times 35 = 247,100\ mg$$

Lethal dosage = 247.1 g of ethanol

DOI: 10.1201/9781003341826-4

PROBLEM 2.1.3
Percentage of a gas in air

The EPA limit for carbon monoxide (CO) is 9 ppm.
What is this quantity as a percentage?

SOLUTION 2.1.3

CO contained in 100 parts of air $= 9 / 1,000,000 \times 100$ parts of CO

Percentage of CO in 100 parts of the air $= 9 / 10,000 = 0.0009\%$

PROBLEM 2.1.4
Decommissioned fossil fuel depots

Near a large decommissioned gasoline storage depot, a 5-L sample of water contained 120 µg of lead. What is this lead concentration in ppb?

SOLUTION 2.1.4

Five liter contains $10^9 \times 5$ µg of water which contains 120 µg of lead.
One liter contains 10^9 µg of water which contains 120/5 µg of lead.
Lead concentration $= 24$ ppb of lead.

PROBLEM 2.1.5
Toxins in water

A recommended safe level of heavy metal cadmium in drinking water is 0.01 ppm. What would be the mass of cadmium in a liter of drinking water (density of water $= 1 \text{ kg} \cdot \text{L}^{-1}$)?

SOLUTION 2.1.5

10^6 mg $= 1$ L and 10^9 µg $= 1$ L

One million parts (ppm) of a liter of water $= 10^6$ mg / L

Therefore, 1 ppm of cadmium $= 1$ mg / L

0.01 ppm of cadmium $= 1$ mg $\times 0.01$

Answer $= 0.01$ mg of cadmium

2.2   CONVERTING TO PPM

The ideal gas occupies 22.414 L per mole, and air is an example of an ideal gas. As the denominator and numerator of every ratio must refer to the same variable, ppm of pollutant gas/air must refer to volume. But a heavier gas than air spreads more slowly than air. Hence, a greater mass of it (than that of air) will occupy 22.414 L per mole, thereby potentially biasing a ppm comparison with air. Our ppm ratio must therefore "convert the mass of a mole of the pollutant gas to that of a mole of an ideal gas."

Using currency exchange as an analogy, where presently the US$ is the world standard, if you have 50 C$50 (Canadian dollars),

Exchange rate being 100 US cents = 120 Canadian cents

Therefore, you really have 50 / 1.2 = US$41.6 = less than US$50

Similarly, gases of the same volume differ in mass. Hence the equation:

$$V_p = M_p / M_W \times 22.414 \text{ L} \cdot \text{mole}^{-1} \qquad\qquad (2.2.1)$$

where

V_p = equivalent volume in liters at standard temperature and pressure (STP)
M_p = mass of the gas
M_W = molar mass of the gas (the currency in the $ analogy above)

Under STP, air occupies $22.214 \text{ L} \cdot \text{mole}^{-1}$. Hence, $22.414 \text{ L} \cdot \text{mole}^{-1}$ = the "k" (constant) for the "ideal" gas.

However, more heat expands the gas volume, while increased pressure decreases volume.

In other words, >heat = multiplication, and >pressure = division:

$$22.414 \text{ L} \cdot \text{mole}^{-1} \times T_2 / 273 \text{ K} \times 101.325 \text{ kPa} / P_2$$

where

T_2 and P_2 are, respectively, the new temperature and pressure.

Parts per million (ppm) compare volumes: that of air with that of pollutant gas, giving rise to:

$$\text{ppm of pollutant} = V_p / V_a + V_p$$

where

• V_a = volume of air in m^3 at the pressure and temperature when the measurement occurred.

For ppm, the above two equations are combined as follows:

$$\left[(M_p / M_W) \times 22.414 \text{ L} \cdot \text{mole}^{-1} \times (T_2 / 273 \text{ K}) \times 101.325 \text{ kPa} / P_2 \right] / V_a \times 1000 \text{ L} \cdot \text{m}^{-3}$$

where

• M_p = the mass of the pollutant under investigation, in micrograms.

But the 10^{-6} conversion factor to micrograms from grams cancels out the 10^6 for ppm. So do several entities as shown below:

$$\left[(M_p \times 10^{-6} \text{ g} / M_W \text{ g mole}^{-1}) \times 22.414 \text{ L} \cdot \text{mole}^{-1} \times (T_2 \text{ K} / 273 \text{ K}) \right.$$
$$\left. \times 101.325 \text{ kPa} / P_2 \text{ kPa} \right] / V_a 10^6 \times \text{L} \cdot \text{m}^{-3}$$

$$\left[(M_p \times 10^{-6}\ \cancel{g} / M_w\ \cancel{g\ mole}^{-1}) \times 22.414\ \cancel{L\ mole}^{-1} \times (T_2\ \cancel{K} / 273\ \cancel{K}) \right.$$

$$\left. \times 101.325\ \cancel{kPa} / P_2\ \cancel{kPa} \right] / V_a\ 10^{-6} \times \cancel{L} \cdot m^{-3}$$

$$= \left[(M_p / M_w) \times 22.414\ L \times (T_2 / 273) \times 101.325 / P_2 \right] / V_a \times 1000\ L \cdot m^{-3}$$

$$= \left[\text{Fraction of } 22.414\,L \text{ of pollutant} \right] / \left[\text{Per cubic meter} \right]$$

$$\text{ppm of pollutant} = V_p / V_a + V_p$$

Normally, it is assumed that $V_a = 1m^3$ (= 1000 L). Hence, the volume of air ($V_a \times$ 1000 L) = $10^3\ L \cdot m^{-3}$.

The value of R, the ideal gas constant, depends on the units chosen for pressure, temperature, and volume in the ideal gas equation. Pressure is commonly measured in one of three units: kPa, atm, or mmHg. Therefore, R can seemingly have three different values.

Calculating R when the pressure is measured in kPa at the volume of 1.00 mole of any gas at STP (standard temperature, 273.15 K and pressure, 1 atm):

$$\text{Volume} = 22.414\,L$$

$$\text{Pressure} = 101.325\,kPa$$

Putting temperature = 273.15 K for temperature into the ideal gas equation and solving for R:

$$R = PV / nT = 101.325\ kPa \times 22.414\ L / 1.000\ mole \times 273.15\ K$$

$$= 8.314\ kPa \cdot L / K \cdot mole$$

This is the value of R that is to be used in the ideal gas equation when the pressure is given in kPa.

- Table 2.1 shows a summary of the possible values of R.
- The correct value of R can be chosen for a given problem.

TABLE 2.1
Values of the Ideal Gas Constant

Unit of P	Unit of V	Unit of n	Unit of T	Value and Unit of R
kPa	L	mole	K	8.314 J/K · mole
atm	L	mole	K	0.08206 L · atm/K · mole
MmHg	L	mole	K	2.36 L · mm Hg · K_{-1} · $mole_{-1}$

NB: The unit for R, when the pressure is in kPa, has been changed to J/K· mole. A kilopascal multiplied by a liter is equal to the SI unit for energy, a joule (J).

PROBLEM 2.2.1
Pollutant gases

What volume is occupied by 3.65 g of oxygen gas at a pressure of 85.6 kPa and a temperature of 21°C? Assume that oxygen is an ideal gas.

SOLUTION 2.2.1

Step 1: Given:

$$P = 85.6\,kPa$$
$$T = 21°C = 294\,K$$
$$Mass\ O_2 = 3.65\,g$$
$$Find: V = L?$$

Step 2: List other known quantities.

$$O_2 = 32\ g/mole$$
$$R = 8.314\ J/K \cdot mole$$

Step 3: Plan the problem
First, determine the number of moles of O_2 from the given mass and the molar mass. Then, rearrange the equation algebraically to solve for V.

$$V = nRT/P$$

Step 4: Calculate the answer.

1. 3.65×1 mole $O_2/32.00$ g $O_2 = 0.1140$ mole O_2
2. Substitute the known quantities into the equation and solve.

$$V = nRT/P$$
$$= 0.1175\ mole \times 8.314\ J/K \cdot mole \times 294\ K/85.6\ kPa$$
$$= 3.28\ L\ of\ O_2$$

3. Work out the difference between the two values

PROBLEM 2.2.2
Gaseous pollutant concentration

As already stated, the conversion of gases between micrograms per cubic meter and ppm is possible because all gases occupy a volume of 22.414 liters per mole at standard temperature and pressure.

$$Ppm = (22.414\ L \cdot mole^{-1}) \times T_2/273\ K \times 101.325\ kPa/P_2$$

At a temperature and pressure of 23°C and 103.823 kPa respectively, a 1-m³ sample of air was found to contain 75 µg of SO_2. What was the concentration of the SO_2 in parts per million?

SOLUTION 2.2.2

$$M_W \text{ of } SO_2 = 32.06 + 2(15.9994) = 64.06 \text{ g} \cdot \text{mole}^{-1}$$

$$23°C + 273 \text{ K} = 296 \text{ K}$$

Concentration of pollutant (the factor of 10^6 for μg & ppm is omitted as they cancel out)

$$\big[(75 \ \mu g / 98.06 \ \text{g} \cdot \text{mole}^{-1}) \times 22.414 \ L \times (296 \ K / 273 \ K)$$

$$\times (101.325 \ kPa / 103.823 \ kPa)\big] / m^3 \times 103 \ L \cdot m^{-3}$$

$$= 0.018 \text{ ppm of } SO_2$$

2.3   WHAT IS SYNCHROTRON TECHNOLOGY?

Without synchrotron technology, it would have been virtually hard to artificially extract carbon directly from plants in dry soils. A synchrotron is an extremely powerful source of X-rays. High energy electrons produce X-rays as they circulate the synchrotron. A synchrotron machine exists to accelerate electrons to extremely high energy and then make them change direction periodically. The resulting X-rays are emitted as dozens of thin beams, each directed toward a beamline next to the accelerator (ESRF 2019).

2.3.1   SYNCHROTRON TECHNOLOGY: ROLE IN AGRICULTURE

Synchrotron light enables the research of the rhizosphere's structure at various scales and with high resolution. It illustrates how atoms and molecules of nutrients and contaminants "walk" in the soil and alter chemically as they interact with other molecules (BSLL 2021). As a result, the activities that occur in the soil can be better understood and regulated, resulting in more efficient and environmentally friendly agricultural outputs (ESRF 2019).

PROBLEM 2.3.1
Calculating the soil organic carbon of a selected depth of topsoil

Soil organic carbon (SOC) is usually reported as a percentage of the topsoil (0–10 cm). How can this value be converted into a meaningful volume?

SOLUTION 2.3.1

Formula:

$$\text{SOC stock in tC/ha} = (SOC\%) \times (\text{mass of soil in each volume}) \qquad (2.3.1)$$

For example, 10,000 m^2 in 1 ha × 0.2 m soil depth × 1.4 g/cm^3 bulk density × 1.2% = 33.6 tC/ha

PRACTICE PROBLEM 2.3.1

A soil has a SOC of 1.3% (0.013). A bulk density of 1.2 g/cm³ would have SOC to a depth of 10 cm (0.1 m) per hectare (10,000 m²) of...? A soil has a SOC of 1.3% (0.013), and a bulk density of 1.2 g/cm³. What is the total weight of SOC between the surface and a depth of 10 cm (0.1 m) per hectare (10,000 m²)?

PROBLEM 2.3.2
Elastic rebound

What is isostatic adjustment?

SOLUTION 2.3.2

Even though the ice retreated long ago, North America is still rising (isostatic readjustment) where the massive layers of ice pushed it down (isostatic adjustment). The US East Coast and Great Lakes regions – once on the bulging edges, or forebulge, of those ancient ice layers – are still slowly sinking from forebulge collapse (isostatic readjustment).

2.4 REMEDIATION BASED ON DAM DEPTH MEASUREMENTS

Flushing of a waterway improves the water quality because it removes toxic substances, and mean depth is important for the following reasons:

- Depth of a waterway is important because it varies indirectly with the time elapsed before the next flushing (referred to as the period of natural flushing).
- Chemically reducing conditions, which increase heavy metal toxicity, vary directly with depth.
- Shallow lakes are generally more productive than deep lakes and mean depth is a quick way of assessing overall depth.
- The depth indicates the potential for waves and mixing events to disrupt bottom sediments, thereby increasing the pollution of purer water in upper levels.
- Depth determines the approximate potential life of the lake due to its complete evaporation

PROBLEM 2.4.1
Rapid depth measurement

How can you rapidly measure the average depth of a waterway?

SOLUTION 2.4.1

$$\text{Mean depth}(z) = \text{volume} / \text{surface area} \qquad (2.4.1)$$

However, if the volume is unavailable and the lake is small, collecting numerous lake depth measurements and averaging them is one option. A dam depth measuring rope can be made from the following:

- 3 mm cord (long enough to be thrown from the bank to the middle of the deepest dam)
- Fishing sinkers, fishing floats, marker pen, plastic drink bottle tops, field collection sheet, pen/clipboard (Industries and Investment NSW 2008)

Measuring Dam Depth

According to Industries and Investment NSW (2008), the measurement entails using some cord, fishing sinkers and floats, a marker pen, and some plastic-colored bottle tops. The cord needs to be long enough to reach the bottom of the biggest dam being measured on the property. One end of the cord is threaded with alternating fishing floats and bottle tops spaced at 1 m intervals and secured with knots on either side. The floats can have the depth written on their side. Sinkers are tied to the end of the cord.

PROBLEM 2.4.2
Calculating lake volume

In an era of climate change, inventories of water resources at the state or district level of administration are required to efficiently plan for conservation and redistribution of water. How can the average volume of a lake be quickly calculated?

SOLUTION 2.4.2

A frustum is the portion of a conical pyramid which remains after its upper part has been cut off by a plane parallel to its base, i.e., the area intercepted between two such planes. The volume of a frustum is the amount of space that is present inside it or the quantity of matter that it can hold (Figure 2.1).

$$\text{Frustum volume: } V = H / 3 \left[S1 + S2 + \sqrt{(S1 \cdot S2)} \right]$$

where
- H = height of the frustum (the distance between the centers of two bases of the frustum)
- $S1$ = area of one base of the frustum
- $S2$ = area of the other base of the frustum

Area at the top (A_{top}) = the area at the top of the layer
Area at the bottom (A_{bottom}) = the area at the bottom of the layer
z = the distance between contour lines
V = the volume of one layer

$$V = z \left[A_{top} + A_{bottom} + \sqrt{(A_{top} \cdot A_{bottom})} \right] / 3$$

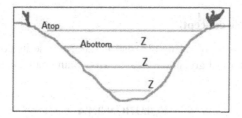

FIGURE 2.1   Mean depth of a lake using frustum values.

PROBLEM 2.4.3
Estimating the area of water as a function of depth

The amount of surface area exposed to the atmosphere helps to determine the potential rate of evaporation of the water as a lake dries out. How can one estimate the potential surface areas of water which will be exposed to evaporation as a natural water body continues to dry out?

SOLUTION 2.4.3

A hypsographic curve = area as a function of depth

Using a bathymetric map of the water body, measure along a selected bathymetric contour interval. Using Figure 2.2 as an example, perform the following:

- Select an appropriate vertical interval.
- Enlarge the map, place an overlay of grid squares on it, and count the squares at each selected vertical interval.
- By interpolation, draw a smooth line linking the points (Figure 2.2).
- Using an appropriate correct interval, join the measured horizontal area with the depth represented (Figure 2.2).

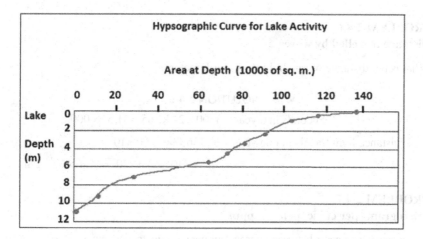

FIGURE 2.2   The hypsographic curve shows the area at sequential depths in a water body.

PROBLEM 2.4.4
Moles in a gas (mole concept)

Hydrogen gas (H_2) is a promising alternative fuel because its combustion does not produce large amounts of greenhouse gases. How many molecules are present in 1 g of hydrogen gas?

SOLUTION 2.4.4

The mole concept is like the decimal currency system, where 100-unit currencies differ in value. Thus, a mole quantity of a substance is Avogadro's number, approximately 6.022×10^{23}. It is the number of atoms contained in 12.0 grams of carbon-12, in 16.0 grams of oxygen, and so on, expressing large quantities. We can determine the number of moles in any chemical reaction given the chemical formula and the mass of the reactants.

One mole of hydrogen gas contains 6.02×10^{23} atoms, which weigh 1 g.

One molecule of hydrogen (H_2) contains two atoms of hydrogen.

Therefore, 1 g of hydrogen contains 6.02×10^{23}) / 2 = 3.01×10^{23} molecules.

PROBLEM 2.4.5
Frequency and absorptivity of molecules

The ability of a greenhouse gas to absorb the energy of electromagnetic waves depends partly on the frequency. What is the wavelength of a radio wave having a frequency of 89 MHz?

SOLUTION 2.4.5

$$\lambda = v/f$$

$$\text{Wavelength} = \text{speed/frequency} = 3 \times 10^5 \, m \cdot s^{-1} / 8.9 \times 10^7 \, Hz$$
$$= 0.00337 \text{ m or } 3.37 \times 10^{-3} \, m$$

PROBLEM 2.4.6
Distance travelled by waves

What is the distance of one light year in kilometers?

SOLUTION 2.4.6

$$\text{Seconds in a year} = 3600 \times 24 \times 365 = 31,536,000$$

$$\text{Distance light travels in one year} = 31,536,000 \times 3.0 \times 10^5$$
$$= 3.154 \times 10^7 \times 3.0 \times 10^5 = 9.46 \times 10^{12} \text{ km}$$

PROBLEM 2.4.7
Micrograms per cubic meter to ppm

The conversion of gases between micrograms per cubic meter and ppm is possible because all gases occupy a volume of 22.414 L per mole at standard temperature and pressure.

An equation is necessary which converts the mass of the pollutant M_p to its equivalent volume V_p in liters at standard temperature and pressure (Davis and Masten 2020).

Thus, under standard atmospheric conditions, how can the molar ratio of a pollutant gas in air can be expressed?

SOLUTION 2.4.7

$$V_p = [M_p / M_w] \times 22.414$$

where
 V_p = volume of pollutant gas
 M_p = mass of pollutant gas
 M_w = molecular weight of pollutant gas

As temperature and pressure (T_2 and P_2) vary directly and indirectly, respectively, with gas volumes, the volume of the pollutant changes as follows:

$$ppm = (22.414 \text{ L} \cdot \text{mole}^{-1}) \times T_2 / 273 \text{ K} \times 101.325 \text{ kPa} / P_2$$

PROBLEM 2.4.8
Gaseous pollutant concentration

At a temperature and pressure of 23°C and 103.823 kPa, respectively, a 1-m^3 sample of air was found to contain 75 µg of SO_2. What was the concentration of SO_2 in parts per million?

SOLUTION 2.4.8

$$M_W \text{ of } SO_2 = 32.06 + 2(15.9994) = 64.06 \text{ g} \cdot \text{mole}^{-1}$$
$$23°C + 273 \text{ K} = 296 \text{ K}$$

Concentration of pollutant (the factor of 10^6 for µg and ppm is omitted as they cancel out, and m^3 is converted into liters)

$$= \left[(75 \text{ µg} / 64.06 \text{ g} \cdot \text{mole}^{-1}) \times 22.414 \text{ L} \times (296 \text{ K} / 273 \text{ K}) \right.$$
$$\left. \times (101.325 \text{ kPa} / 103.823 \text{ kPa}) \right] / m^3 \times 10^3 \text{Lm}^{-3}$$
$$= 0.027 \text{ ppm of } SO_2$$

REFERENCES

BSLL (2021) Brazilian Synchrotron Light Laboratory (BSLL). Retrieved November 7, 2022, from https://lnls.cnpem.br/sirius-en/synchrotron-light-and-its-benefits/
Davis and Masten (2020) Principles of Environmental Engineering & Science. 4th edition. ISBN10: 1259893545 I ISBN13: 9781259893544

Encyclopedia.com (2018) Gasohol. Oxford University Press. https://www.encyclopedia.com/
     science-and-technology/chemistry/organic-chemistry/gasohol#:~:text=Gasohol%20
     is%20a%20term%20used%20for%20the%20mixture,promoted%20as%20a%20
     means%20of%20reducing%20corn%20surpluses
ESRF (2019) What is a synchrotron? European Synchrotron Radiation Facility. Retrieved
     November 7, 2022, from https://www.esrf.fr/home/education/what-is-the-esrf/what-is-
     synchrotron-light.html
Industries and Investment NSW (2008) Measuring Dam Depth. https://calculator.agriculture.
     vic.gov.au/fwcalc/information/field-measurement-guide

3 Sequestration by Forests

MAIN FACTORS – DETERMINANTS OF SEQUESTRATION IN FOREST TREES:

- Growth rate (high)
- Bulk density (high)
- Geographical distribution (wide)
- Invasiveness (strong)
- Drought tolerance (high)
- Heat tolerance (resistance to fire)
- Commercial value (high-value timber is always preserved)
- The difference between inbound (photosynthesis) and outbound fluxes (respiration and mineralization), or the net absorption flux (Figure 3.1).

3.1 FATE OF CARBON IN TREES

Primary forests have a large carbon pool and a low sequestration rate. A plantation has a smaller carbon pool and a high sequestration rate. The important fact is that primary forests store a great quantity of carbon, so destroying them for other uses will release large quantities of carbon dioxide into the atmosphere. Comparing the alternatives of (A) maintaining primary forest and (B) conversion of a primary forest to a plantation), B would release more carbon into the atmosphere, making a greater contribution to climate change.

PROBLEM 3.1.1
Carbon sequestration

What are some attributes of one named efficient carbon sequestering tree?

SOLUTION 3.1.1

For effective carbon sequestration, the selected tree species exhibits the following characteristics: high growth rate, a variety of reproductive methods, invasiveness, and rapid acquisition of biomass. As a hardy perennial tropical evergreen, *Spathodea campanulata* (Flame of the Forest or African tulip tree) of the family Bignoniaceae grows aggressively in a wide range of tropical sub-climates and thrives in many soil types (Polunin 1987). Achieving a growth rate of up to 160 cm per year in wet conditions (ISSG 2006), *S. campanulata* is among the fastest growing tropical evergreens, being often the first large tree to colonize wastelands (Chin 1989). It is rapidly invasive (ISSG 2010), especially in wet tropical fertile conditions as it propagates readily from seeds (wind-dispersed), suckers (from rapidly extending roots), or cuttings, the species having been nominated as among the 100 "World's Worst Invaders" (Pagad 2010) and being an invasive species in many tropical areas.

DOI: 10.1201/9781003341826-5

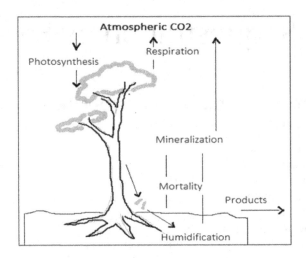

FIGURE 3.1   Inbound and outbound carbon in trees.

PROBLEM 3.1.2
Burning forest trees for fuel

Two cooking fuels used often by backpackers are kerosene and dried wood. Which of these fuels emit more CO_2 into the atmosphere?

SOLUTION 3.1.2

At 15.5°C (59.9°F or 288.65 K) and standard atmospheric pressure, kerosene weighs 800 grams per liter.

$$2\ C_{12}H_{26}(l) + 37\ O_2(g) \rightarrow 24\ CO_2(g) + 26\ H_2O(g);\ \Delta H° = -7513\ kJ\quad (3.1.1)$$

$$[288 + 52] + 1184 \rightarrow 288 + 768 + 46H_2O$$

$$\text{Moles of kerosene in 1 liter} = 800\ /\ 170 = 4.7$$

Kerosene has an 84.7 percent carbon content, or 720.7 grammes of carbon per liter.

$$\text{1 mole of kerosene combusted produces 12 moles of } CO_2$$

$$\text{4.7 moles of kerosene combusted produce } 12 \times 4.7 \text{ moles of } CO_2 = 2481.6\ g\ CO_2$$

Ans. = 2481.6 g of CO_2 forms when 1 liter (800 g) of kerosene combusts.

For 800 g of oven-dry wood, which, chemically is largely carbohydrate, the combustion is:

$$C_6H_{12}O_6 + 6O_2 \rightarrow 6CO_2 + 6H_2O + energy$$

$$[72 + 12 + 96] + 192 \rightarrow [72 + 192] + 6H_2O + energy$$

$$\text{192 g of carbohydrate (wood) produces 264 g of } CO_2$$

$$1 \text{ gram of wood produces } 264 / 192 \text{ g of } CO_2$$

$$800 \text{ g of wood produces } 264 / 192 \times 800 \text{ g of } CO_2$$

Ans. = 1,100 g of CO_2 comes from the combustion of 800 g of wood.

NB: Dry wood therefore emits only 44% of the CO_2 emitted by kerosene. However, 340 g of kerosene release 7513 kJ. Assuming a regular hardwood used typically for firewood has a net calorific value of 4.06 kWh / kg at air-dry (i.e., a moisture level of 20%), and:

$$0.000278 \text{ Kilowatt hours} = 1 \text{ Kilojoule}$$

$$1 \text{ kWh} = 1 / 0.000278 \text{ kJ} = 3597.1 \text{ kJ}$$

$$4.06 \text{ kWh} \times 3597.1 \text{ kJ} \times 0.8 = 11,683.38 \text{ kJ}$$

$$800 \text{ g of hardwood releases } 11,683.38 \text{ kJ}$$

As such wood exceeds the thermal capacity of kerosene by 35.6%, while kerosene emits more than twice the amount of CO_2, wood is the better fuel for restricting global warming.

3.2 WHAT IS CARBON SEQUESTRATION?

3.2.1 KEEPING CARBON IN CHECK

In aggrading natural ecosystems and well-managed agroecosystems, a portion of the added carbon is converted into soil organic matter. Conversely, intensive tillage, erosion, inadequate nutrient management, and conversion of forests and grasslands to agriculture can lead to carbon losses of up to 50% from that present in native ecosystems. Carbon dioxide is the most produced greenhouse gas. Carbon sequestration is the process of capturing and storing atmospheric carbon dioxide. It is one method of reducing the amount of carbon dioxide in the atmosphere to reduce global climate change.

3.2.2 CALCULATING THE AMOUNT OF CO_2 SEQUESTERED IN A TREE PER YEAR

The rate of carbon sequestration, which is greatest in the younger stages of tree growth, between 20 and 50 years depends on the growth characteristics of the tree species, the conditions for growth where the tree is planted, and the density of the tree's wood.

A yearly rate can lead to a rough estimate of the amount of CO_2 sequestered in a tree depending on the tree's age according to the following process:

1. Determine the total weight of the tree.
2. Determine the dry weight of the tree.
3. Determine the weight of carbon in the tree.
4. Determine the weight of carbon dioxide sequestered in the tree.
5. Determine the weight of CO_2 sequestered in the tree per year.

3.2.3 Determining the Total (green) Weight of the Tree

Based on tree species in the Southeast United States, for example, the algorithm to calculate the weight of a tree is as follows (Clark et al. 1986):

W = aboveground weight of the tree in pounds

D = diameter of the trunk in inches

H = height of the tree in feet

For trees with D < 11 in., $W = 0.25D^2 H$.

For trees with D > = 11 in., $W = 0.155D^2 H$.

As the root system weighs about 20% as much as the aboveground weight of the tree, we can determine the total green weight of the tree by multiplying the aboveground weight of the tree by 120%.

3.2.4 Determine the Dry Weight of the Tree

This is based on a table by Birdsey (1992) where average weights for one cord of wood for different temperate tree species exist. As the combined average of all the tree species is 72.5% dry matter and 27.5% moisture, the dry weight of the tree can be calculated by multiplying the weight of the tree by 72.5%.

3.2.5 Determine the Weight of Carbon in the Tree

As the average carbon content is generally 50% of the tree's total volume (UNM 2019), multiply the dry weight of the tree by 50% to find the weight of carbon in the tree.

3.2.6 Determine the Weight of Carbon Dioxide Sequestered in the Tree

CO_2 is composed of one molecule of carbon and two molecules of oxygen.

The atomic weight of carbon is 12.001115.

The atomic weight of oxygen is 15.9994.

The weight of CO_2 is $C + 2 \times O = 43.999915$.

The ratio of CO_2 to C is 43.999915/12.001115 = 3.6663.

Therefore, to determine the weight of carbon dioxide sequestered in the tree, multiply the weight of carbon in the tree by 3.6663.

3.2.7 Determine the Weight of CO_2 Sequestered in the Tree Per Year

Divide the weight of carbon dioxide sequestered in the tree by the age of the tree.

PROBLEM 3.2.1
Average amount sequestered by a tree

As stated above, the higher the growth rate, the faster the carbon sequestration of trees. In just 5 years, a teak (*Tectonia grandis*) tree can grow to about 15 ft tall with a trunk about 8 inches in diameter. How much CO_2 would it have sequestered?

$$W = 0.25D^2H = 0.25(8^2)(15) = 240 \text{ lbs. green weight above ground}$$

240 lbs. $\times$ 120% = 288 lbs. green weight (roots included)

288 lbs. $\times$ 72.5% = 208.8 lbs. dry weight

208.8 lbs. $\times$ 50% = 104.4 lbs. carbon

104.4 lbs $\times$ 3.6663 = 382.8 lbs. CO_2 sequestered

382.8 lbs / 5 years = 76.44 lbs. CO_2 sequestered per year

3.3 CARBON SEQUESTRATION IN RIPARIAN FORESTS: A GLOBAL SYNTHESIS AND META-ANALYSIS

Forests near bodies of water or riparian forests support biodiversity, benefit the local ecosystem in numerous ways (Figure 3.2), and have the potential for fast removal of carbon dioxide from the atmosphere through carbon sequestration. Riparian forests hold, on average, 68–158 Mg C / ha (USGS 2021), with the highest values in relatively warm and wet climates (USGS 2021). Increasing the scale and speed of

FIGURE 3.2 Biological carbon sequestration.

Source: USGS (2021).

riparian forest restoration would benefit the ecosystem through atmospheric carbon dioxide removal.

PROBLEM 3.3.1
Sequestration in riparian forests

How much carbon on average can a small riparian tree sequester over a set period (ignoring density differences among tree species)?

SOLUTION 3.3.1 (AFTER UNM 2019)

Tree details: 6-year-old tree (time period)
    4.5 meters tall / 14.76 feet (height)
    20 cm / 7.8-inch-wide tree trunk (width)
Step 1: Calculate the weight above ground of the tree (lbs.).
    Weight above ground (lbs.) = $0.25\ D^2H$, D = diameter (in)., H = height (ft)
    Thus
    $0.25\ (7.8^2)\ (14.76) = 224.5$ lbs.
Step 2: Calculate the green weight of the tree (weight of the tree while alive).
    The value 1.2 is used to account for the average root system accounting for, on average, 20% of a tree's total weight.
    Green weight (lbs.) = weight above ground (lbs.) × 1.2
    Thus
    224.5 lbs. × 1.2 = 269.4 lbs.
Step 3: Calculate the dry weight of the tree.
    The value 0.725 is used to account for the average 72.5% dry matter and 27.5% moisture.
    Dry weight (lbs.) = green weight × 0.725
    Thus
    269.4 lbs. × 0.725 = 195.3 lbs.
Step 4: Calculate the carbon content weight of the tree.
    The value 0.5 is used to account for carbon content in trees generally being 50% of the dry weight total volume.
    Carbon weight = dry weight × 0.5
    Thus
    172.5 × 0.5 = 86.2 lbs.
Step 5: Calculate the weight of sequestered carbon dioxide of the tree.
    Weight of CO_2 within tree is determined by the average ratio of CO_2 to C, 44:12 = 3.67, and thus 3.67 is used as the multiplicative value.
    Carbon dioxide sequestered = 3.67 × carbon content weight
    Thus
    86.2 lbs × 3.67 = 316 lbs. of carbon dioxide sequestered over a 5-year period.

Conclusion

It can be reasonably estimated that a 6-year-old riparian tree can sequester a total of 316 lbs of carbon dioxide over the course of its life, highlighting the importance of replanting riparian forests to assist with carbon sequestration.

PROBLEM 3.3.2
Effects of climate on riparian sequestration

"Riparian forests hold on average 68–158 Mg C/ha in biomass at maturity, with the highest values in relatively warm and wet climates" (USGS 2021). Using the information provided, calculate the average biomass of a sample area of a riparian forest in a warm, wet climate.

SOLUTION 3.3.2

Time: during August 2020 (wet period)
 Correlation coefficient for vegetation (average weight) (CF) = 1000.5 g

$$\text{Height (H)} = 20 \text{ m} / 400 \text{ m}^2$$
$$\text{Width (W)} = 5 \text{ m} / 25 \text{ m}^2$$
$$\text{Breadth} = 7 \text{ m} / 249 \text{ m}^2$$

Step 1) Measure the height (H) of a sample area
Step 2) Calculate the cover

$$\text{Cover} = W(\text{width}) \times B(\text{breadth})$$
$$\text{Cover} = 25 \text{ m}^2 \times 49 \text{ m}^2 = 1225 \text{ m}^2$$

Step 3) Calculate the biomass by multiplying height × cover × weight

$$\text{Biomass} = H \times C \times CF$$
$$= 400 \text{ m}^2 \times 1225 \text{ m}^2 \times 1000.5 \text{ g}$$
$$\text{Average biomass} = 490,245,000 \text{ g} / \text{m}^2$$

 Therefore, the average biomass of a sample area of a riparian forest is approximately 490,245,000 g/m².

PROBLEM 3.3.3
Carbon sequestration in moist to dry climates

What will the biomass density of mature closed forest growing in moist to dry climates be [Hint: volume over bark (VOB) = measured from stump to top of bole (FAO 1998)]?

SOLUTION 3.3.3 (AFTER FAO 1998)

Forest details:

$$\text{Broadleaf forest with a VOB} \left(\text{volume over bark, m}^3/\text{ha}\right) = 300 \text{ m}^3/\text{ha}$$
$$\text{Weighted average wood density WD} = 0.65 \text{ t}/\text{m}^3$$

Calculations

$$\text{Aboveground biomass density}\,(t\,/\,ha) = VOB \times WD \times BEF \qquad (3.3.1)$$

where
  WD = volume-weighted average wood density
  BEF = biomass expansion factor

Step 1

$$\text{Calculate biomass of VOB} = 300\ m^3\,/\,ha \times 0.65\ t\,/\,m^3 = 194\ t\,/\,ha$$

Step 2
  Calculate the BEF
  NB: BEF includes the canopy biomass extrapolated above the bole (the section between the ground and the first branch). According to FAO (1998), the BEF method used here rests on existing volume per ha data and suits secondary to mature closed forests only, growing in moist to dry climates.
  BV > 190 t/ha, and therefore BEF = 1.74
  BV = biomass of inventoried volume in t/ha, calculated as the product of VOB/ha (m^3/ha) and wood density (t/m^3)
Step 3
  Calculate aboveground biomass density (Equation 3.3.1) = 1.74 × 300 × 0.65 = 338 t/ha

PROBLEM 3.3.4
Role of forest fires

Fires and nitrogen are important to the germination of ectomycorrhizal-associated trees. However, fire-suppressive management practices have given arbuscular mycorrhizal-associated trees an advantage. What is the product of the combustion of nitrogen?

SOLUTION 3.3.4

$$N_2 + O_2 \rightarrow 2NO \qquad (3.3.2)$$

PROBLEM 3.3.5
Micorrhyzae and nitrogen fixation

Mycorrhizae are fungi that symbiotically exist in soil with roots. Two main types, ectomycorrhizae and endomycorrhizae, respectively, are externally associated with the plant root and form their associations within the cells of the host. How do ectomycorrhizae fix nitrogen so that plants can use it?

SOLUTION 3.3.5

$$N_2 + 8H^+ + 8e^- \rightarrow 2NH_3 + H_2$$

Biological nitrogen fixation (BNF) occurs when an enzyme called nitrogenase (Mikola 1986) converts atmospheric nitrogen to ammonia. The reaction for BNF results in N_2 gaining electrons (see above equation) and is thus termed a reduction reaction.

3.4 CARBON FIXATION STRATEGIES: NATURAL AND ARTIFICIAL CO_2 FIXATION

The creation of an efficient catalyst is a crucial first step in the construction of a synthetic photosynthesis system that uses sunlight to transform carbon dioxide into organic molecules.

3.4.1 BASIS

Plants are known as purifiers of the air as they help to absorb and remove greenhouse gases from the atmosphere, where through photosynthesis plants are known to give off oxygen and absorb CO_2 emissions.

PROBLEM 3.4.1
Role of photosynthesis

    a. Write the balanced equation for photosynthesis.
    b. How many grams of carbon dioxide will yield 36 g of glucose?

SOLUTION 3.4.1

Calculations

$$\text{Step 1: } CO_2 + H_2O + \text{Light energy} \rightarrow C_6H_{12}O_6 + O_2 \qquad (3.4.1)$$

Balanced, the above equation will be:

$$\text{Step 2: } 6\,CO_2 + H_2O + \text{Light energy} \rightarrow C_6H_{12}O_6 + O_2 \qquad (3.4.2)$$

Ratio of carbon (CO_2) to glucose ($C_6H_{12}O_6$) = 1:6

$$\text{Mass of } C_6H_{12}O_6 = 36 \text{ g}$$
$$\text{Grams of } CO_2 = ?$$
$$\text{Mass of 1 mole of glucose} = 180 \text{ g/mole}$$
$$\text{Molar mass of } CO_2 = 44.1 \text{ g/mole}$$

Step 1
Calculate the number of moles in $C_6H_{12}O_6$

$$\text{Number of moles} = \text{Mass of substance / mass of 1 mole}$$
$$\text{Number of moles is } C_6H_{12}O_6 = 36 \text{ g} / 180 \text{ g}$$
$$\text{Number of moles} = 0.2 \text{ moles of glucose}$$

Step 2

To get the numbers of moles of CO_2, multiply the moles of glucose by 6 to equate the ratio

$$\text{Moles of } CO_2 = 0.2 \text{ moles} \times 6$$
$$\text{Moles of } CO_2 = 1.2 \text{ moles}$$

Step 3

To find grams of carbon dioxide, multiply the moles of CO_2 by CO_2's molar mass.

$$\text{Grams of } CO_2 = 1.2 \text{ moles} \times 44.1 \text{ g} / \text{mole}$$
$$\text{Grams of } CO_2 = 52.92 \text{ g}$$

Therefore, 52.92 g of carbon dioxide will yield 36 g of glucose.

The calculations given show that if a plant requires 36 g (0.079 lbs) of glucose for development, it would need 52.92 g (0.17 lbs) of CO_2 to yield such an amount, thus removing 52.92 g (0.17 lbs) of CO_2 from the air.

Application:

- Using figures found in the calculation to explore synthetic photosynthesis.
- According to the National Resource Defense Council (NRDC 2015), the average human exhales 2.3 lb of CO_2 per day.
- By converting grams to lbs, it is shown that 0.17 lbs CO_2 will yield 0.079 lbs of glucose.

This calculation therefore not only shows how the photosynthetic requirements of a plant can aid in the removal of CO_2 from the air but also suggests a starting point for its sequestration, for example, using a future technique to collect and fix CO_2 gas artificially from sunlight.

PROBLEM 3.4.2
Forest attributes required for CO_2 sequestration

How can the forest sector mitigate climate change?

SOLUTION 3.4.2

- Creating plantations
- Developing agroforestry
- Reducing deforestation
- Reducing emissions caused by forest activities
- Producing biomaterials and bioenergy

3.5   CHANGES BETWEEN FOREST AND CROPLAND

PROBLEM 3.5.1
Evaluating carbon stocks in the forest

How is the carbon in forests estimated?

SOLUTION 3.5.1

Estimation of acres of US forest preserved from conversion to cropland comes from the Inventory of U.S. Greenhouse Gas Emissions and Sinks: 1990–2019, based on the methodology for estimating carbon stored in US forests (EPA 2020). On that basis, the carbon stock density of US forests in 2019 was 200 metric tons of carbon per hectare (or 81 metric tons of carbon per acre) (EPA 2021) and composed of the five carbon pools: aboveground biomass (54 metric tons C/hectare), belowground biomass (11 metric tons C/hectare), dead wood (10 metric tons C/hectare), litter (13 metric tons C/hectare), and soil carbon, which includes mineral soils (90 metric tons C/hectare) and organic soils (21 metric tons C/hectare).

The Inventory of U.S. Greenhouse Gas Emissions and Sinks:

According to the USDA Natural Resource Inventory and the DayCent biogeochemical model (EPA 2021), estimates of soil carbon for 1990–2019 based on the IPCC guidelines indicate that the average carbon stock change is equal to the carbon stock change due to removal of biomass from the outgoing land use (i.e., forestland) plus the carbon stocks from one year of growth in the incoming land use (i.e., cropland), or the carbon in biomass immediately after the conversion minus the carbon in biomass prior to the conversion plus the carbon stocks from one year of growth in the incoming land use (i.e., cropland) (IPCC 2006). They averaged carbon stock in annual cropland biomass (in the USA) after one year at 5 metric tons C per hectare and the carbon content of dry aboveground biomass as 45 percent (IPCC 2006). Therefore, the carbon stock in cropland after one year of growth is estimated to be 2.25 metric tons C per hectare (or 0.91 metric tons C per acre).

The average carbon content for soils of several types was estimated at (for high-activity clay, low-activity clay, sandy soils, and histosols for all climate regions in the United States) 40.43 metric tons C/hectare (EPA 2021). As carbon stock change is time dependent, the default time for transition to equilibrium after a change in land use is 20 years (after the last change) for soils in cropland systems (IPCC 2003. As it is assumed that the change in equilibrium soil carbon occurs incrementally, during those 20 years will be annualized over to represent the annual change in mineral and organic soils.

Organic soils also emit CO_2 when drained. Emissions from drained organic soils in forestland and drained organic soils in cropland vary based on the drainage depth and climate (IPCC 2006). The Inventory of U.S. Greenhouse Gas Emissions and Sinks: 1990–2019 estimates emissions from drained organic soils using US-specific emission factors for cropland and IPCC (2003) default emission factors for forestland (EPA 2021).

The annual change (i.e., after the changed land use) in emissions from 1 ha of drained organic soil can be calculated as the difference between the emission factors for forest soils and cropland soils. The emission factors for drained organic soil on temperate forestland are 2.60 metric tons C/hectare/year and 0.31 metric tons C/hectare/year (EPA 2021, IPCC 2014), and the average emission factor for drained organic soil on cropland for all climate regions is 13.17 metric tons C/hectare/year (EPA 2021).

The IPCC (2006) guidelines indicate that there are insufficient data to provide a default approach or parameters to estimate carbon stock change from dead organic matter pools or belowground carbon stocks in perennial cropland (IPCC 2006).

REFERENCES

Aguinaldo GT (2016) How to calculate air emissions (for AQMD AERs and other purposes). Envera Consulting. https://enveraconsulting.com/how-to-calculate-air-emissions/

Chin WY (1989) A Guide to the Wayside Trees of Singapore. Singapore Science Centre, pp. 140–141. Singapore. https://eresources.nlb.gov.sg/printheritage/detail/d28b09c8-ab46-46ab-8636-effc229d86b7.aspx Accessed December 6 2019.

EPA (2020) AVERT, U.S. national weighted average CO2 marginal emission rate, year 2019 data. U.S. Environmental Protection Agency, Washington, DC.

IPCC (2003) Intergovernmental Panel on Climate ChangeGood Practice Guidance for Land Use, Land-Use Change and Forestry. https://www.ipcc.ch/site/assets/uploads/2018/03/GPG_LULUCF_FULLEN.pdf. Accessed January 14, 2023

ISSG (2006) International Global Invasive Species Data Base. Issg.org/database/species/ecology.asp?si=75. Accessed December 18, 2017

Mikola PU (1986) Relationship between nitrogen fixation and mycorrhiza. *Mircen Journal* 2, 275–282. https://doi.org/10.1007/BF00933493

NRDC (2015) Do we exhale carbon? May 19, 2015 Brian Palmer https://www.nrdc.org/stories/do-we-exhale-carbon

Pagad S (2010) Invasive species specialist group (ISSG) of the IUCN Species Survival Commission. https://www.iucn.org/our-union/commissions/group/iucn-ssc-invasive-species-specialist-group. Accessed June 11, 2014.

Polunin I (1987) Plants and Flowers of Singapore. Times Editions, p. 124.

USGS (2021) Riparian Forests. Forest and Rangeland Ecosystem Science Center November 13, 2017. Riparian Forests | U.S. Geological Survey (usgs.gov). Riparian Forests | U.S. Geological Survey (usgs.gov). Accessed March 6, 2023.

4 CO_2 Sequestration by Water

MAIN POINTS: DETERMINANTS OF SEQUESTRATION IN WATER

- Salinity
- Depth
- pH
- Temperature
- Density
- Total dissolved solids (TDS)

4.1 DEEP OCEAN SINK

Preuss (2001) concluded that the oceans are one of the most promising places to sequester carbon, as they currently take up a third of the carbon emitted by human activity, roughly two billion metric tons each year, and that the amount of carbon that would double the load in the atmosphere would increase the concentration in the deep ocean by only two percent.

At the Department of Energy's Center for Research on Ocean Carbon Sequestration (DOCS), direct injection, which would pump liquefied carbon dioxide a thousand meters deep or deeper, either directly from shore stations or from tankers trailing long pipes at sea, was being studied (Preuss 2001).

Manahan (2005) also, supports the approach of injecting carbon dioxide from combustion into deep ocean regions, which he states to be the only viable alternative for sequestering carbon dioxide, though it remains an unproven technology on a large scale. However, discrepancies lie in the envisioned oceanic pH that would result. Thus, he predicts a slight increase in ocean water pH which, though only of the order of a tenth of a pH unit, could adversely affect many of the organisms that live in the ocean, while Preuss (2001) cites a slight increase in acidity which could be buffered if the carbonic acid species produced by the sequestered CO_2 dissolves calcium carbonate sediments. As temperatures rise, carbon dioxide leaks out of the ocean like a glass of aerated tonic water going flat on a warm day. Though carbonate gets used up and must be restocked by upwelling of deeper waters, which are rich in carbonate dissolved from limestone and other rocks, the warmer the surface water becomes, the harder it is for winds to mix the surface layers with the deeper layers (NASA 2009). Normally, in the center of the ocean, wind-driven currents bring cool waters and fresh carbonate to the surface, with the relatively low-CO_2 water taking up yet more carbon to be in equilibrium with the atmosphere, while the old water carries the carbon it has captured into the depths of the ocean (NASA 2009). However, as the ocean settles into layers or stratifies, the surface water saturates with CO_2 (NASA 2009).

DOI: 10.1201/9781003341826-6

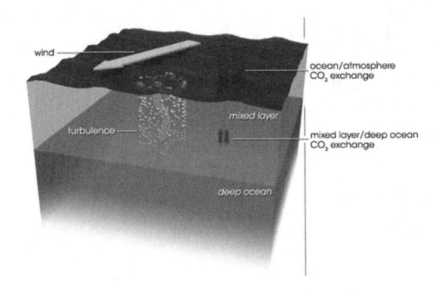

FIGURE 4.1   Turnover of CO_2 from deep oceans.

Saturation with CO_2 acidifies the ocean. A declining pH increases rates of dissolution of calcium carbonate and diminishes the amount of free carbonate ions in the
water (as $CO3^{2-}$ cannot exist in water without linking up with a cation, whereupon
most salts, such as sodium carbonate, are practically insoluble in water). The relative proportions of the different carbon compounds in seawater are dependent on pH
(Figures 4.1 and 4.2). As pH declines, the amount of available carbonate declines,
so there is less available for organisms to incorporate into their shells and skeletons
because acidification both dissolves existing shells and makes it harder for shell formation to occur (NASA 2009).

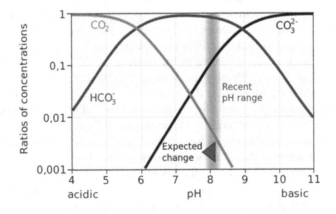

FIGURE 4.2   Turnover of CO_2 from deep oceans due to acidity concentration.

PROBLEM 4.1.1
CO$_2$ from atmosphere to ocean

What is the evidence for the decreasing ability of the oceans to sequester CO$_2$ from the atmosphere?

SOLUTION 4.1.1

According to Figure 4.3, which shows decreasing pH during the last 30 years, solubilization of carbon dioxide has increased ocean acidity, and based on Henry's law, the ratio of a gas increases in the air above the water as the concentration in the water increases.

PROBLEM 4.1.2

The ocean takes up carbon dioxide through photosynthesis by plant-like organisms (phytoplankton), as well as by simple chemistry: carbon dioxide dissolves in water. It reacts with seawater, creating carbonic acid (Riebeek 2008), which releases hydrogen ions, which then combine with carbonate in seawater to form bicarbonate (Figure 4.2), a form of carbon that doesn't escape the ocean easily (Riebeek 2008). What would be the effect of an increase in water temperature on the rate of CO$_2$ sequestration by water bodies?

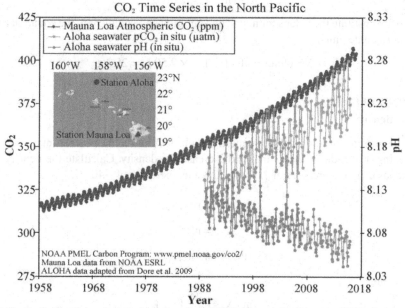

Data: Mauna Loa (ftp://aftp.cmdl.noaa.gov/products/trends/co2/co2_mm_mlo.txt) ALOHA (http://hahana.soest.hawaii.edu/hot/products/HOT_surface_CO2.txt)
Ref: J.E. Dore et al, 2009. Physical and biogeochemical modulation of ocean acidification in the central North Pacific. Proc Natl Acad Sci USA 106:12235-12240.

FIGURE 4.3 Time series of carbon dioxide and ocean pH at Mauna Loa, Hawaii.

Source: www.pmel.noaa.gov

SOLUTION 4.1.2

CO_2 (aq) $\rightarrow$ CO_2 (g) equilibrium constant K_H (Henry's constant) increases with T. Hence, the sequestration rate would be lowered, and more would be left in the atmosphere where it can contribute to additional warming (positive feedback).

PROBLEM 4.1.3
Calculating the mass concentration of a solution

The concentration of one or more dissolved solutes affect the sequestration of CO_2 in water. Determine the mass of solid sodium hydroxide, NaOH(s), needed to make 2.50 L of an aqueous solution of sodium hydroxide, NaOH(aq), with a mass concentration of 10.00 g L^{-1}.

SOLUTION 4.1.3

Mass of solute = mass concentration $\times$ volume of solution

$$M(NaOH(s)) = m/v \times v(NaOH(aq))$$

Mass concentration of the solution (include units) m/v = 10.00 g L^{-1}

Step 4. Identify the volume of the solution (include units)

$$V(solution) = 2.50 \text{ L}$$

Step 5. Substitute the values for mass concentration and volume into the equation and solve (include units)

$$M(solute) = 10.00 \text{ g L}^{-1} \times 2.50 \text{ L} = 25.0 \text{ g}$$

PROBLEM 4.1.4
CO_2 density

Efficient transportation before sequestration of CO_2 in the deep ocean may require freezing it to reduce the volume and increase the density. Calculate the density of gaseous CO_2 at a pressure of 1.42 atm and a temperature of $-40°$ C.

SOLUTION 4.1.4

$$PV = nRT \tag{4.1.1}$$

$$PV = (m / M) RT$$

Rearrange:

$$m / V = P / RT(M)$$

$$m / V = d = P / RT(M)$$

$$d = P / RT(M)$$

$$M = \text{grams per mole} = 44 \text{ g for } CO_2$$

$d = 1.42$ atm $/ (0.082$ L $\cdot$ atm $\cdot$ mole$^{-1} \times 233$ K^{-1}) 44 g $\cdot$ mole$^{-1} = 0.001689$ or

Density of gaseous CO$_2$ at a pressure of 1.42 atm and a temperature of $-40°C$

$$= 1.6 \times 10^{-3} \text{ g} \cdot \text{L}^{-1}.$$

PROBLEM 4.1.5
CO$_2$ partial pressure

Burnt fossil fuels vary greatly in the proportion of CO$_2$ released, and which the ocean absorbs. A solid hydrocarbon is burned in air in a closed container, producing a mixture of gases having a total pressure of 4.27 atm. Analysis of the mixture shows it to contain 0.360 g of water vapor, 0.835 g of carbon dioxide, 0.355 g of oxygen, 3.970 g of nitrogen, and no other gases. What are the mole fraction and partial pressure of carbon dioxide in this mixture?

SOLUTION 4.1.5

n total $= nH_2O \quad\quad +nO_2 \quad\quad\quad +nCO_2 \quad\quad +nN_2$
$\quad\quad = 0.36 / 18 \quad +0.835 / 44 \quad +0.355 / 32 \quad +3.87 / 28$

$= 0.188$ mole
X_{CO2} (mole fraction of CO$_2$) in mixture $= 0.019/0.188 = 0.101$ mole
$P_{CO2} = X_{CO2} \times (P$ total$) = 0.101$ mole $\times 4.27$ atm
$P_{CO2} = 0.431$ atm

REFERENCES

Manahan S (2005) Environmental Chemistry. Taylor and Francis, Boca Raton, Fla. Pp 355–363.
NASA Earth Observatory (2009) The Ocean's Carbon Balance https://earthobservatory.nasa.gov/features/OceanCarbon
Preuss P (2001) Climate change scenarios compel studies of ocean carbon storage. https://www2.lbl.gov/Science-Articles/Archive/sea-carb-bish.html
Riebeek H (2008) The Ocean's Carbon Balance. NASA Earth Observatory. https://earthobservatory.nasa.gov/features/OceanCarbon. Accessed December 16, 2022.

5 Sequestration by Soils

MAIN POINTS – DETERMINANTS OF SEQUESTRATION IN SOILS:

- Age of soil
- Depth
- Water content
- Mineral richness, e.g., iron, aluminum
- pH
- Bulk density

5.1 DEEP SOIL ADVANTAGES

According to Kramer and Chadwick (2018), a Washington State University researcher found that one-fourth of the carbon held by soil is bound to minerals as far as 6 ft below the surface. Professor Murphy of the University of Western Australia corroborates this finding by stating that there is potential for soils between 10- and 30cm below ground to store more carbon under certain conditions and that his findings suggest that these layers could theoretically store twice as much carbon as they do currently (phys.org 2018). Another researcher placed that figure at up to five times as much. The need to investigate management practices that can increase the amount of carbon stored in soil at this depth is paramount, he asserts. The discovery opens a new possibility for dealing with the element as it continues to warm Earth's atmosphere. They used new data from soils around the world to describe how water dissolves organic carbon and takes it deep into the soil, where it is physically and chemically bound to minerals. The likely explanation is energy-starved microbes; they may be the force that causes huge amounts of carbon to be stored in deep soils, according to a Dartmouth College study (phys.org 2018). The lack of this energy source at depth denies microbes the energy they need to efficiently decompose dead roots. The research finds that less food energy at depth makes it more difficult to decompose deposits of organic carbon, creating an underground storehouse for the climate-destabilizing chemical element. Therefore, to drive back global warming to below 2°C, carbon storage in soils plays an important part, since soils can store a notable share of human CO_2 emissions (Doetterl et al. 2018).

5.2 SEQUESTRATION IN IMPOVERISHED SOILS

But for sequestration to occur, the soil organic matter must first be present, and some mined out and/or highly drained tropical soils with low water retention do not support the requisite vegetation. The SOC delivery system depends on water to leach carbon from roots, fallen leaves, and other organic matter near the surface and carry it deep into the soil, where it will attach to iron- and aluminum-rich minerals prone to form strong bonds (phys.org (2018) while also tying up phosphorus (Figures 5.1a). Hence, drought-tolerant economically viable tree species such as the *Cajanus cajan*

DOI: 10.1201/9781003341826-7

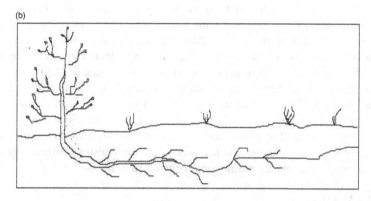

FIGURE 5.1 (a) A strong bond occurs between PO_4^{3-} and Al^{3+}, thereby preventing the availability of P in many laterites and bauxite soils. (b) The sideways-growing root system of the legume *C. cajan*, which releases phosphorus bound to iron and aluminum oxides, helps in soil fertility.

(pigeon or gungo pea) can be important facilitators of soil carbon sequestering in such water-deficient phosphorus-unavailable soils (Figure 5.1b).

This is because *C. cajan* dissolves and releases tightly bound phosphorus from iron and aluminum phosphates in soil by exuding piscidic acid, also known as (2R,3S)-piscidate, belonging to the class of organic compounds known as phenyl-propanoic acids, in the root zone. Phenylpropanoic acids contain a benzene ring conjugated to a propanoic acid (Figure 5.2).

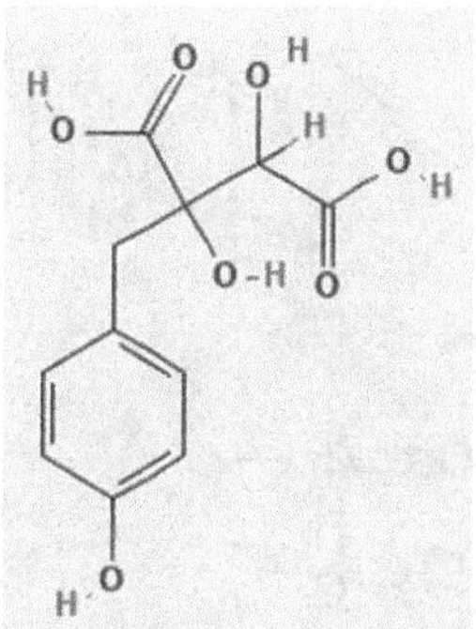

FIGURE 5.2 Chemical structure of piscidic acid.

Source: nih.gov
https://pubchem.ncbi.nlm.nih.gov/compound/Piscidic-acid#section=2D-Structure

5.3   MEASURING OXIDIZABLE SOIL ORGANIC MATTER

Soil organic carbon is a measurable component of soil organic matter. The Walkley-Black chromic acid wet oxidation method (Walkley & Black 1934) determines oxidizable soil organic carbon in the soil as organic matter can be oxidized, i.e., solubilized, by 1 N potassium dichromate ($K_2Cr_2O_7$) solution. Ten (10) milligrams of potassium dichromate is mixed with 1 g of soil. The term chromic acid usually applies to a mixture made by adding concentrated sulfuric acid to a dichromate, which may contain a variety of compounds, including solid chromium trioxide (Figure 5.3). The heat generated when two volumes of H_2SO_4 are mixed with one volume of the dichromate assists the reaction. Some dichromate (i.e., chromic acid) will remain unreacted, and it is titrated with ferrous ammonium sulfate $(NH_4)_2$ Fe $(SO_4)_2 \cdot (6H_2O)$, an amphoteric substance, i.e., one which can dissolve in water to develop an acid or a basic service. The titer relates inversely to the amount of C present in the soil sample. In other words, the more ferrous sulfate required to achieve titration, the smaller the amount of carbon present.

PROBLEM 5.3.1
Determining soil carbon percentage

Calculate the average mean percentage of soil organic carbon for three replicates 1, 2, and 3 using ferrous ammonium sulfate (FAS) and 0.5 g of air-dried soil.

SOLUTION 5.3.1

Walkley-Black formula (WBF) for soil organic carbon (%) (Griffin et al. 2013) is

$$WBF = 10(B-T) / B \times 0.003 \times 100 / S \qquad (5.3.1)$$

where
   B = volume (mL) of ferrous ammonium sulfate required to titrate the blank
   T = volume (mL) of ferrous ammonium sulfate required to titrate the soil samples
   S = weight of the soil (g)

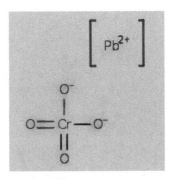

FIGURE 5.3   Chromic acid H_2CrO_4: the H^+ ions (not shown) are attached to the O^- sites. Chromic acid forms when an acid reacts with a chromic salt.

Source: https://www.nlm.nih.gov/pubs/techbull/ja22/ja22_pubchem.html

Titration of Blank

R	Burette Initial Reading (mL)	Burette Final Reading (mL)	Volume of Fas (mL)
R1	0	10.3	10.3
R2	10.3	20.8	10.5
R3	20.8	30.9	10.1

Volume of ferrous ammonium sulphate = Burette final reading – Burette initial reading
Mean value (B) of the volume of FAS = (10.3 + 10.5 + 10.1)/3 = 10.3 mL

Titration of Soil Samples

R	Burette Initial Reading (mL) (X)	Burette Final Reading (mL) (Y)	Volume of Fas (mL) (T = Y – X)
R1	20.6	26.7	6.1
R2	26.7	32.4	5.7
R3	32.4	38.4	6

Volume of Fas (T) = Burette final reading (Y) – Burette initial reading (X)

Calculation of Soil Organic Carbon %

R	Burette Initial Reading (mL)	Burette Final Reading (mL)	Volume of Fas (mL) (T = Y – X)	Blank Reading (mL) (B)	B – T (mL)
R1	20.6	26.7	6.1	10.3	4.2
R2	26.7	32.4	5.7	10.3	4.6
R3	32.4	38.4	6	10.3	4.3

Organic Carbon % = 10(B – T)/B × 0.003 × 100/S

R	10(B – T)/B	0.003	100/S	10(B – T)/B × 0.003 × 100/S
R1	4.08	0.003	200	2.45
R2	4.47	0.003	200	2.68
R3	4.17	0.003	200	2.50

100/S (weight of the soil) = 100/0.5 g = 200
R1: 10(B – T)/B = 10(4.2)/10.3 = 4.08
R2: 10(B – T)/B = 10(4.6)/10.3 = 4.47
R3: 10(B – T)/B = 10(4.3)/10.3 = 4.17
Soil organic carbon percentage is
R1: 10(B – T)/B × 0.003 × 100/S
    10(4.2)/10.3 × 0.003 × 100/0.5 = 2.45
R2: 10(B – T)/B × 0.003 × 100/S
    10(4.6)/10.3 × 0.003 × 100/0.5 = 2.68
R3: 10(B – T)/B × 0.003 × 100/S
    10(4.3)/10.3 × 0.003 × 100/0.5 = 2.50
Average mean value of soil organic carbon percentage in the samples analyzed is

$$(2.45 + 2.68 + 2.50) / 3 = 2.54\%$$

Therefore, the average mean percentage of soil organic carbon for three replicates in the 0.5 g sample of air-dried soil is 2.54%.

5.4   CALCULATING CARBON MASS IN A SINGLE STRATUM

Organic carbon that is found in soil is crucial for the functionality of an ecosystem, while the increase of organic carbon storage contributes to lessening climate change.

PROBLEM 5.4.1
Formulas for total carbon

How is total carbon for a stratum calculated?

SOLUTION 5.4.1

The formula (Civil Engineering Information 2022 in use is:

$$C_T = C_F \times D \times V$$

where
   C_T = total carbon for the layer in metric tons
   C_F = the fraction of carbon (percentage carbon divided by 100)
   D = density
   V = volume of the soil layer in cubic meters

PROBLEM 5.4.2
Carbon in a single stratum

Randomly selected plots in a stratum of a 12-ha field average 1.8% carbon, bulk density = 1.20, and the depth of sampling is 15 cm. What is the total carbon for the layer?

SOLUTION 5.4.2

First, find the volume by multiplying square meters in a hectare by the depth. Since there are 10,000 m^2 in a hectare, the volume will be $12 \times 10,000 = 120,000$. Then 120,000 $m^2 \times 0.15$ m = 18,000 m^3.

$$C_T = C_F \times D \times V$$

$0.18 \times 1.2 \times 18,000 = 388.8$ tons of carbon for the layer.
This equates to 32.4 tons of carbon per hectare.

5.5   PRECISE MEASURING OF SOC

The precision and usefulness of soil organic carbon (SOC) monitoring can be significantly increased by combining field-delineated spatial units, remote and close range monitoring of land cover, crops, and surface residues, soil sampling, and simulation modeling (Griffin et al. 2013).

PROBLEM 5.5.1
Estimating bulk density to calculate SOC stocks

How much SOC is in a paddock to a depth of 10 cm? How much soil organic carbon is in a paddock given that dry soil weight was 500 g, tube diameter 7 cm (3.5 cm radius), depth of soil is 10 cm, and 1.5% carbon value?

SOLUTION 5.5.1

Step 1: estimate bulk density

A bulk density (BD) estimate is required to calculate soil organic carbon stocks in tons of carbon per hectare.

A tube 7 cm in diameter (3.5 cm radius) driven into a depth of 10 cm has a volume of $\pi r^2 h$:

$$3.14 \times (3.5 \times 3.5) \times 10 = 385 \text{ cm}^3$$

Step 2: amount of organic carbon

$$SOC = Carbon \times BD \times Depth$$

The amount of organic carbon to 10 cm depth in soil with a carbon value of 1.5% and bulk density of 1.3 g/cm^3 is:

10,000 m^2 in 1 ha × 0.1 m soil depth × 1.3 g/cm^3 bulk density × 1300 t/ha soil

15 × 1,300,000 = 1.9 tons per carbon per hectare

1.5 × 1.3 × 10 = 19.5 t C/ha

Therefore, there will be 19.5 t C/ha of stock carbon in the paddock.

5.6   KEEPING CARBON IN CHECK

In aggrading natural ecosystems and well-managed agroecosystems, a portion of the added carbon is converted into soil organic matter. Conversely, intensive tillage, erosion, inadequate nutrient management, and conversion of forests and grasslands to agriculture can lead to carbon losses of up to 50% from that present in native ecosystems (Griffin et al. 2013).

5.7   HOW TO STORE MORE CARBON IN SOIL DURING CLIMATE CHANGE

Microbes that are found within moist soils process organic inputs and store the organic carbon from the soil more efficiently than in drier soils. Therefore, knowing the soil's water content prior to watering can reduce the wastage of excess irrigation in an era of climate change. Water content in soil is often defined as the percentage of water that is present within a soil mass on a weight/weight basis according to the following procedure (Mahajan 2022).

W = the water content in the soil

W1 = weight of the container

W2 = weight of the container + wet or moist soil

W3 = weight of the container + dry soil

1. Dry the soil in a heat-resistant container, preferably a beaker, in an oven at a temperature range of 105°C to 110°C for 24 hours.

2. Find the percentage of the water content in the specific soil sample using
   the following equations:

$$W2 - W3 = \text{water weight in g}$$
$$W2 - W1 = \text{wet soil weight in g}$$
$$W = (W2 - W3) \times 100 / W2 - W1$$
$$= \text{water percentage in soil}$$

PROBLEM 5.7.1
Soil water content

A moist soil sample has a weight of 135 g. The container that the wet sample was
placed in weighed 15 g. When placed in the oven and dried, the soil sample had a
weight of 95 g. What is the percentage of soil moisture by weight?

W1 = 15 g
W2 = 150 g
W3 = 110 g
W = (W2 – W3) × 100/W2 – W1
W = (150 g – 110 g) × 100/150 g – 15 g
W = 40 g × 100/135 g
W = 4000 g/135 g
W = 29.63%

Therefore, the water weight by percentage that was found is 29.63% of the wet
soil sample.

Adjusting for gravel content:

If there is gravel in the soil sample, laboratory results will need to be adjusted as this
is taken out before carbon analyses.

So, if SOC was 1.5% but the soil had 25% gravel (by volume), then 1.5 – (1.5 ×
0.25) = 1.1%.

5.8   WATER CONTENT AND ADVANTAGES OF VOLUME UNITS

Based on location, environments and soils can become wetter or dryer.

Soil moisture calculations in percentages on a mass (weight) basis have been commonly used, but this does not give a true picture of soil-moisture relationships because
two soils may have similar moisture content on a weight basis but not on a volume basis.

Calculations on a volume basis are more meaningful and practical because water
is retained in the soil within a given volume and plant roots also absorb moisture
from a volume of soil.

PROBLEM 5.8.1
Water content by volume

Low water contents in soils restrict carbon sequestration. For Problem 5.8.1, determine the following:

a. The total depth of water presently contained in the top 35 cm
b. The depth to which 25 mm (1 inch) of irrigation would wet this uniform soil
c. The available water the soil contains in the top 35 cm when the soil is at field capacity

Given information:

Present water content – 18%
   Water content at field capacity – 24%
   Percentage of water at permanent wilting point – 10%
   Bulk density (BD) of 0–35 cm depth surface soil – 1.3 g/cc

SOLUTION 5.8.1

a. Depth of water (d_w) = bulk density soil / bulk density water × depth of soil (ds)

$$= 1.3 / 1 \times (18 / 100) \times 35 \text{ cm}$$
$$= 8.2 \text{ cm}$$

So, the depth of total water present in the top 35 cm of soil is 8.2 cm.

b. To calculate the depth of wetting (or saturation, d_s) by a 25 mm (1.1 inch) irrigation, the following equation (after Colorado State University Extension 2022) is substituted:

$$25 \text{ mm} = 1.3 \times \left[(23 - 18) / 100 \right] \times ds$$
$$25 \text{ mm} = 1.3 \times (5 / 100) \times ds$$
$$ds = 25 \text{ mm} / 0.065$$
$$= 384.6 \text{ mm}$$

So, the soil will be wetted to a depth of 384.6 mm.

c. To calculate the total possible plant-available water in the top 35 cm, when the soil is wetted, subtract permanent wilting percentage (PW) from field capacity (because the water in soil below wilting point is tightly bound and cannot be extracted by roots). Field capacity (FC) occurs when all the air spaces are filled. So, the plant available water (d_w) is:

$$dw = BD \times (FC - PW) \times ds$$
$$= 1.3 \times \left[(24 - 10) / 100 \right] \times 35 \text{ cm}$$
$$= 1.3 \times 0.14 \times 35 \text{ cm}$$
$$= 6.37 \text{ cm}$$

So, the top 30 cm of soil contains 5.46 cm of available water.

PRACTICE PROBLEM 5.8.2
Role of grassed surfaces

How do increased grassed surfaces decrease the accumulation of carbon monoxide and carbon dioxide from heavy traffic in the air of large cities?

SOLUTION 5.8.2

The oxygen released by photosynthesis of grassed surfaces reacts with carbon monoxide from vehicles to eliminate carbon monoxide by breakdown into carbon dioxide as follows:

$$2CO + O_2 \rightarrow 2CO_2, \text{ etc.}$$

PROBLEM 5.8.3
Temperature and CO_2 losses from soils

To reduce losses of CO_2 already sequestered in soil, what are the ideal temperatures for tilling the soil?

SOLUTION 5.8.3

It has been shown that the peaks in all emissions from soil including those of CO_2 occurred after tillage on hot days, so it is recommended to avoid tilling on hot days. By comparing carbon storage in places with different average temperatures, researchers found that for every 10°C of increase in temperature, average carbon storage (across all soils) fell by more than 25% (Hartley et al. 2021).

REFERENCES

Civil Engineering Information (2022) How to Calculate the Water Content of Soil (7 methods) https://www.civilengineeringinformation.com/2022/06/How-to-Calculatethe-Water-Content-of-Soil-7-Methods.html

Colorado State University Extension (2022) Measurement of soil moisture. https://extension.colostate.edu/disaster-web-sites/measurement-of-soil-moisture/

Doetterl S, Berhe AA, Arnold C et al. (2018) Links among warming, carbon and microbial dynamics mediated by soil mineral weathering. Nature Geosci 11, 589–593. https://doi.org/10.1038/s41561-018-0168-7

Hoyle FC, Murphy DV & Griffin E (2013) 'Soil organic carbon', in Report card on sustainable natural resource use in Agriculture, Department of Agriculture and Food, Western Australia. In N Schoknecht, D Bicknell, J Ruprecht, F Smith, & A Massenbauer (Eds.), *Report card on sustainable natural resource use in agriculture* (pp. 78–91). Department of Agriculture and Food, Western Australia.

Hartley IP, Hill TC, Chadburn SE et al. (2021) Temperature effects on carbon storage are controlled by soil stabilisation capacities. Nat Commun 12, 6713. https://doi.org/10.1038/s41467-021-27101-1

Kramer MG, Chadwick OA (2018) Climate-driven thresholds in reactive mineral retention of soil carbon at the global scale. Nature Clim Change 8, 1104–1108. https://doi.org/10.1038/s41558-018-0341-4

Phys.org (2018) New research unravels the mysteries of deep soil carbon. https://phys.org/news/2018-11-feet-approach-global.html

Pontikes Yiannis, Dimitri Boufounos, Dimitris Fafoutis (2006) Environmental aspects on the use of Bayer's process bauxite residue in the production of ceramics. Adv Sci Technol 45, 2176–2181. doi: 10.4028/www.scientific.net/AST.45.2176

Walkley AJ, Black IA (1934) Estimation of soil organic carbon by the chromic acid titration method. Soil Sci 37, 29–38.

6 Carbon Footprint

6.1 CALCULATION OF CARBON FOOTPRINT

A material with a high thermal mass such as concrete, is often colder when in the ground than when not in the ground. When the surroundings are hot, concrete floors can absorb heat and feel cooler than surrounding objects. If used to refrigerate foods, they could substantially reduce global CO_2. The author of this book, whose refrigerator has been continuously unplugged for the last 10 years, currently uses the concrete floor of a small vacant room to cool and preserve all fresh fruits and vegetables for periods >7 days without substantial deterioration. His monthly home electricity cost averages US$5.00. Before that, the cost was five to six times that amount.

The three main economic sectors that use fossil fuels are electricity/heat, transportation, and industry. The first two sectors, electricity/heat and transportation, produced nearly two-thirds of global carbon dioxide emissions in 2010 (IEA 2012). The calculation of carbon footprint is the standard way of measuring and reporting the environmental impact that a building, land, a structure, or a retail location has on the environment. The carbon footprint of a company, a building, land, a structure, or a retail location is measured in tons of CO_2 per year. The carbon footprint calculator function bases calculations on the following three environmental scopes:

Scope 1 (direct) energy emissions
Scope 2 (indirect) energy emissions
Scope 3 (indirect) travel emissions for vehicles that are not owned by the company including the following components:
- On-site energy production and other industrial activities
- Area of facilities and percent of occupancy
- Facility energy use such as electricity, gas, coal, oil, and solar
- Corporate travel such as plane, rail, vehicle
- Corporate waste

PROBLEM 6.1.1
CO_2 from fuels

How much carbon dioxide is produced when different fuels are burned?

SOLUTION 6.1.1

Different fuels emit different amounts of CO_2 in relation to the energy they produce when burned. To analyze emissions across fuels, the carbon content of the fuel must be multiplied by the ratio of the molecular weight of carbon dioxide (44) to the molecular weight of carbon (12), i.e., 44/12 = 3.7.

 DOI: 10.1201/9781003341826-8

Carbon dioxide emissions from burning a fuel can be described as

$$QCO_2 = Cf / Hf \times MCO_2 / Mc$$

where

QCO_2 = specific CO_2 emissions (kgCO$_2$/kWh)
 Cf = specific carbon content in the fuel (kgc/kgfuel)
 Hf = specific energy content in the fuel (kWh/kgfuel)
 Mc = molecular weight content carbon (kg/kmole carbon)
MCO_2 = molecular weight carbon dioxide (kg/kmole CO_2)

PROBLEM 6.1.2
Carbon footprints of homes

Adapting one's lifestyle and thinking more about one's actions can have a big impact on global warming and reducing or limiting greenhouse gas emissions. How does one calculate the carbon footprint of a home?

SOLUTION 6.1.2

First, identify the sources that produce CO_2 emissions and put in place the measures needed to reduce them. Okin (2017) calculated energy consumption using the following equation:

$$E^a Total = E^a Individual^N \qquad (6.1.1)$$

where

$E^a Total$ = the total energy consumed annually
$E^a Individual^N$ = the per capita annual consumption
 N = the number of individuals

$E^a Individual$ was calculated separately for humans, dogs, and cats.

Even though more than 60% of US households have pets (APPA 2016), these consumers of agricultural products are rarely included in calculations of the environmental impact of dietary choices (Okin 2017). For example, he calculated that meat-eating by dogs and cats creates the equivalent of about 64 million tons of carbon dioxide a year, which has about the same climate impact as a year's worth of driving from 13.6 million cars. Giving details for the USA, he found that dogs and cats consume about 19% ± 2% of the amount of dietary energy that humans do (203 ± 15 PJ yr^{-1} vs. 1051 ± 9 PJ yr^{-1}) and 33% ± 9% of the animal-derived energy (67 ± 17 PJ yr^{-1} vs. 206 ± 2 PJ yr–1-1), while producing about 30% ± 13%, by mass, as much feces as Americans (5.1 ± Tg yr^{-1}-vs. 17.2 Tg yr^{-1}), and through their diet, constitute about 25–30% of the harmful environmental impacts of food production from animals in terms of the use of land, water, fossil fuel, phosphate, and biocides. Offering a four-pronged potential solution, he mentioned the following:

• Reduce the rate of cat and dog ownership
• Replace it with ownership of pets that offer similar health and emotional benefits

- Simultaneous, industry-wide efforts to reduce overfeeding
- Find sources of protein other than animal-based foods

6.2 DECENTRALIZED POWER DELIVERY

The cost of solar installations is declining as utility bills currently increase. Many specialists in power consumption costs feel that it is now appropriate to switch from a centralized to a distributed style of power delivery (NREL 2022). Every ton of solid reactant carbon releases 3.7 tons of carbon dioxide gas products into the atmosphere (NREL 2022).

PROBLEM 6.2.1
Decentralization of power provision

What is the quantity of CO_2 emissions given off by electricity usage of my community in a recent time period?

SOLUTION 6.2.1

The cost of solar installations is declining because there is an 85% cost decline in the module price (Ramdas et al. 2021).

6.3 IMPACT OF SMARTPHONES CHARGED

The 24-h energy consumed by a common smartphone battery, including the amount of energy needed to charge a fully depleted smartphone battery and maintaining that full charge throughout the day, is 14.46 Wh (DOE 2020). According to Ferreira et al. (2011), the average time required to recharge a smartphone battery completely is 2 h. The power consumed when the phone is fully charged and the charger is still plugged in, also known as maintenance mode power, is 0.13 W (DOE 2020).

PROBLEM 6.3.1
Determination of carbon emissions per smartphone

How were the carbon emissions per smartphone determined?

SOLUTION 6.3.1

By multiplying the energy use per smartphone charged by the national weighted average carbon dioxide marginal emission rate for delivered electricity, carbon dioxide emissions per smartphone charged were determined. The national weighted average carbon dioxide marginal emission rate for delivered electricity in 2019 was 1562.4 lbs CO_2 per megawatt-hour, which accounts for losses during transmission and distribution (EPA 2020).

PROBLEM 6.3.2
Smartphone CO_2 emissions from the USA

How many tons of CO_2 do charging smartphones place in the atmosphere every year?

SOLUTION 6.3.2

To obtain the amount of energy consumed in charging the smartphone, subtract as follows:

24-hour energy consumed – Energy consumed in "maintenance mode"
[14.46 Wh – (22 h × 0.13 W)] × 1 kWh / 1000 Wh = 0.012 kWh / phone charged
0.012 kWh / charge × 1562.4 pounds CO_2 / MWh delivered electricity × 1 MWh / 1000 kWh × 1 metric ton / 2204.6 lbs = 8.22 × 10^{-6} metric tons CO_2 / smartphone charged
10^6 phones produce 8.22 × 10^{-6} × 10^6 metric tons CO_2 = 8.22 metric tons.
10^8 phones = 8.22 × 10^2 metric tons CO_2, and 10^8 × 2 phones = 1.62 × 10^3 tons of CO_2 (based on 200 million users, i.e., 2 × 10^8 smartphones).

That is, 1620 tons of CO_2 were placed in the atmosphere by charged smartphones in the USA.

PRACTICE PROBLEM 6.3.1
Smartphone CO_2 emissions from countries

Using information from the above problem, how much CO_2 does the world's population put into the atmosphere per year?

SOLUTION 6.3.1

Extrapolating from figures for the USA of > 1600 tons of CO_2 p.a. produced from charging smartphones, a world population of 8,000,000,000 in 2022, and a report from the World Bank saying that 75% of the world's population own smartphones (RFE/RL (2012):

CO_2 from 10^8 × 2 phone charges = 1,600 tons of CO_2 per day

CO_2 from 6 × 10^9 phone charges = 1,600 × (6 × 10^9 / 10^8 × 2) tons of CO_2 per day

= 48,000 tons of CO_2

48,000 tons of CO_2 × 365 day · yr^{-1} = 1.752 × 10^7 tons

That is, all the world's smartphones currently put at least 17,520,000 tons of CO_2 in the atmosphere every year.

PROBLEM 6.3.3
Potential impact of people-powered smartphones

How can humans significantly reduce the impact of charging smartphones on global warming?

SOLUTION 6.3.3

Just as mechanically powered (hand-squeezed) flashlights are considered a green technology, the potential invention of smartphones charged in this manner could significantly reduce the amount of CO_2 placed in the atmosphere. Moreover, smartphone batteries contain heavy metals and toxic chemicals which end up in the environment. If mechanical hand-squeezed charging were to be fully practiced, the smartphone batteries would last longer (actually, there would be no need for such batteries), the physical exercise occurring would improve the health of millions (e.g., by charging the phones while walking), thereby reducing tons of CO_2 from fossil fuels currently used in hospitals.

6.4   REMEDIATION BY CHANGING THE DURATION OF CHARGE

PROBLEM 6.4.1
Avoiding long charging

Power consumed while charging a phone is 0.13 Watts (DOE 2020). How much power is avoided by changing the charge mode?

SOLUTION 6.4.1

According to Crookes (2021), An Android phone turned on, has apps and other processes running in the background and increasing the time taken to charge the device. He recommends powering down the phone before charging it.

If it is desired to keep the handset on during charging, activating airplane mode disables wireless technologies such as cellular data networks, Wi-Fi, and Bluetooth — each of which draw on the Android phone's battery even while the device charges. To cut charging time by as much as 25%, Crookes (2021) suggests:

- first swipe down from the top of the screen to open the Quick Settings panel.
- Tap the Airplane mode icon (or, if it is hidden, swipe left to view more icons).
- Alternatively, launch the Settings app and navigate to Network and Internet.
- Toggle the Airplane mode switch

6.5   BIOCHAR REMEDIATION OF CARBON FOOTPRINTS

PROBLEM 6.5.1
Re-purposing food wastes

How can food waste decrease carbon footprints in winter months?

SOLUTION 6.5.1

Role of home fireplaces

Considering that millions of tons of carbon dioxide move into the atmosphere annually by the decomposition of food wastes, pyrolysis of food waste in home fireplaces is a favorable alternative. With the oxygen supply opening restricted to save fuel, many wood-burning fireplaces continue to burn throughout the night during the

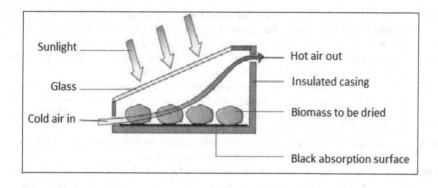

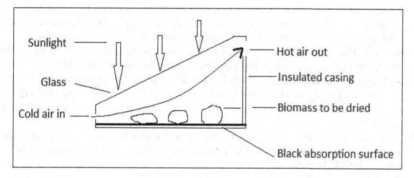

FIGURE 6.1 Solar powered passive biomass waste dryer.

cold season. This potential in large urban centers outside the tropics to hold back CO_2 from the atmosphere can be effective. Compared to microbial decomposition of organic matter, food waste pre-dried with solar power (Figure 6.1) and loaded into the anoxic conditions of a low-oxygen home fireplace lessens the CO_2 released while producing volatile gases for fuel.

Drying biomass in a parked car

To reduce the water content of the biomass prior to the biochar process, drying can be easily achieved without cost if the biomass to be bio-charred is first placed on the dashboard of a car parked in sunlight. Othoman et al. (2018) showed that the temperature inside the car exceeds the ambient temperature by 60% and that the difference in temperature inside and outside the car could reach approximately 25°C. A window periodically slightly opened at the top will release the extracted moisture from the high vapor pressure developed inside the vehicle.

REFERENCES

APPA (American Pet Products Association) (2016) 2015–2016 APPA National Pet Owners Survey. American Pet Products Association, Greenwich, CT.

DOE (2020) Compliance Certification Database. Energy Efficiency and Renewable Energy Appliance and Equipment Standards Program EXIT EPA WEBSITE.https://www.regulations.doe.gov/news/new-compliance-certification-database

EPA (2020) AVERT, U.S. national weighted average CO_2 marginal emission rate, year 2019 data. U.S. Environmental Protection Agency, Washington, DC.

Ferreira D, Dey AK, Kostakos V (2011) Understanding human-smartphone concerns: A study of battery life. Pervasive Computing, pp. 19–33. Lecture Notes in Computer Science, vol 6696. Springer, Berlin, Heidelberg. https://doi.org/10.1007/978-3-642-21726-5_2 doi: 10.1007/978-3-642-21726-5_2.

IEA (International Energy Agency) (2012) CO_2 Emissions from Fuel Combustion 2012. Paris: Organisation for Economic Co-operation and Development.

NREL (2022) From the Bottom Up: Designing a Decentralized Power System Cross-Discipline Team Envisions the Grid of the Future. https://www.nrel.gov/news/features/2019/from-the-bottom-up-designing-a-decentralized-power-system.html

Okin Gregory S (2017) Environmental impacts of food consumption by dogs and cats. PloS One (August 2). https://doi.org/10.1371/journal.pone.0181301

Othoman MA, Fouzi MSM, Nordin A (2018) Assessment of thermal comfort in a car cabin under sun radiation exposure. Eng Appl New Mater Technol 85, 469–479.

Ramdas A, Feldman D, Ranasamy V, Fu R., Desai J, Margolis R (2021) U.S. Solar Photovoltaic System and Energy Storage Cost Benchmark: Q1 2020. National Renewable Energy Laboratory. Retrieved December 19, 2022, from https://www.nrel.gov/docs/fy21osti/77324.pdf

RFE/RL (2012) Report says 75 percent of world's population have mobile phones https://www.rferl.org/a/report-says-75-percent-of-worlds-population-have-mobile-phones/24648234.html

7 The Chemistry of CO_2 Fixation

MAIN POINTS:

- Formic acid is a CO_2 converter
- CO_2 sequestration in the sea
- Potential for synthetic CO_2 fixation

7.1 FORMIC ACID AS A CO_2 CONVERTER

Formate dehydrogenases are a set of enzymes that catalyzes the oxidation of formate to carbon dioxide, a family of enzymes that has attracted attention as inspiration or guidance on methods for carbon dioxide fixation, relevant to global warming (Amao 2018). There are two pathways:

NAD (nicotinamide adenine dinucleotide)-dependent reaction

$$\text{Formate} + \text{NAD}^+ \rightleftharpoons CO_2 + \text{NADH} + \text{H}^+ \tag{7.1.1}$$

Cytochrome-dependent reaction

$$\text{Formate} + 2 \text{ ferricytochrome b1} \rightleftharpoons CO_2 + 2 \text{ ferrocytochrome b1} + 2\text{H}^+ \tag{7.1.2}$$

NADPH represents nicotinamide adenine dinucleotide phosphate, while nicotinamide adenine dinucleotide (NAD) is an oxidizing coenzyme central to metabolism and found in all living cells (Nelson & Cox 2005). Formate dehydrogenase (FDH) is a catalyst that accelerates the reaction of converting carbon dioxide (CO_2), which accounts for the largest share of radiative forcing (Figure 7.1), into formic acid (Sato & Amao 2020). However, until recently, the details of how this happened were unclear. They found that in solution, CO_2 is only converted to formic acid when the carbon dioxide ratio is high. Formic acid did not decrease when the carbon dioxide solution was converted to bicarbonate and carbonate. They believe that this discovery will guide the development and design of catalysts that will help bring about an artificial photosynthesis system that efficiently converts carbon dioxide into organic molecules (Sato & Amao 2020).

PROBLEM 7.1.1
Mole ratios of CO_2 fixation

a. What is the mole ratio between carbon dioxide and formic acid ($HCOO^-$)?
b. How many moles of formic acid will result if 50 g of carbon dioxide is used?

DOI: 10.1201/9781003341826-9

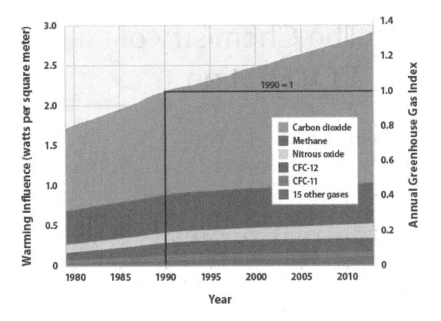

FIGURE 7.1 Of the 20 greenhouse gases, carbon dioxide accounts for by far the largest share of radiative forcing since 1990, and its contribution continues to grow at a steady rate.

Source: US EPA 2021.

SOLUTION 7.1.1

a. $CO_2 + NADH \rightarrow HCOO^- + NAD^+$ (7.1.3)

This is a reduction reaction (i.e., gaining electrons).

The carbon dioxide ratio to the formic acid is 1:1

1 mole of carbon dioxide will yield 1 mole of formic acid ($HCOO^-$).

b. Known – a mass of 50 g

Unknown – number of moles in carbon dioxide gas

Solution is the number of moles of carbon dioxide; it can be calculated using the following formula:

$$\rightarrow \text{Mass / molar mass}$$

$$\text{Molar mass of carbon dioxide} = 44 \text{ g / mole}$$

Therefore, keeping the values in the given formula to find the number of moles in each mass of carbon dioxide:

Number of moles = 50/44

Number of moles = 1.14

Number of moles in a given mass of carbon dioxide is 1.14.

The molar ratio is 1:1.

Therefore, 1.14 moles of CO_2 (i.e., 50 g) will yield 1×1.14 moles of formic acid.

Therefore 50 g of carbon dioxide will yield 1.14 moles of formic acid.

7.2   CO$_2$ SEQUESTRATION IN THE SEA

PROBLEM 7.2.1
CO$_2$ sequestration in seawater

Baking soda is a pure form of bicarbonate. How many moles of sodium bicarbonate (NaHCO$_3$) are in 227 g of NaHCO$_3$?

SOLUTION 7.2.1

M = 227 g of NaHCO$_3$
  n =?
  M = Mass/molar mass
  Finding molar mass of NaHCO$_3$:
  M of Na + MH + MC + 3MO:

$$22.99\,g\,/\,mole + 1.01\,g\,/\,mole + 12.01\,g\,/\,mole + 3(16\,g\,/\,mole)$$
$$= 84.01\,g\,/\,mole$$
$$n = Mass\,/\,molar\;mass$$
$$= 227\,g\,/\,84.01\,g\,/\,mole$$
$$= 2.70\;mole$$

7.3   CO$_2$ POTENTIAL FOR SYNTHETIC FIXATION

CO$_2$ has a reduction target set on a global scale. It is possible to repurpose CO$_2$ emissions, but to utilize it as a raw material and transform it into effective substances is another important issue. The creation of an efficient catalyst is a crucial first step in the construction of a synthetic photosynthesis system that uses sunlight to transform carbon dioxide into organic molecules.

PROBLEM 7.3.1
Photosynthetic equation as the basis for synthetic fixation

Write the balanced equation for photosynthesis. How many grams of carbon dioxide will yield 36 g of glucose?

SOLUTION 7.3.1

Calculations

Step 1:

$$CO_2 + H_2O + Light\;energy \rightarrow C_6H_{12}O_6 + O_2$$

The reactant side only has one carbon atom, whereas the product side contains six. Put 6 in front of the carbon dioxide molecule as you attempt to balance the equation above, and the resulting equation will be:

Step 2:

$$6CO_2 + H_2O + \text{Light energy} \rightarrow C_6H_{12}O_6 + O_2$$

Six carbon atoms are present on each side currently. Hydrogen and oxygen must be balanced as the last two atoms. Only 2 hydrogen atoms are present on the reactant side, but 12 are present on the product side.

Place six hydrogen atoms in front of the water molecule on the reactant side to balance the quantity of hydrogen atoms. The partially balanced photosynthetic formula is represented by the following after this stage:
Step 3:

$$6CO_2 + 6H_2O + \text{Light energy} \rightarrow C_6H_{12}O_6 + O_2$$

At this point, the ratios of carbon and hydrogen atoms on either side of the photosynthesis equation are equal. The quantity of oxygen atoms must therefore be balanced as the last stage.

Calculate the amount of oxygen atoms present on the reactant side with care. For example, 12 oxygen atoms from carbon dioxide (6 CO_2) and 6 from water (6 H_2O) combine to create a total of 18 oxygen atoms. Eight atoms total – six from the glucose molecule ($C_6H_{12}O_6$) and two from the oxygen molecule (O_2) – are present on the product side.

Put six in front of the oxygen molecule to make up for the missing atoms on the product side:
Step 4:

$$6CO_2 + 6H_2O + \text{Light energy} \rightarrow C_6H_{12}O_6 + 6O_2$$

The above represents the balanced equation for photosynthesis, where the organic component glucose is created. It demonstrates how, in the presence of light energy, six molecules each of carbon dioxide and water combine to create one molecule of glucose and six molecules of oxygen.

The grams of CO_2 yielding 36 g of glucose:

Ratio of carbon (CO_2) to glucose ($C_6H_{12}O_6$) = 1:6

$$\text{Mass of } C_6H_{12}O_6 = 36\,g$$
$$\text{Grams of } CO_2 = ?$$
$$\text{Mass of 1 mole of glucose} = 180\,g\,/\,\text{mole}$$
$$\text{Molar mass of } CO_2 = 44.1\,g\,/\,\text{mole}$$

Step 1
Calculate the number of moles in $C_6H_{12}O_6$

$$\text{Number of moles} = \text{Mass of substance} / \text{Mass of 1 mole}$$
$$\text{Number of moles is } C_6H_{12}O_6 = 36\,g\,/\,180\,g$$
$$\text{Number of moles} = 0.2 \text{ moles of glucose}$$

Step 2
To get the number of moles of CO$_2$, multiply the moles of glucose by six to equate the ratio

$$\text{Moles of } CO_2 = 0.2 \text{ moles} \times 6$$
$$\text{Moles of } CO_2 = 1.2 \text{ moles}$$

Step 3
To find grams of carbon dioxide, just multiply the moles of CO$_2$ by CO$_2$'s molar mass.

$$\text{Grams of } CO_2 = 1.2 \text{ mole} \times 44.1 g / \text{mole}$$
$$\text{Grams of } CO_2 = 52.92 g$$

Therefore, 52.92 g of carbon dioxide will yield 36 g of glucose.

The calculations given show that if a plant requires 36 g (0.079 lbs) of glucose for development, it would need 52.92 g (0.17 lbs) of CO$_2$ to yield such amounts, thus removing 52.92 g (0.17 lbs) of CO$_2$ from the air.

Application

According to the National Resource Defense Council (NRDC 2015), the average human exhales 2.3 lbs of CO$_2$ per day. Converting grams to lbs, it is shown that 0.17 lbs CO$_2$ will yield 0.079 lbs of glucose. This calculation, therefore, shows us how the photosynthetic requirements of a plant can aid indirectly in the removal of CO$_2$ from the air.

7.4 FORMATE DEHYDROGENASE, CARBONATE, AND BICARBONATE

Formate dehydrogenase reduces carbon dioxide to formic acid only when the carbon dioxide ratio is large (Amao 2018). When the carbon dioxide in the solution was changed to bicarbonate and carbonate, it was not reduced to formic acid. Formic acid did not decrease when the carbon dioxide solution was converted to bicarbonate and carbonate.

Therefore, to convert CO$_2$ most efficiently to formic acid, the formate hydrogenases must be applied before large dissolutions of carbonate-bearing materials ensue.

PROBLEM 7.4.1
Calculating dissolved CO$_2$

Suppose that the current average atmospheric concentration of CO$_2$ is 387 ppm, that is, 387×10^{-6} atm (although it is higher at present). Henry's law constant for CO$_2$ is $k_H = 0.142 \times 10^{-4}$ atm. What is the molar concentration $[CO_2]$?

What is the concentration of dissolved carbon dioxide in an aqueous solution?

SOLUTION 7.4.1

The greater the partial pressure of the gas, the greater its solubility in the liquid.

Henry's law is a gas law that states that the amount of dissolved gas in a liquid is directly proportional to its partial pressure above the liquid.

Concentration of carbon dioxide in all surface water is calculated thus:

- Atm = standard atmosphere pressure
- ppm = parts per million
- M = molar
- P denotes the partial pressure of the gas in the atmosphere above the liquid.
- 'k_H' is Henry's law constant of the gas.
- C is concentration of the gas in the liquid.

$$C = k_H P_{gas}$$
$$[CO_2] = ?$$
$$= 3.87 \times 10^{-4} \text{ atm} \left(0.142 \times 10^{-4} \text{ atm}\right)$$
$$= 0.549 \times 10^{-8} \text{ M}$$

Concentration of CO_2 in the water $= 5.49 \times 10^{-9}$ M

Conclusion

There is 5.49×10^{-9} M of carbon dioxide dissolved in water to form bicarbonate and carbonate. When the carbon dioxide in the solution was changed to bicarbonate and carbonate, it was not reduced to formic acid because water lowered the ratio of carbon dioxide. A change in the ratio will affect the pH of the fluid.

REFERENCES

Amao Y (2018) Formate dehydrogenase for CO_2 utilization and its application. J CO_2 Util 26, 623–641. doi: 10.1016/j.jcou.2018.06.022

Hester KC et al. 2008. Unanticipated consequences of ocean acidification: A noisier ocean at lower pH. Geophys Res Lett 35, L19601.

Nelson D L, Cox M M (2005) Principles of Biochemistry (4th ed.). W. H. Freeman, New York.

NRDC (2015) Do we exhale carbon? www.nrdc.org/stories/do-we-exhale-carbon

Sato R, Amao Y (2020) Can formate dehydrogenase from Candida boidinii catalytically reduce carbon dioxide, bicarbonate, or carbonate to formate? New J Chem. doi: 10.1039/D0NJ01183E

US EPA (2021) Climate Change Indicators: Climate Forcing. https://www.epa.gov/climate-indicators/climate-change-indicators-climate-forcing

8 CO$_2$ Capture

MAIN DEVICES: CARBON CAPTURE

Five basic systems for capturing CO_2 from fossil fuels and/or biomass (Allam et al. 2018):

- Capture from industrial process streams
- Post-combustion capture
- Oxy-fuel combustion capture
- Pre-combustion capture
- Removal by natural phenomena

8.1 CARBON CAPTURE SYSTEMS

EPA (2022) defines carbon dioxide capture and sequestration as a set of technologies that can potentially greatly reduce CO_2 emissions from new and existing coal- and gas-fired power plants, industrial processes, and other stationary sources of CO_2. They found that a CCS project might capture CO_2 from the stacks of a coal-fired power plant before it enters the atmosphere, transport the CO_2 via pipeline, and inject the CO_2 deep underground at a carefully selected and suitable subsurface geologic formation, such as a nearby abandoned oil field, where it is securely safely.

CO_2 capture systems require an increase in energy use for their operation because the energy and resource requirement for CO_2 capture (which includes the energy needed to compress CO_2 for subsequent transport and storage) is typically much larger than for other emission control systems, as evaluated from the "systems" perspective (Allam et al. 2018). They postulate that, in general, the CCS energy requirement per unit of product can be expressed in terms of the change in net plant efficiency (η) when the reference plant without capture is equipped with a CCS system:

$$\Delta E = \left(\eta_{ref} / \eta_{ccs} \right) - 1 \qquad (8.1.1)$$

where
ΔE = the fractional increase in plant energy input per unit of product
η_{ref} = net efficiency of the reference plant
η_{ccs} = net efficiency of the capture plant

PROBLEM 8.1.1
CCS energy requirement

To produce each bicycle, a factory normally uses 15 kWh of electrical energy. After fitting a CCS, the factory requires 18 kWh to make one bicycle. What is the change in plant efficiency?

DOI: 10.1201/9781003341826-10

SOLUTION 8.1.1

$$\eta_{ref} = 15 \text{ kWh}$$

$$\eta_{ccs} = 18 \text{ kWh}$$

$$\Delta E = ?$$

$$\Delta E = (15 \cancel{\text{kWh}} / 18 \cancel{\text{kWh}}) - 1$$

Change in plant efficiency $= -0.17$

8.2 INCREMENTAL PRODUCT COST

Reacting coal with controlled amounts of oxygen and/or steam at high temperatures produces syngas (hydrogen and carbon monoxide) which also contains impurities such as carbon dioxide, methane, and water vapor. Gasification sequentially involves four stages: drying, low-pyrolysis, combustion, and gasification reactions (CSIRO 2022). Synthetic natural gas is close as possible in composition and properties to natural gas (Cui et al. 2021).

PROBLEM 8.2.1
Gasification process

Sometimes producing a new fuel uses more energy than the original fuel itself. For example, the oxidative manufacture of methanol will be exothermic, so burning the methanol produced (without a catalyst) will yield less heat than would be produced by burning the methane from which it was produced (LibreTexts Chemistry 2022). How can coal gasification reduce global CO_2?

SOLUTION 8.2.1

Having about the same energy density by volume, CO and H_2 both are very clean burning as they only need to take on one oxygen atom, in a simple step, to arrive at the normal end states of combustion, CO_2 and H_2O. Hence the emissions from engines run on producer gas are relatively clean.

Gasification aims to break down carbonaceous material into the simple fuel gases of H_2 and CO— hydrogen and carbon monoxide. In the absence of oxygen, a combusting fuel cannot produce CO_2. Gasification removes energy-rich gases from coal before they are oxidized, mimicking a burning match, where pyrolysis, combustion, and cracking (of some form of solid carbonaceous material), sequentially occur.

Gasification isolates these separate processes, to pipe the resulting gases elsewhere. In the early stages of heating, fragments of the original biomass break off with heat into a combination of solids, liquids, and gases. Charcoal remains. The gases and liquids, referred to as volatiles, are collectively called tars.

Thus, in review, pyrolysis is the application of heat to biomass in the absence of air/oxygen.

Oxygen is unnecessary because pyrolysis merely releases compounds already produced; new compounds are not produced. The volatiles in the biomass are evaporated off as tar gases, and the fixed carbon-to-carbon chains are what remains—otherwise known as charcoal.

Cracking

Tar gases condense into sticky tar that will rapidly foul the valves of an internal combustion engine. Cracking breaks down large complex molecules such as tar into lighter gases by exposure to heat. Complete combustion only occurs when combustible gases thoroughly mix with oxygen.

Reduction

Reduction, the reverse of combustion, removes oxygen atoms off combustion products of hydrocarbon (HC) molecules, to return the molecules to forms that can burn again. Combustion releases heat, producing water vapor and carbon dioxide as waste products. Reduction is the removal of oxygen from these waste products at high temperature to produce combustible gases.

Reduction in a gasifier occurs by passing CO$_2$ or water vapor across a bed of red-hot charcoal (C), where the carbon in the hot charcoal has such a high affinity for oxygen that it strips the oxygen off water vapor and carbon dioxide and redistributes it to as many single bond sites as possible. The oxygen is more attracted to the bond site on the C than to itself, thus no free oxygen can survive in its usual diatomic O$_2$ form.

All available oxygen will bond to available C sites as individual O until all the oxygen is gone. When all the available oxygen is redistributed as single atoms, reduction stops. Through this process, CO$_2$ and H$_2$O are reduced by carbon to produce two CO and H$_2$ and CO respectively. Hence, gasification in this manner minimizes the CO$_2$ released into the atmosphere.

8.3 REMEDIATION OF FOREST FIRES

If fossil fuel combustion rates were reduced by a factor of 25 and deforestation is reduced, carbon dioxide levels could be kept permanently below 500 parts per million (ppm). Carbon dioxide capture, utilization, and sequestration (or storage) (CCUS) is increasingly becoming a core supporting technology component of clean coal projects, such as coal gasification facilities, to reduce the overall environmental impact of coal utilization.

PROBLEM 8.3.1
Rapidly locating forest fires

Forest fires place thousands of tons of carbon dioxide in the sky every year. But early and precise locations of such fires can expedite the extinguishing process. Triangulation is the process of pinpointing the location of an object by taking bearings to it from two or three remote points. How can fire rangers quickly find the exact location of a distant but visible forest fire?

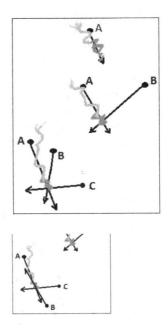

FIGURE 8.1 To locate forest fires, compass bearings are taken from two or three towers in a triangulation process. Ranger A takes a bearing on a distant fire. He notifies Ranger B who then takes a bearing from Ranger B's location. Where the two bearings cross marks the location of the fire. A third bearing may be necessary (from C) if B had been located at too large an angle from A.

SOLUTION 8.3.1

Forest fire lookout towers use triangulation to locate spot fires in the following manner: a ranger at Tower A (Figure 8.1) sees smoke in the distance and takes a bearing to it on his compass. But because the fire could be anywhere along that line, the fire could be close or as far away as the eye can see. So, locating the fire requires more information.

Hence, the ranger would radio Tower B and declare the general direction of the fire. Tower B would then find the fire from his position by taking a bearing (Figure 8.1). The fire occurs where the two bearings cross. By looking at a map, firefighters could be efficiently dispatched to extinguish it.

Though two readings are all that is usually necessary, a fire occurring directly between two towers or off in a direction that causes their bearings to be similar will require a third bearing (C).

8.4 BIOCHAR AND AVOIDANCE OF FOREST FIRES

A fire needs three things: heat, fuel, and oxygen (Frontier Fire 2020). Forest fires occurring in the summer season place thousands of tons of carbon dioxide into the atmosphere. Therefore, the removal of fuel from forest floors using biochar in the winter (less heat) can greatly reduce the initiation of forest fires in the ensuing summer. According to Baranovskiy and Kirienko (2022), drying, pyrolysis, and ignition of forest fuel are of particular importance in predicting forest fires, such that the knowledge of these process patterns allows for both improving the forecast accuracy and developing a set of measures to slow down or stop these processes, which will avoid ignition.

PROBLEM 8.4.1
Ignition of forest fires

What is the role and source of gases in igniting forest fires, and how can biochar preempt such gases?

SOLUTION 8.4.1

A study of forest pyrolysis showed that the gasification process is a kind of "conductor" to the ignition of the material since during decomposition gaseous volatile substances that support combustion are released into the atmosphere (Baranovskiy & Kirienko 2022). Pyrolysis thermochemically decomposes forest biomass into solid, liquid, and gaseous fractions coming from three main polymers, cellulose, hemicellulose, and lignin, in varying proportions, as well as extractive and mineral substances (Sharma et al. 2015).

Release of most gaseous components in typical forest biomass occurs at the lowest temperatures of decomposition. However, for mainly ligneous material the vapor phase dominates at the highest temperatures. Hence, a low smoldering biochar in the cooler seasons in forests can remove a large proportion of biomass gases, thereby reducing the chances of starting forest fires in the ensuing hot seasons.

PROBLEM 8.4.2
Setting biochar temperatures

Heat is a major factor in the spreading of forest fires. How can biochar production reduce heat in forests?

SOLUTION 8.4.2

As stated above, forest biomass contains three main polymers: cellulose, hemicellulose, and lignin. Forest fuel pyrolysis will proceed according to its main components' decomposition:

- Category 1: Cellulose decomposition includes three steps with corresponding temperature ranges: below 300°C, intermediate decomposition products of active cellulose or anhydrous cellulose are mainly formed (Banyasz et al. 2001). At 300–390°C, active decomposition occurs during depolymerization reactions. Also, about 80% of volatile compounds and anhydro-oligosaccharides are formed (Lu et al. 2011).
- Category 2: At 380–800°C, charring occurs, where hemicellulose promotes the binding of cellulose microfibrils to the cell wall, a substance of low thermal

stability and a different structure and composition (Baranovskiy & Kirienko 2022). Hemicellulose decomposition occurs at 200–350°C with main peaks at temperatures of 290°C and 310°C. The degradation of hemicellulose copies that of cellulose, where dehydration occurs at low temperatures (<553 K) with further depolymerization at higher temperatures (Hameed et al. 2019).

- Category 3: The decomposition of major parts of lignin, the complex three-dimensional polymer of phenylpropane units, occurs at temperatures of 200–450°C (Collard & Blin 2014), with dehydration dominating at relatively low temperatures and different lignin monomers occurring at higher temperatures (Baranovskiy & Kirienko 2022). Their report shows monomer decomposition happening at temperatures higher than 700°C, after which the released products enter the vapor phase. As lignin is more stable than the previous components, the proportion of char and aromatics formed is also higher (Demirba (2000).

Therefore, the major components of the forest floor organic materials, when known, can be first segregated and preferentially charred at the required temperatures of decomposition. In other words, biocharring highly lignified biomass in the same heap with lower order polysaccharides requiring lower temperatures unnecessarily increases the heat among forest trees and the likelihood of forest fires. In contrast, a selective, sequential biochar procedure limits the intensification of heat.

8.5   TORREFACTION AND CARBON CAPTURE

Solar dehydration of food waste preceding low-temperature pyrolysis

Biochar addition is heralded to prolong the longevity of inorganic carbon against release into the atmosphere by preserving the recalcitrant carbon fractions. But the high water content typical of plant materials increases the startup energy used in biochar kilns, thereby enlarging the amount of startup fuels used.

PROBLEM 8.5.1
Solar energy in carbon capture

How can solar energy decrease greenhouse gas release during biochar production?

SOLUTION 8.5.1

Solar energy can reduce the carbon footprint of food waste biochar. In a recent unpublished study (Harris 2023), the author incorporated four food wastes to examine the physical effect of water content on the time required for biochar production from food wastes. Fresh 1 kg lots of banana peels, corn cobs, bean pods, and carrot peels were dehydrated on the dashboard (DS) of a closed automobile at midday with an outside temperature of 25–27°C. Samples of various sizes (Figure 8.2) were spread out in a single layer on wire trays which facilitated airflow.

The samples were weighed before and after the dehydration periods of 1, 2, and 3 hours or until there was no further change in weight. For the duration of observations,

FIGURE 8.2 Food waste biochar. Solar powered dehydrated banana peels on the dashboard of an enclosed automobile at an outside temperature of 25°C which produced a dashboard temperature of 40°C. Note the small change in size between dashboard dehydration (center) and that of low-temperature biochar or torrefaction (left image) and the similar blackening of the dehydrated and torrefaction samples.

Photo by Sarah Marshall, South Australia.

a cloudless sky prevailed. The ambient relative humidity oscillated between 60% and 70%. The samples were then subjected to pyrolysis at <250°C in a sliding-door oxygen-free kiln, along with control samples of non-DS treated fresh food waste. This process, carried out under atmospheric pressure and in the absence of oxygen, is torrefaction.

Results

The following results occurred after 2 hours, when temperatures in the car reached 40°C:

- Weight losses observed in DS samples were >90%; i.e., only 0.1 of 1.0 kg remained for every sample (Figure 8.2)
- Pyrolysis times to produce biochar (<250 °C) were 3 minutes for DS samples, and 50 minutes for non-DS samples. The shortened biochar period for DS samples is due to the large removal of water by solar energy preceding torrefaction in the kiln.
- With such a low water content, there would have been no need for applying startup fuel; the DS biomass would be its own startup fuel, thereby avoiding the need for any fossil fuel in the process.

- Given enough fresh DS, its evaporation would have significantly dropped the temperature inside the car, thereby reducing the initial need for air conditioning in the car, thus saving fuel.

PROBLEM 8.5.2
Food waste carbon capture

Decomposing food and crop wastes are one of the serious environmental problems of the world (Gholz 1987; Popp et al. 2014), releasing thousands of tons of CO_2 into the atmosphere yearly. What single procedure can substantially decrease these problems?

SOLUTION 8.5.2

Such materials accumulate in water sources, resulting in problems such as unpleasant odors, eutrophication, high levels of biological oxygen demand, and chemical oxygen demand (Kanu & Achi 2011; Pardeshi & Vaidya 2015). Therefore, the sustainable conversion of waste into useful products is needed. Producing biochar from agricultural waste is a practical solution for handling these massive amounts of garbage and lowering greenhouse gas emissions from burning (Islam et al. 2019). Subjecting such food waste first to solar dehydration would improve carbon capture by decreasing the startup fuel required for torrefaction. Additionally, such solar drying done in stationary automobiles parked while owners are at work should avoid the need for fabricating solar dryers.

PROBLEM 8.5.3
Biochar carbon capture and greenhouse gas emissions

Compared with other methods of incineration, what is the biochar contribution to greenhouse emissions?

SOLUTION 8.5.3

Torrefaction and conventional biochar produce fewer greenhouse gas emissions than other incineration technologies (Bridgewater 2003; Verma et al. 2012).

PROBLEM 8.5.4
Biochar as a fuel

What attributes account for biochar's efficiency as a fuel?

SOLUTION 8.5.4

Low-temperature biochar production (torrefaction) causes combustible aliphatic volatile materials to become trapped in the pores rather than escaping the structure, such that the volatiles in such a low-temperature biochar act as fuel when heated, enabling the combustion of the biochar (or charcoal) (Suman & Gautam 2017). However, all the volatiles are lost when biochar is produced at higher temperatures (>500°C), leaving only a carbon skeleton (Wei et al. 2019). Lam et al. (2018) also used banana peel and orange peel in the slow pyrolysis experiment at a heating rate of 10°C · min to produce biochar as an adsorbent in the treatment of palm oil mill effluent.

PROBLEM 8.5.5
Mechanics of low-temperature pyrolysis

How can low-temperature pyrolysis occur?

SOLUTION 8.5.5

Torrefaction removes moisture and superficial volatiles from biomass, leaving bio-coal (Li et al. 2015). The low-temperature thermal method between 200°C and 400°C under atmospheric pressure and the absence of oxygen produced six biochar types (Wu et al. 2012).

PROBLEM 8.5.6
Activated charcoal advantages

What is activated charcoal (AC) and how is it made?

SOLUTION 8.5.6

ACs have a very porous structure with a large internal surface area ranging from 500 to 3895 m^2 · g^{-1} (Patil & Kulkarni 2012). Activation is a key step to enhance the pore structures. There are two types of activation methods, namely, physical activation and chemical activation. In the activation process, an oxidizing gas will remove more reactive carbon species forming pores and vessels.

PROBLEM 8.5.7
Kinds of pyrolyzed carbon

What differentiates the two kinds of pyrolyzed carbon, and what is the significance of the differences?

SOLUTION 8.5.7

According to Ryan and Yoder (1997), non-graphitizable carbon is a non-graphitic carbon that cannot be transformed into graphitic carbon, whereas graphitizable carbon is a non-graphitic carbon that can be converted into graphitic carbon by the process of heat treatment and is soft, non-porous and has a high density, with microstructures arranged in a preferential direction. Contrastingly, they describe non-graphitizable carbon as hard, porous, of low density, with very disordered, porous microstructures.

Conclusion

Biochar in several forms facilitates not only carbon capture but also several other necessary requirements for human existence and survival (Figure 8.3).

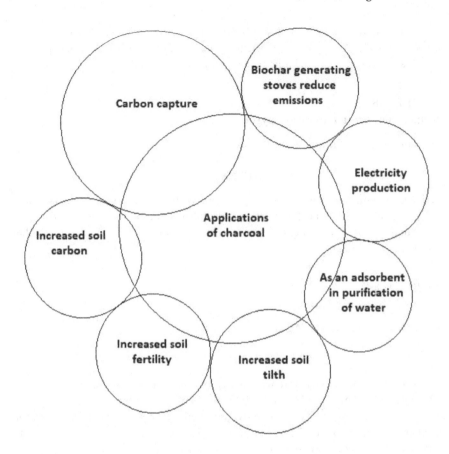

FIGURE 8.3   Carbon capture and other functions of biochar.

REFERENCES

Allam R, Bolland O, Davison J, Feron P, Goede F (2018) IPCC special report on carbon dioxide capture and storage. https://www.ipcc.ch/site/assets/uploads/2018/03/srccs_chapter3-1.pdf

Banyasz JL, Li S, Lyons-Hart J, Shafer KH (2001) Gas evolution and the mechanism of cellulose pyrolysis. Fuel 80, 1757–1763.

Baranovskiy NV, Kirienko V (2022) Forest fuel drying, pyrolysis and ignition processes during forest fire: A review. https://www.researchgate.net/publication/357710183_Forest_Fuel_Drying_Pyrolysis_and_Ignition_Processes_during_Forest_Fire_A_Review

Bridgewater AV (2003) Renewable fuels and chemicals by thermal processing of biomass. Chem Eng J 91(2–3), 87–102. https://doi.org/10.1016/S1385-8947(02)00142-0

Collard FX, Blin JA (2014) Review on pyrolysis of biomass constituents: Mechanisms and composition of the products obtained from the conversion of cellulose, hemicelluloses and lignin. Renew Sustain Energy Rev 38, 594–608.

Comino Theo (2018) Understanding net and gross generation. https://www.agl.com.au/thehub/articles/2018/06/understanding-net-and-gross-generation

CSIRO (2022) Coal gasification. https://www.csiro.au/en/work-with-us/ip-commercialisation/hydrogen-technology-marketplace/Coal-gasification

Cui X, Song G, Yao A, Wang H, Wang L, Xiao, J (2021) Technical and economic assessments of a novel biomass-to-synthetic natural gas (SNG) process integrating O_2-enriched air gasification. Process Saf Environ Prot 156, 417–428.

Demirba A (2000) Mechanisms of liquefaction and pyrolysis reactions of biomass. Energy Convers Manag 41, 633–646.

EPA (2022) Overview of greenhouse gases. https://www.epa.gov/ghgemissions/overview-greenhouse-gases

Frontier Fire Protection (2020) Three things a fire needs. https://www.frontierfireprotection.com/3-things-fire-needs/

Gholz HL (1987) Agroforestry: Realities, Possibilities and Potentials. Martinus Nijhoff Publishers in cooperation with ICRAF, Boston, MA. Springer Dordrecht. Heidelberg.

Hameed S, Sharma A, Pareek V, Wu H, Yu Y (2019) A review on biomass pyrolysis models: Kinetic, network and mechanistic models. Biomass Bioenergy 123, 104–122.

Islam M, Halder M, Siddique MAB et al. (2019) Banana peel biochar as alternative source of potassium for plant productivity and sustainable agriculture. Int J Recycl Org Waste Agric 8(Suppl 1), 407–413. https://doi.org/10.1007/s40093-019-00313-8

Kanu I, Achi OK (2011) Industrial effluents and their impact on water quality of receiving rivers in Nigeria. J Appl Technol Environ Sanit 1, 5–86.

Su Shiung Lam, Rock Keey Liew, Chin Kui Cheng, Nazaitulshila Rasit, Chee Kuan Ooi, Nyuk Ling Ma et al. (2018) Pyrolysis production of fruit peel biochar for potential use in treatment of palm oil mill effluent. J Environ Manage 213, 400–408. https://doi.org/10.1016/j.jenvman.2018.02.092

Li MF, Li X, Bian J, Chen CZ, Yu YT et al. (2015) Effect of temperature and holding time on bamboo torrefaction. Biomass Bioenergy 83, 366–372.

Lu Q, Yang X, Dong C, Zhang Z, Zhang X, Zhu X (2011) Influence of pyrolysis temperature and time on the cellulose fast pyrolysis products: Analytical Py-GC/MS study. J Anal Appl Pyrolysis 92, 430–438.

Pardeshi DS, Vaidya S (2015) Physico-chemical assessment of Waldhuni River Ulhasnagar (Thane, India): A case study. Int J Curr Res Acad: Rev 3, 234–248.

Patil SB, Kulkarni SK (2012) Development of high surface area activated carbon from waste material. Int J Adv Eng Res Stud 1, 109–113.

Popp J, Lakner Z, Harangi-Rákos M, Fári M (2014) The effect of bioenergy expansion: Food, energy, and environment. Renew Sust Energ Rev 32, 559–578. https://doi.org/10.1016/j.rser.2014.01.056.

Ryan MG, Yoder BJ (1997) Hydraulic limits to tree height and tree growth. BioScience 47, 235–242.

Sharma A, Pareek V, Zhang D (2015) Biomass pyrolysis—A review of modelling, process parameters and catalytic studies. Renew Sustain Energ Rev 50, 1081–1096.

Su Shiung L, Rock Keey L, Chin Kui C, Nazaitulshila R, Chee Kuan O, Nyuk Ling M et al. (2018) Pyrolysis production of fruit peel biochar for potential use in treatment of palm oil mill effluent, Journal of Environmental Management, Volume 213, 2018, Pages 400–408, ISSN 0301-4797 https://doi.org/10.1016/j.jenvman.2018.02.092.

Suman S, Gautam S (2017) Pyrolysis of coconut husk biomass: Analysis of its biochar properties. Energy Sources, Part A: Recovery, Util Environ Eff 39(8), 761–767. https://doi.org/10.1080/15567036.2016.1263252

Verma M, Godbout S, Brar SK, Solomatnikova O, Lemay SP (2012) Biofuels production from biomass by thermochemical conversion technologies. Int J Chem Eng Jan 2012, 1–18.

Wei J, Tu C, Yuan G, Liu Y, Bi D, Xiao L et al. (2019) Assessing the effect of pyrolysis temperature on the molecular properties and copper sorption capacity of a halophyte biochar. Environ Pollut 251, 56–65.

Wu KT, Tsai CJ, Chen CS, Chen HW (2012) The characteristics of torrefied microalgae. Appl Energy 100, 52–57.

Section II

Other Potent Greenhouse Gases

9 Methane

9.1 INCREASED ATMOSPHERIC METHANE

Due to the chemical bonds within its molecule, methane is much more efficient at absorbing heat than carbon dioxide, but it can only last about 10–12 years in the atmosphere before it gets oxidized mainly by atmospheric OH⁻ radicals to water and carbon dioxide. Carbon dioxide lasts for centuries, but the amount of methane in the atmosphere has multiplied since the industrial revolution, growing from an estimated 722 parts per billion (ppb) in 1750 to more than 1800 ppb in 2020 (EPA 2022; Figure 9.1).

9.1.1 HISTORICAL GROWTH OF METHANE

Since the industrial revolution, human sources of methane emissions have grown, mainly through fossil fuel production and intensive livestock farming such that these two sources are responsible for 60% of all human methane emissions, the former, including combustion of fossil fuels, creating 33% (100 million tons per year) of human methane emissions (Bousquet et al. 2006). Other sources include landfills and waste (16%), biomass burning (11%), rice agriculture (9%), and biofuels (4%) (Bousquet et al. 2006).

9.1.2 ROLE OF TERMITES AND RUMINANTS

Termites, a significant natural source of methane, eat cellulose but rely on microorganisms in their gut to digest it, during which methane gets produced, thereby creating 12% of natural methane emissions (Bousquet et al. 2007). According to their calculations, each termite produces small amounts of methane daily, but when multiplied by the world population of termites, their emissions add up to a total of 23 million tons of methane per year.

PROBLEM 9.1.1
The process of methanogenesis

What are the sequential biochemical pathways producing methane?

SOLUTION 9.1.1

The synergistic process using microorganisms to produce methane includes four basic phases of anaerobic digestion, which are (1) hydrolysis, (2) acidogenesis, (3) acetogenesis, and (4) methanogenesis (Clifford 2020).

Phase 1: Hydrolysis
First, hydrolysis reactions break down carbohydrates (cellulose, starch, and simple sugars) using water and enzymes:

$$\text{Biomass} + H_2O \rightarrow \text{Monomers} + H_2 \tag{9.1.1}$$

DOI: 10.1201/9781003341826-12

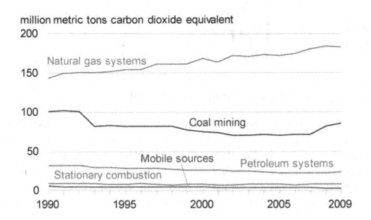

FIGURE 9.1 US methane emissions from energy sources, 1990–2009.

Source: www.eia.gov

Phase 2: Acidogenesis

Gut organisms (several bacteria, protozoa, and fungi), through anaerobic digestion using exoenzymes (cellulosome, protease, etc.), convert soluble monomers to small organic compounds, such as short chain (volatile) acids (propionic, formic, lactic, butyric, succinic acids) (cellulosomes are highly efficient nanomachines that play a fundamental role during the anaerobic deconstruction of complex plant cell wall carbohydrates (Bras et al. 2012), being large cell-surface bound multienzyme complexes that synergistically degrade plant cell wall polysaccharides):

$$C_6H_{12}O_6 + 2H_2 \rightarrow 2CH_3CH_2COOH + 2H_2O \qquad (9.1.2)$$

or ketones (glycerol, acetone) and alcohols (ethanol, methanol), such as

$$C_6H_{12}O_6 \rightarrow 2CH_3CH_2OH + 2CO_2 \qquad (9.1.3)$$

Phase 3: Acetogenesis

Acetogenic bacteria attack the products of acidogenesis; the products from acetogenesis include acetic acid, CO_2, and H_2. The following are some reactions that occur during acetogenesis:

$$CH_3CH_2COO^- + 3H_2O \rightarrow CH_3COO + H^+ + HCO_3^- + 3H_2 \qquad (9.1.4)$$

$$C_6H_{12}O_6 + 2H_2O \rightarrow 2CH_3COOH + 2CO_2 + 4H_2 \qquad (9.1.5)$$

$$CH_3CH_2OH + 2H_2O \rightarrow CH_3COO^- + 2H_2 + H^+ \qquad (9.1.6)$$

$$2HCO_3^- + 4H_2 + H^+ \rightarrow CH_3COO^- + 4H_2O \qquad (9.1.7)$$

Phase 4: Methanogenesis

The last phase of anaerobic digestion, methanogenesis, entails several reactions using the intermediate products from the other phases, with the main product being methane:

$$2CH_3CH_2OH + CO_2 \rightarrow 2CH_3COOH + CH_4 \text{ (ethanoic acid plus methane)} \quad (9.1.8)$$

$$CH_3COOH \rightarrow CH_4 + CO_2 \text{ (ethanoic acid broken to methane and } CO_2) \quad (9.1.9)$$

$$CH_3OH \rightarrow CH_4 + H_2O \text{ (methanol broken to methane and water)} \quad (9.1.10)$$

$$CO_2 + 4H_2 \rightarrow CH_4 + 2H_2O \quad (9.1.11)$$

PROBLEM 9.1.2
CO_2 from burning cubic feet of methane

How much carbon dioxide is produced by combustion of 1000 cubic feet of a gas under the assumption it is pure methane?

SOLUTION 9.1.2

Steps:

1. Identify the gas law
2. Identify the STP
3. Convert 1 cubic foot to liters
4. Determine how many moles of methane are in a set volume at STP
5. As 1 mole $CO_2 = 44$ g, determine how many pounds of carbon dioxide are produced
6. Convert the grams of CO_2 to lbs

According to the ideal gas law

If P = pressure, v = volume, n = moles of gas, T = temperature, and R = constant, then at STP:

$$T = 273\,K / 0°C$$

$$P = 1 \text{ atm } (14.7 \text{ psia, pound per square inch absolute})$$

$$14.7 \text{ psi} = 760 \text{ torr} = 760 \text{ mm mercury}$$

$$\text{If 1 cubic foot} = 0.0283165 \text{ cubic meters} (m^3)$$

$$1m^3 = 1000 \text{ liters} (L)$$

$$\text{Hence, 1 cubic foot} = 28.31685 \text{ L} (0.0283165 \times 1000)$$

Therefore, if 1 mole of gas occupies 22.4 L at STP.

28,316.85 L of methane (at STP) will contain 1264.145 moles of methane.

Completely burning methane produces carbon dioxide and water, depicted as:

$$CH_4 + 2O_2 \rightarrow CO_2 + 2H_2O \quad (9.1.12)$$

Each mole of methane burnt produces 1 mole of carbon dioxide.

$$\text{Mass of moles of } CO_2 = 1264.145 \times 44 \text{ or } 55,622.38\,g$$

$$1 \text{ lb} = 454\,g$$

Therefore, 55,622.38 g/454 = 122 lbs.

Hence, after combustion of 1000 cubic feet while at STP, pure methane will result in the production of 122 lbs of carbon dioxide.

9.2 EMISSION FACTORS

The characteristics of the fuels and raw materials utilized in each process determine the CO_2 emission factor for that process. The term "carbon dioxide equivalent" (CO_2 eq) refers to the quantity of CO_2 emissions which, multiplied by the respective GWP (global warming potential), would produce the same radiative intensity as a given quantity of a greenhouse gas or "bundle" of greenhouse gases, based on the various atmospheric retention times of the gases (GOV.UK 2021).

PROBLEM 9.2.1
GHG emission from the burning of diesel in stationary source

Proportionally, the following products are released when diesel is burnt:

$$CO_2 - 0.00265$$

$$CH_4 - 0.00000036$$

$$N_2O - 0.000000021$$

If 100 L of diesel are burnt, what mass of methane and nitrous oxide are released? What are the energy changes for methane combustion in the air?

SOLUTION 9.2.1

Data: 100 L of diesel consumption

Calculation

Step 1

$$CO_2 \text{ emission} = \text{Fuel burnt in volume unit} \times CO_2 \text{ emission factor}$$
$$\text{(tons per volume unit)}$$

$$CO_2 \text{ emission} = 100 \times 0.00265$$

Step 2

$$CH_4 \text{ emission} = \text{Fuel burnt in volume unit} \times CH_4 \text{ emission factor}$$
$$\text{(tons per volume unit)}$$

$$CH_4 \text{ emission} = 100 \times 0.00000036$$

Step 3

$$N_2O \text{ emission} = \text{Fuel burnt in volume unit} \times N_2O \text{ emission factor}$$
$$\text{(tons per volume unit)}$$

$$N_2O \text{ emission} = 100 \times 0.000000021$$

$$\text{Total GHG emission (in tCO}_2 \text{ eq)} = (\text{CO}_2 \text{ emission}) + (\text{CH}_4 \text{ emission} \times 21)$$
$$+ (\text{N}_2\text{O emission} \times 310)$$

$$\text{Total GHG emission (in tCO}_2 \text{ eq)} = 0.265299393 + (0.000035819 \times 21)$$
$$+ (0.00000215 \times 310)$$

$$\text{Total GHG emission (in tCO}_2 \text{ eq)} = 0.2667$$

Combustion of methane to produce carbon dioxide

Methane is an odorless, colorless, tasteless gas that is lighter than air. When methane burns in the air, it has a blue flame. In sufficient amounts of oxygen, methane burns to give off carbon dioxide and water. When it undergoes combustion, it produces a great amount of heat; hence, it is a very useful fuel source.

$$\text{CH}_4 \Delta\text{H} = -74.6$$

$$\text{CO}_2 \Delta\text{H} = -393.51$$

$$\text{H}_2\text{O}\Delta\text{H} = -285.83$$

Step 1: Write a balanced equation

The problem says that methane combines with oxygen to form carbon dioxide and water, so the formula would be:

$$\text{CH}_4 + 2\text{O}_2 \rightarrow \text{CO}_2 + 2\text{H}_2\text{O} \qquad (9.2.1)$$

Step 2: Calculate the enthalpy change

$$\Delta\text{H} = -(-74.6) + (-393.51) + 2(-285.83)$$

$$\Delta\text{H} = 74.6 - 393 - 571.66$$

$$\Delta\text{H} = -890.57 \, \text{kJ}$$

9.3   CLATHRATE SOURCES OF METHANE

A clathrate is like an almost-finished jigsaw puzzle. A compound, usually polymeric, consisting of a lattice, traps molecules, thereby completely enveloping the guest molecule (Figure 9.2). As the clathrate contains holes of an appropriate size, particular molecules often constitute the "best fit." In the case of methane hydrate, frozen water provides that rigid lattice. The source of the methane is often decaying vegetation. While locked in the ice, the methane is safe from escape to the atmosphere. But a warming planet can melt ice. The occurrence of methane hydrates in marine sediments enriched in organic carbon is controlled by temperature, pressure, and ionic strength (Baird & Cann 2013). Hence, organic-carbon-enriched hemipelagic sediment substances in temperate and polar latitudes at depths that maintain an ice-like consistency are important sources. It has been found that 1 m³ of gas hydrate is equivalent in energy to 164 m³ of methane gas under normal conditions (Baird & Cann 2013). Therefore, if gas hydrates could replace coal, the impact on global warming would be reduced substantially. The challenge, however, lies in the technical

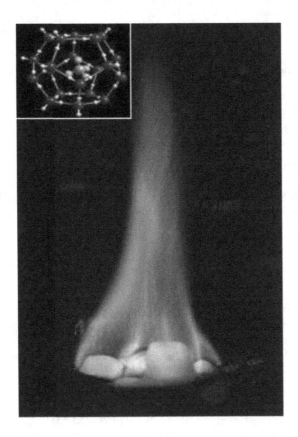

FIGURE 9.2 "Burning ice". Methane from a clathrate, released by heating, burns, and water drips. Inset: clathrate structure.

Source: United States Geological Survey.

difficulties of recovering methane from clathrates, the structure therein consisting of an open framework with nanocages or voids, which are filled by guest atoms.

9.4   CLATHRATE EUSTASY: THE POTENTIAL FOR RAPID FALLS IN SEA-LEVEL FALL

Besides carbon dioxide, other greenhouse gases are methane, water vapor, nitrous oxide, CFCs, and ozone. Methane is produced naturally when vegetation is burnt, digested, or rotted in the absence of oxygen. However, clathrates present a potential for slowing down sea-level rise despite their potential role in sea-level fall having not been appreciated, as stated by Bratton (1999), who cites recent estimates of the total volume occupied by gas hydrates in marine sediments as varying by 20-fold, from 1.2×10^{14} to 2.4×10^{15} m^3. He reports that using a specific volume change on the melting of –21%, dissociation of the current global inventory of hydrate would result in a decrease of submarine hydrate volume of 2.4×10^{13} to 5.0×10^{14} m^3. The release of free gas bubbles present beneath hydrates would increase these volumes by 1.1–2.0 × 10^{13} m^3. The combined effects of hydrate melting and subhydrate gas release would cause, conservatively, a global sea-level fall of 10–146 cm. Such a mechanism may offset some future sea-level rise associated with thermal expansion of the oceans and could also explain anomalous sea-level drops during ice-free periods such as the early Eocene, the Cretaceous, and the Devonian (Bratton 1999).

PROBLEM 9.4.1
Clathrate melting: potential for sea-level change

As the observed density of the methane clathrate is around 0.9 g · cm^{-3}, it will float to the surface of the sea or of a lake unless it is bound in place by sediments or anchored in some other way. With a 25-fold heat trapping capacity as CO_2, and a melting point of 18°C, exceeding that of ice, such potentially floating clathrates could enhance the global greenhouse effect. What is the mass of methane trapped in 1 ton (1000 kg) of methane hydrate?

SOLUTION 9.4.1

$$CH_4 \cdot 6H_2O \rightarrow CH_4 + 6H_2O \qquad (9.4.1)$$

$$12 + 4 + 6(18) \rightarrow 16\,g \text{ of } CH_4$$

$$1g \text{ of } CH_4 \cdot 6H_2O \rightarrow 16/124\,g \text{ of } CH_4$$

$$\text{Therefore, } 1 \text{ ton} \left(10^6\,g\right) \text{of } CH_4 \cdot 6H_2O \rightarrow 0.129 \times 10^6\,g \text{ of } CH_4$$

1 ton of methane clathrate releases 0.129 ton of methane.

9.5   NON-CLATHRATE SOURCES OF METHANE

The second major source of atmospheric methane is the cattle feedlot and the rapid conversion of forestlands to pasture. Though this problem may be an ethical one based on the culinary proclivities of humans for beef cattle, there may be indirect pathways to counteract atmospheric methane from pasture cattle. Cattle ranching is now the biggest cause of deforestation in the Amazon, and nearly 80% of deforested areas in Brazil are now used for pasture. Hence, the rate of forest clearing can be decreased if the efficiency of these pastures increases. Pasture conservation, in pragmatic terms, is, therefore, one of the most important aspects of slowing deforestation in the Amazon. Indeed, in the quest to reduce the enhanced greenhouse effect, the curtailing of pasture expansion by deforestation applies not only to the Amazon but also to all forests. One way to achieve this goal is to improve the efficiency of cattle pastures. The third major source of methane is paddy (lowland rice) fields, but wetlands sources are very large.

PRACTICE PROBLEM 9.5.1
Longevity, source, and significance of major atmospheric gases

Fill in the blanks in Table 9.1.

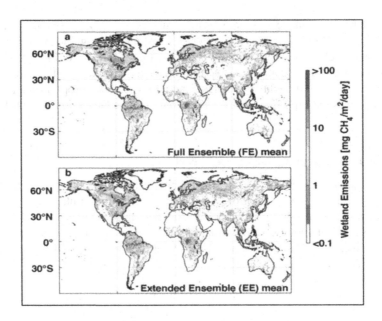

FIGURE 9.3   Methane emissions from wetlands.

Source: https://daac-news.ornl.gov/

TABLE 9.1
Summary of Some Important Atmospheric Gases

Gas	Name	Natural Source	Anthrop. Source	Conc. Range	Lifetime	Sinks	Significance
OH							
O_2							
O_3 Tropospheric							
O_3 Stratospheric							
CO_2							
CH_4							
N_2O							
NO_2							
NO							
CFCs							
HCs (VOCs)							
RCHO							
SO_2							
CO							
H_2S							
SO_3							

9.6   METHANE REDUCTION BY ADDITIVES

Methane-reducing feed additives and supplements inhibit methanogens in the rumen, subsequently reduce enteric methane emissions, and are most effective when grain, hay, or silage is added to the diet, especially in beef feedlots and dairies (DPI 2022).

Methane-reducing feed additives and supplements can include synthetic chemicals, natural supplements, and compounds, such as tannins and seaweed, fats, and oils. But synthetic chemicals, such as antibiotics, are not recommended based on legislative restrictions and human health concerns about using antibiotics as growth promoting agents in livestock (DPI 2022).

Though methane inhibitors consisting of natural compounds have not been widely commercialized, feeding one type of seaweed at 3% of the diet has resulted in up to 80% reduction in methane emissions from cattle (DPI 2022), while fats and oils as a group show the most potential for practical application to farming systems and have shown methane emission reductions of 15–20% (DPI 2022).

9.6.1   BENEFITS FROM USING FEED ADDITIVES OR SUPPLEMENTS TO REDUCE METHANE EMISSIONS

Carbon benefits

There are two approved methodologies under the Emissions Reduction Fund (ERF) for using feed additives or supplements to reduce methane emissions and claim carbon credits (DPI 2022).

* Reducing greenhouse gas emissions by feeding nitrates to beef cattle
* Reducing greenhouse gas emissions through feeding dietary additives to milking cows

PROBLEM 9.6.1
Mechanism of nitrates

How can the addition of nitrates to the diet of ruminants decrease the production of methane? With the aid of one or more equations, construct a real-world example of the process, showing a step-by-step calculation of the solution.

SOLUTION 9.6.1

The use of dietary additives is currently approved only for grazing milking cows and includes the addition of eligible additives to increase the fat content of the diet to reduce methane emissions. Adding nitrates to the diet at a specified rate optimizes rumen fermentation and changes the pathway of hydrogen to produce ammonia rather than methane (DPI 2022). This can have the dual effect of reducing methane emissions while improving or maintaining animal performance. It is recommended that producers seek specialist advice before using this option because overdosing can result in nitrate poisoning (DPI 2022).

PROBLEM 9.6.2
Remediation of methane

What is one rudimentary step to curtail anthropogenic methane emissions outside of wetlands?

SOLUTION 9.6.2

An aerial methane survey determined that a fraction of point sources is responsible for more than a third of California's methane emissions. CEC (2019) reports that NASA, through a methane survey contract with two state agencies, flew remote sensing equipment over selected portions of the state and identified hundreds of methane point sources including "super-emitters," sources responsible for an outsized proportion of the total methane released into the atmosphere.

The survey found that just 10% of the point sources produced 60% of the total methane emissions detected. CEC (2019) put forth the possibility that statewide, these relatively few super-emitters are responsible for about a third of California's total methane emissions. Because such large, hitherto hidden sources are revealed in California, it is reasonable to conclude that similar or more urgent conditions prevail in less modern locations, as the use of such aerial tools promises success at locating such sites outside of California.

REFERENCES

Baird C, Cann M (2013) Environmental Chemistry. WH Freeman & Company, New York.
Bratton JF (1999) Clathrate eustasy: Methane hydrate melting as a mechanism for geologically rapid sea-level fall. Geology, Woods Hole Coastal and Marine Science Center. USGS Publications Warehouse, Woods Hole Coastal and Marine Science Center. https://www.usgs.gov/publications/clathrate-eustasy-methane-hydrate-melting-mechanism-geologically-rapid-sea-level-fall
CEC (2019) Aerial methane survey finds a fraction of point sources responsible for more than a third of California's methane emissions. https://ww2.arb.ca.gov/es/news/aerial-methane-survey-finds-fraction-point-sources-responsible-more-third-californias-methane
Clifford C (2023) Alternative Fuels from Biomass Sources: Anaerobic Digestion. https://www.e-education.psu.edu/egee439/node/727
D'Alessandro Deanna M, Smit Berend, Long Jeffrey R (2010) Carbon dioxide capture: Prospects for new materials. Ang Chem Int Ed 49(35): 6058–6082.
DPI (2022) Managing livestock to reduce methane emissions. New South Wales Department of Primary Industries. www.dpi.nsw.gov.au/dpi/climate/Carbon-and-emissions/emissions-reduction-pathways/livestock-industries/methane_emissions
EPA (2022) Overview of greenhouse gases. https://www.epa.gov/ghgemissions/overview-greenhouse-gases
FAO (2017) Banana facts and figures. https://www.fao.org/economic/est/est-commodities/oil-crops/bananas/bananafacts/en/#.ZDXwe3tBzrc
GOV.UK (2021) Methods of calculating greenhouse gas emissions. https://www.gov.uk/government/publications/methods-of-calculating-greenhouse-gas-emissions
IPCC (2007) Summary for Policymakers, in Climate Change (2007) The Physical Science Basis. Contribution of Working Group I to the Fourth Assessment Report of the Intergovernmental Panel on Climate Change. Cambridge University Press, Cambridge, United Kingdom and New York, NY, USA.

10 Nitrous Oxide

10.1 SOURCES OF NITROUS OXIDE

As is the case when fixing any problem, knowing the extent and sources of the nitrous oxide problem becomes necessary here. Hence, this section examines such sources. Nitrous oxide (N_2O), one of the natural components of Earth's atmosphere, and one of several types of nitrogen oxides existing in the environment, N_2O, NO, NO_2, N_2O_3, N_2O_4, N_2O_5 (Skalska et al. 2010), was controlled by dynamic equilibrium before the Industrial Revolution, when the atmospheric concentration of nitrous oxide stayed in a safe range of levels because of natural sinks. Today, nitrous oxide levels are higher than at any other time during the last 800,000 years (Schilt et al. 2010), having more than doubled over the last 150 years (IPCC 2007) due to agriculture, fossil fuel combustion, and industrial processes, which together account for 77% of all human nitrous oxide emissions in the atmosphere (Kroeze & Rasmussen 1992). Other sources include biomass burning (10%), atmospheric deposition (9%), and human sewage (3%) (Denman et al. 2007).

10.1.1 THE OCEANS

Sinking particles, such as fecal pellets, produce much of the nitrous oxide in the oceans. This arises from microbial activity as they provide the anaerobic conditions necessary for denitrification, a process that creates nitrous oxide as a by-product:

$$CH_4 + 2O_2 \rightarrow CO_2 + 2H_2O \ (\Delta H = -891 \ kJ / mole, \text{ at STP conditions}) \qquad (10.1.1)$$

10.1.2 ATMOSPHERIC CHEMICAL REACTIONS

Chemical reactions in the atmosphere produce a significant amount of nitrous oxide emissions. According to Denman et al. (2007), the atmosphere acts as a source of nitrous oxide through the oxidation of ammonia, a naturally occurring gas in the atmosphere, which creates 5% of emissions. The oceans, manure from wild animals, and aging and rotting plants form the most important natural sources of ammonia in the air, where the oxidization of ammonia from natural sources creates 600,000 tons of nitrous oxide per year (Denman et al. 2007).

10.1.3 THE ROLE OF FUELS

According to Gong et al. (2018), some fuels contain nitrogen compounds, which are often tolerated in oils and refined products because fuel specifications do not directly limit the nitrogen content of transportation fuels. Most immobile sources of NO_2 include power plants using coal, with nitrogen in coal typically taking the form of aromatic structures such as pyridines (DOE 2022), the burning of which oxidizes the intrinsic nitrogen, creating nitrous oxide emissions. But in comparison to double-bonded O_2,

DOI: 10.1201/9781003341826-13

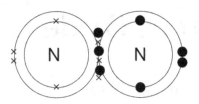

FIGURE 10.1   The triple bond within the nitrogen molecule.

the greater stability of the strong triple N–N bond of elemental nitrogen (Figure 10.1) requires more intense energy, which is the reason almost all mobile emissions are formed when NO_x (i.e., both nitric oxide (NO) and nitrogen dioxide (NO_2) form at temperatures exceeding 500°C in internal combustion engines, both of which can react to form nitrous oxide (N_2O), which contributes to global warming.

10.1.4   ROLE OF HUMAN ACTIVITIES: IMPACT ON OZONE LAYER

Its concentrations having increased for the last two centuries and more rapidly so during the last 50 years, nitrous oxide is a major scavenger of stratospheric ozone, with an impact comparable to that of CFCs (Ravishankara et al. 2009), produced largely by denitrifying bacteria stimulated by hypoxia or low oxygen levels in the soil and by nitrogen in applied manure and the decomposing biomass from the cover crops often exceeding the soil's capacity to hold it. Indeed, globally, about 40% of total N_2O emissions come from human activities (IPCC 2007). Hence, nitrous oxide emissions increase as both legume and manure input simultaneously increase (Flatley 2022). When soils are warm and moist, with a ready supply of reactive nitrogen, nitrous oxide may be produced quickly (Flatley 2022).

Nitrous oxide is produced on a large scale by carefully heating ammonium nitrate to around 250°C, when it decomposes into nitrous oxide and water vapor (Cao et al. 2015). Both natural and anthropogenic sources contribute to nitrous oxide emissions. Agriculture (nitrous oxide fertilizers, soil cultivation), livestock dung, biomass or fossil fuel burning, and industrial operations are also important human sources. Recent studies indicate that the increasing emissions and concentrations are largely associated with agricultural activities (Kroeze and Mosier 2000).

PROBLEM 10.1.1
The role of fertilizers

Do inorganic fertilizers produce more nitrous oxide than organic ones?

SOLUTION 10.1.1

Calculation

Scenario (Flatley 2022): Two samples are taken, sample 1 from an organic dairy manure and sample 2 from an inorganic fertilizer, and sample 1 was taken from a bag weighing 200 lbs and had a percentage description of 40% nitrogen, while sample 2 was taken from a bag weighing the same amount but a different nitrogen percentage of 60%. Which sample produces the highest amount of nitrous oxide for a year?

Equation to produce nitrous oxide:

$$2N_2(g) + O_2(g) \rightarrow 2N_2O(g) \tag{10.1.2}$$

NO reacts with a radical generated from HCN or NH_3 to create nitrous oxide from nitrogen-containing species; the reactions are (Cao et al. 2015)

$$NCO + NO \rightarrow N_2O + CO \text{ and } NH + NO \rightarrow N_2O + H \tag{10.1.3}$$

To calculate the pounds of nitrogen in a bag of fertilizer, multiply the weight of the bag by the percent nitrogen.

Sample 1 (legume manure): 200 lbs times 40 equals 8000 lbs of nitrogen
Sample 2 (inorganic fertilizer): 200 lbs times 60 equals 12,000 lbs of nitrogen
Now we need to find out the emission of nitrous oxide from both samples.
As 1 kg = 2.205 lbs, multiply total lbs N_2O-N emissions by 2.205 to convert to kg N_2O-N and then divide by 1000 to convert to metric tons (t N_2O-N).

$$\text{Sample } 1: 8000 \times 2.205 / 1000 = 17.64 \text{ t}(N_2O\text{-N})$$
$$\text{Sample } 2: 12,000 \times 2.205 / 1000 = 26.46 \text{ t}(N_2O\text{-N})$$
$$\text{As formula weight of } N_2O = 44\,g, \text{the } N_2 \text{ mole} = 14\,g$$
$$\text{Ratio of } N_2O \text{ to } N_2 = 44 / 28\,g = 1.5711$$

Multiply t N_2O-N by 1.5711 to convert to t N_2O/year (1 t N_2O-N = 1.5711 t N_2O).

$$\text{Sample } 1: 17.64 \times 1.5711 = 27.71 \text{ t } N_2O$$
$$\text{Sample } 2: 26.46 \times 1.5711 = 41.57 \text{ t } N_2O$$

Conclusion

Inorganic fertilizer creates more nitrous oxide than legume manure according to the scenario and calculation above.

PRACTICE PROBLEM 10.1.1

In the approved methodology for feeding nitrates to beef cattle, nitrate salt licks are fed to animals previously fed urea and are potentially applicable outside of feedlots (Australian Government 2022). You are commissioned to manufacture nitrate salt licks. How will you reduce the carbon footprint of the typical manufacturing and production process?

SOLUTION PLAN 10.1.1

A. Describe the present process (with credible references)
B. Show proposed modifications with explanations and calculations including equations
C. Based on your findings, write a question for future research

10.2   NITROUS OXIDE IN PERMAFROST

Nitrogen is the most common gas in the atmosphere. Nitrous oxide, or laughing gas, is a more powerful warming agent compared to carbon dioxide. Much of this gas is trapped. Thawing permafrost is common in Tundra like the Yedoma domain. Yedoma usually forms in lowlands or stretches of land with rolling hills where ice wedge polygonal networks are present, in stable relief features with accumulation zones of poor drainage, severe cold and arid continental climate zones resulting in scanty vegetation cover, and amid intense periglacial weathering processes (Strauss et al. (2017). Within it is contained approximately 40 Gt (gigatons) of nitrogen in its first 20 m, 90% (37 Gt) of this nitrogen being frozen and three-fourths of that 90% stored deeper than 3 m (Zimov et al. 2006), and this 3 m is the active layer. Regions with excessively high ice content like the Yedoma (Figure 10.2) are known to contain long-frozen organic matter and carbon. This thawing can result in the release and movement of these organic materials. The Yedoma domain only represents 12% of the northern permafrost regions. Little is known about the nitrous oxide found in thawing permafrost, but this could have a highly significant impact on climate change.

PROBLEM 10.2.1
Calculating nitrogen in permafrost

Considering that permafrost covers roughly 22.8 million square kilometers (or 8.8 million square miles) in Earth's Northern Hemisphere according to National Geographic (2022) and the Yedoma is 12% of these regions, the volume of potential nitrous oxide in these areas can be calculated. (A) How much nitrogen exists there? (B) How much nitrogen exists per km^2 of the permafrost?

SOLUTION 10.2.1

If other permafrost areas contain similar proportions of nitrogen, nitrogen in first 20 m of Yedoma is approximately 41.2 Gt (Strauss et al. 2022).

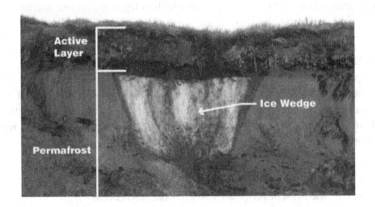

FIGURE 10.2   Nitrous oxide stored in active layer in the Tundra.

Source: Photo by Benjamin Jones, USGS (public domain).

Percentage of permafrost in the Northern Hemisphere that is just the Yedoma: 12% (Strauss et al. 2017).

Total nitrogen in the Northern Hemisphere permafrost (NHP):

12% of NHP nitrogen = 41.2 Gt

1% of Northern Hemisphere N = 41.2 Gt / 12 = 3.43 Gt

100% of Northern Hemisphere N = 100 × 3.43 = 343.33 Gt

At up to roughly 20 m depth = 343.33 Gt of nitrogen for the entire permafrost

Calculation 2: How much nitrogen do we have per km² of the permafrost?

Total permafrost in the Northern Hemisphere = 22.8 million km²
   Nitrogen for the permafrost up to roughly 20 m depth: 343.33 Gt
   22.8 million km² = 22,800,000
   343.33 Gt/22,800,000 km² = 0.0000150583 Gt · km⁻²
   There is 0.0000150583 Gt of nitrogen · km⁻².
   Considering that 1 Gt = 10^9 tons, that is, 15,058 tons of nitrogen · km⁻², such a potential increase in the nitrous oxide content of the atmosphere could be detrimental if more of the nitrous oxide is released and the cycle continues.

PROBLEM 10.2.2
Measuring of half-lives of gases in the atmosphere

The half-life is the time it takes for 50% of the initial influx of a material to be removed. In the case of a 10% per annum decline, it works out that the half-life is about 6 years. After another 6 years (t_2), another 50% of what was there after t_1 is removed and so on (Figure 10.3). In general, after five half-lives have expired, very little is left of the original material. The rate of N_2O photolysis varies spatially and temporally throughout the stratosphere, peaking near the equator at altitudes between 30 and 35 km and at noon when the solar UV radiation is most intense (Butenhoff & Khalil 2007).

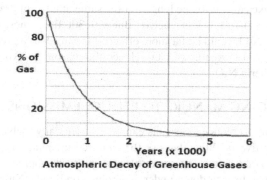

Atmospheric Decay of Greenhouse Gases

FIGURE 10.3 Schematic diagram of exponential photolytic destruction of atmospheric gases. The half-life of each gas occurs when 50% of the remaining material has been removed.

PROBLEM 10.2.3
Half-life of atmospheric gases

If in the atmosphere there are 500 g each of N_2O, CH_4, and CO_2, how long will it take 450 g to decay*?

*NB: Ln = the natural logarithm. This is not to the base 10 but to base e. To find the natural logarithm of "6," for example, on the calculator, press 6 followed by the Ln key = 1.791715.

SOLUTION 10.2.3

Half-life of N_2O = 132 years (Iowa State University 2019)

$$K = \ln \text{ of 2 divided by } t1/2 \text{ (i.e., divided by the half-life)}$$

For N_2O

$$K = 0.6931 / 48180 \text{ (}Half\text{-}life \text{ of } N_2O \text{ in days)}$$

$$K = .0000143$$

$$\text{Ln } (A_f/A_o) = K_t$$

$$\text{Ln } (50/500) = .0000143(t)$$

$$-2.3026 / -.0000143 = 161{,}000 \text{ days} = 441 \text{ years}$$

Half-life of CO_2 = 120 years (Iowa State University 2019)

$K = 0.6931/43{,}800$ (*half-life of CO_2 in days*)
$K = 0.00001$
$\text{Ln}(A_f/A_o) = K_t$
$\text{Ln}(50/500) = 0.00001582(t)$
$-2.3026/-0.0000158 = 145{,}734$ days or approx. 399.27 years

Conclusion

The fact that it takes nitrous oxide longer for the same number of grams of molecules to degrade suggests that it is less reactive. But with 300 times more intrinsic heating power than CO_2, N_2O would be far more effective than CO_2 as a greenhouse gas had it had the atmospheric volume of CO_2. Hence the reasons to expeditiously reduce levels of atmospheric N_2O.

10.3   MANAGING MANURE TO REDUCE EMISSIONS

Managing manure to reduce emissions can be economically viable for larger enterprises or cooperative facilities that use the captured methane to generate heat and electricity. Livestock urine and manure are significant sources of methane and nitrous oxide when broken down under anaerobic conditions. Nitrous oxide is produced during the nitrification–denitrification of the nitrogen contained in livestock waste. Anaerobic conditions often occur where manure is stored in large piles or settlement ponds to deal with waste from large numbers of animals managed in a confined area (for example, dairy farms, beef feedlots, piggeries, and poultry farms). Hence, methane having 25 times-, and nitrous oxide, with 300 times the global warming potential of carbon dioxide, are sleeping giants of global warming.

PROBLEM 10.3.1
Agriculture and nitrous oxide

How does agriculture increase nitrous oxide?

SOLUTION 10.3.1

Many farmers use nitrogen-based fertilizers, legumes, or animal manures to enrich their soil with nitrogen and help crops and pastures flourish. About 1% of nitrogen in the soil, from any source, is lost as nitrous oxide. The excess nitrogen left in the soil from the application of rapidly acting inorganic fertilizers and via animal urine, dung, and legumes provides soil bacteria with energy for metabolic processes such that microbial actions of nitrification and denitrification produce nitrous oxide which is then released into the atmosphere. When grazing ruminant livestock eat nitrogen-rich pastures or crops, they use only a fraction of the nitrogen consumed to support the production of meat and milk. They excrete most of it in urine and dung, which creates very concentrated nitrogen patches in the soil. For comparison, a urine patch can contain the equivalent of up to 1000 kg N per hectare, while fertilizer application rates are typically 30–50 kg N per hectare (although there may be several applications per year).

PROBLEM 10.3.2
Mass of nitrogen in a commercial bag of fertilizer

Calculate how much nitrogen is in a bag of nitrogen-based fertilizer.

SOLUTION 10.3.2

To calculate the pounds of nitrogen in a bag of fertilizer, multiply the weight of the bag by the percent nitrogen (this is the first number in the N-P-K designation on the front of the bag). This indicates the pounds of nitrogen in the bag. Then divide the pounds of nitrogen by the area the bag states it will cover to get the pounds of nitrogen per 1000 ft².
Note: Fertilizer bags usually exist for 5000 or 10,000 ft² areas of effectiveness.
For example:

One 19 lbs bag of nitrogen-based fertilizer with an analysis of 26-4-12 (N-P-K) covering 5000 ft².

$$(wf \times Np) / Al$$

where
Wf = weight of nitrogen-based fertilizer
Np = percentage of nitrogen
Al = area of land the fertilizer will cover (as printed on the bag of fertilizer)

19 lbs × (26/100) = 4.94 lbs. Total nitrogen in the bag is:
4.94 lbs nitrogen in the bag ÷ 5000 ft² bag = 0.98 lbs of nitrogen/1000 ft².

PROBLEM 10.3.3
Importance of nitrogen in carbon sequestration

Why is nitrogen vital for carbon dioxide uptake in plants?

SOLUTION 10.3.3

It's vital because it is a major component of chlorophyll, the compound by which plants use the energy from sunlight to produce starch. It is also a major component in amino acids, the building blocks of proteins, without which plants will wither and die. Nitrogen also plays a critical role within plants to ensure energy is available when and where the plants need it to optimize yield. This nutrient is even present in the roots as proteins and enzymes that help regulate water and nutrient uptake.

PROBLEM 10.3.4
C:N ratio in soils

What is the carbon-to-nitrogen ratio within soil?

SOLUTION 10.3.4

The carbon-to-nitrogen ratio in soil is the ratio of the mass of carbon-to-nitrogen. A C:N ratio of 12:1 means there are 12 units of carbon (C) for each unit of nitrogen (N) in the soil. This ratio can have a substantial impact on the functioning of the soil, such as crop residue decomposition, particularly residue-cover on the soil and crop nutrient cycling (predominantly N). Soils with a carbon-to-nitrogen (C:N) ratio of 24:1 have the optimum ratio for soil microbes to stimulate the release of nutrients like N, phosphorus, and zinc to crops, thereby influencing the amount of soil-protecting residue cover that remains on the soil (Schultheis et al. 2020).

PROBLEM 10.3.5
Nitrogen use efficiency and nitrous oxide

Nitrogen use efficiency in soil determines largely any leftover nitrogenous residue which can be converted to the greenhouse gas N_2O. Calculate nitrogen use efficiency on-farm if the grain yield is 10 tons · ha^{-1}, grain protein is 13.5% and the soil nitrogen supply before any nitrogen applications is 80 kg N · ha^{-1} (soil mineral nitrogen plus estimate of soil nitrogen mineralization plus spring crop nitrogen), organic manure N application is 40 kg N · ha^{-1}, and the inorganic nitrogen applications is 220 kg N · ha^{-1}.

SOLUTION 10.3.5

Nitrogen use efficiency is used to demonstrate the relationship between a soil's total nitrogen input and its nitrogen output (Gillbard 2021).

Step 1:
Nitrogen output

- Grain yield at 100% dry matter (DM) = 10 tons $\cdot$ ha^{-1}
- Grain protein at 100% DM = 13.5%
- Grain nitrogen = 13.5/5.7 = 2.37% N

Step 2:

$$\text{Total nitrogen output} = \text{Grain yield} \times \text{Grain N percentage}$$
$$\text{Total nitrogen output} = 10,000 \times 2.37\%$$
$$\text{Total nitrogen output} = 237 \text{ kg N / ha}$$

Step 3:
Nitrogen input

- Soil nitrogen supply before any nitrogen applications = 80 kg N $\cdot$ ha^{-1} (soil mineral nitrogen plus estimate of soil nitrogen mineralization plus spring crop nitrogen)
- Organic manure N application = 40 kg N $\cdot$ ha^{-1}
- Inorganic nitrogen applications = 220 kg N $\cdot$ ha^{-1}

Step 4:

$$\text{Total nitrogen supply} = 80 + 40 + 220$$
$$\text{Total nitrogen supply} = 340 \text{ kg N / ha}$$

Step 5:

$$\text{Nitrogen use efficiency} = \text{Nitrogen output / nitrogen input}$$
$$\text{Nitrogen use efficiency} = 237 / 340 = 70\%$$

Conclusion

A nitrogen use efficiency value below 50% suggests a nitrogen surplus (Schultheis et al. 2020), thereby increasing the risk of nitrogen leaching and nitrous oxide. Conversely, very high nitrogen use efficiency values like 90% indicate mining of soil nitrogen stock, which causes depletion of the soil mineral nitrogen supply. Therefore, to restrict N_2O emissions from nitrate leaching, the optimal goal is 70–80% nitrogen use efficiency.

PROBLEM 10.3.6
Mycorrhizae and nitrous oxide

Trees with arbuscular mycorrhizal fungi have accelerated nitrogen cycles. In forests where these trees predominate, sticks and leaves decompose more quickly, releasing more nitrogen into the soil (CAFÉ 2015). If subsequent agricultural practices are improperly managed, excess nitrous oxide can occur in the environment. What is the nitrogen concentration and uptake in grain if dry matter yield is 7000 kg · ha^{-1} and nutrient concentration for nitrogen is 2%?

SOLUTION 10.3.6

To calculate the uptake of any nutrient (CAFÉ 2015):

$$N\% \text{ in dry matter of grain or straw in kg / ha} \div 100 =$$
$$\text{Uptake in kg / ha in grain or straw} \tag{10.3.1}$$

where
   N% = nutrient, in this case nitrogen
   Grain or straw = examples of agriculture by-product/plant
   kg · ha^{-1} = kilogram per hectare

$$2 \times 7000 \div 100 = 140 \text{ kg} \cdot \text{ha}^{-1}\text{ha nitrogen in grain}$$

Conclusion

Planning for nutrient management aids in reducing the amount of plant nutrients that pollute rivers. Nutrients can dissolve in soil water and enter surface or groundwater through leaching or runoff if they are not managed properly (OMAFRA 2016). If the nitrogen concentration uptake amount is what is required by the plant, then the levels of nitrogen would not be in excess, so there would not be a risk of pollution (including from N_2O) and the plant itself would have been receiving the necessary nutrients required for development. Conversely, if the nitrogen is in excess, possible pollution is a risk that needs to be addressed. Therefore, means by which nitrogen reduction could be enabled would have to be determined and practiced.

PROBLEM 10.3.7
Enthalpy of combustion

The heat of a reaction is the amount of heat emitted or absorbed in the process. Nitric oxide gas, NO (g), can be oxidized in air to produce nitrogen dioxide gas, NO_2 (g), in an exothermic (producing heat) reaction. What is the enthalpy change per gram of nitric oxide in the following reaction? Colorless nitric oxide combines with oxygen to form nitrogen dioxide.

SOLUTION 10.3.7

Step 1: Write a balanced equation.
   The problem says that nitric oxide combines with oxygen to form nitrogen dioxide, so the formula would be:

$$NO + O_2 \rightarrow NO_2$$

The balanced equation is:

$$2NO + O_2 \rightarrow 2NO_2 \tag{10.3.2}$$

Step 2: List the enthalpy change of formation for each molecule in the equation.
These numbers can be found on a standard thermodynamic value chart:

$$NO = 90.2 \text{ kJ / mole}$$
$$O_2 = 0 \text{ kJ / mole}$$
$$NO_2 = 33.1 \text{ kJ / mole}$$

Step 3: Calculate the enthalpy change of the entire reaction.
To do this, we use the following equation:

$$\Delta H \text{ reaction} = \sum \Delta H \text{ products} - \sum \Delta H \text{ reactants}$$

In this case, the equation would be:

$$\Delta H \text{ reaction} = 2(NO_2) - \left[2(NO) + O_2\right]$$

Now we plug in the listed enthalpy changes for each molecule.

$$\Delta H \text{ reaction} = 2 \text{ mole} (33.1 \text{ kJ/mole}) - \left[2 \text{ mole} (90.2 \text{ kJ/mole}) + 0 \text{ kJ/mole}\right]$$
$$\Delta H \text{ reaction} = 66.2 \text{ kJ} - 180.4 \text{ kJ}$$
$$\Delta H \text{ reaction} = -114.2 \text{ kJ}$$

The enthalpy change in this reaction is −114.2 kJ when 2 moles of nitric oxide react with 1 mole of oxygen to produce 2 moles of nitrogen dioxide.

Step 4: Convert to enthalpy change per gram of nitric oxide.
The molar mass of nitric oxide is 30.01 g and the balanced equation indicates 2 moles of nitric oxide. A conversion chart is required.

$$-114.1 \text{ KJ / 2 mole NO} \times 1 \text{ mole NO / 30.01 g NO}$$
$$= -114.1 \text{ kJ / 60.02 g NO}$$
$$= -1.90 \text{ kJ} \cdot \text{g}^{-1} \text{ NO}$$

Conclusion

Beneficial purposes of cover crops are the sustainability of their production system because cover crops help in the reduction of erosion, increasing soil carbon, fixing nitrogen against leaching (with the potential for releasing N_2O), increasing water infiltration, and grazing for livestock.

10.4 REMEDIATION OF NITROUS OXIDE

According to Skalska et al. (2010), one of many proposals for NO_x emission reduction is the application of an oxidizing agent which would transform NO_x to higher nitrogen oxides with higher solubility in water, where they determined the rate constant of NO_x gases. However, though the results of their studies were a good basis for the determination of kinetics of NO reaction with ozone, no results were shown for that of N_2O, the latter potentially being the subject of another study.

PROBLEM 10.4.1
Burning and N_2O

How much nitrous oxide does burning biomass place annually into the atmosphere?

SOLUTION 10.4.1

According to Denman et al. (2007), a substantial amount of nitrous oxide is caused by biomass burning, i.e., burning of living and dead vegetation, which accounts for 10% of human-caused emissions. In these fires, some of the nitrogen in the biomass and surrounding air is oxidized, creating nitrous oxide emissions. While natural wild-fires can contribute to this, they say, the great majority of biomass burning is caused by human beings, where large open fires are mainly used by humans to destroy crop waste and clear land for agricultural or other uses, thereby creating 700,000 tons of nitrous oxide per year. Incorporating this biomass otherwise by shallow burial in the soil can sequester it as soil organic carbon.

PROBLEM 10.4.2
Dealing with vapors

Vapors from fertilizers and smoke from fossil fuel or biomass burning contain reactive nitrogen gases. These gases eventually fall out of the atmosphere because either rain washes them out or they get attached to dust and pollen that settle to the ground. Atmospheric deposition produces thousands of tons of nitrous oxide per year. How can such vapors be decreased?

SOLUTION 10.4.2

Curtailing excess fertilizer

Excess nitrous oxide signifies excess fertilizer applications and excess fertilizers are unnecessary. Therefore, moderately decreasing fertilizers where there is excess will decrease such vapors.

Monitoring field workers

The very nature of the fieldwork remuneration system encourages the farmworker to act speedily, often at the unwitting expense of quality work.

The following story, though anecdotal and unproven, signifies the importance of monitoring the actions of such workers because the cost of savings can vary directly with the resulting environmental damage. According to a fertilizer company (1st Products 2022), a farmer called a fertilizer company to come and apply 500 lbs · ac^{-1} to one of his fields before he bedded for corn. He received a call from the spreader truck as he was making his last round and the spreader operator asked him what was to be done with the rest of this fertilizer. Several weeks later, he could see the variation in all his fields behind a spreader truck: tall cotton, then short cotton, i.e., the trucker failed to spread the material evenly.

The farmer then purchased fertilizer hoppers and from that day forward started seeing the savings. He mounted three hoppers on his six-row strip till and three hoppers on a toolbar for side dressing. He applied his P andK at 150 lbs/ac while laying off the rows behind the subsoil shank of the strip till and broadcasted 200 lb $\cdot$ ac^{-1}. The 150 lb $\cdot$ ac^{-1} produced 200 lbs $\cdot$ ac^{-1} more cotton, and this was just one trial. Now he applies fertilizer on all 1000 acres with hoppers. He shared, "After I adapted this method across the entire farm, I saved \$70,000 in one year and had better yields."

10.5 REMEDIATION OF NITROUS OXIDE USING BIOCHAR

As reported by Ameloot et al. (2016), during denitrification, specialized microorganisms reduce nitrate (NO_3^-) via intermediates to N_2O and finally to N_2. Prerequisites for N_2O release during denitrification are oxygen-depleted conditions, high availability of NO_3^- and a C source as an electron donor (energy supplier).

Manipulations likely to promote labile C bioavailability, here either by glucose addition or by soil particulate OM disclosure after disruption of soil aggregates, resulted in the most prominent biochar-induced N_2O emission reductions.

Ameloot et al. (2016) further report that denitrifier activity and N_2O emissions increase relatively abruptly at moisture conditions of at least 70% water-filled pore space (WFPS). As the enzyme N_2O reductase has a high sensitivity for O_2, in soils approaching full water saturation complete denitrification to N_2 occurs and the ratio of N_2O/N_2 approaches zero.

PROBLEM 10.5.1
Biochar: effects on nitrous oxide

Under what environmental conditions does biochar promote, or decrease, nitrous oxide?

SOLUTION 10.5.1

Dong et al. (2020) added biochar to calcareous soils in 0 and 1% (w/w) amounts and moisture was maintained at 70% WFPS during an incubation period. The results revealed that biochar reduced the emissions of soil-produced N_2O by 37–47% and those of injected N_2O by 23–44%. The addition of glucose solution strongly increased N_2O emissions, while biochar reduced total N_2O emissions by as much as 64–81% and those of injected N_2O alone by 29–51%. Differences between low-fertility and high-fertility soils in the apparent N_2O emission mitigation by biochar were relatively small but tended to be larger for the low-fertility soil. The results suggest that biochar addition can suppress the production of N_2O in soil and simultaneously stimulate the reduction of N_2O to N_2.

Biochar not very effective under rice

After Petter et al. (2016) applied biochar to soil as treatments consisting of fertilization with 100 kg N ha^{-1} split into two applications, 60% at sowing and 40% at

45 days after crop emergence, combined with four doses of biochar (0, 8, 16 and 32 Mg ha^{-1}, with four replications, they observed a significant correlation of the application of biochar with moisture retention (r = 0.94, N$_2$O emission (r = 0.86, and soil pH (r = 0.65, and N$_2$O emissions showed a positive correlation (p < 0.05) with soil moisture (r = 0.77 and pH (r = 0.66. Thus, the highest N$_2$O emissions were observed shortly after N fertilization and in the treatments with 32 Mg ha^{-1} of biochar. Despite the higher N$_2$O emissions from the application of 32 Mg ha^{-1} of biochar, the emission factor was lower (0.81%) than the maximum recommended by the IPCC. The higher N$_2$O emissions with the application of biochar are offset by more efficient use of N and consequently the possibility of reduction of applied doses.

Irrigation regimes

Differential irrigation regimes in rice production have also been shown to have a significant impact on nitrous oxide (N$_2$O) emissions. Nitrous oxide emissions have been shown to increase with soil water content, particularly more than 60% WFPS (Bateman & Baggs 2005).

Lunga et al. (2021) report that practices involving any degree of field drainage, such as the rice irrigation practices of alternate wet and dry, mid-season drainage, mid-season aeration, and intermittent flooding, have four times greater N$_2$O emissions than from a continuous, full-season-flood rice production system.

Biochar failure to suppress nitrous oxide

Coleman et al. (2019) report that contradictory outcomes are often attributed either to differences in biochar properties or to mechanisms by which biochar affects N and P cycling, which are in turn dependent on environmental conditions and can affect N cycling in complex ways, such as indirectly by altering the microbial community structure and function and directly by altering the soil C:N ratio. Differences in biochar feedstock (e.g., lignocellulose, manure) and pyrolysis process parameters (e.g., temperature, oxygen content) result in differences in biochar properties (e.g., pH, surface area, nutrient content, and surface charge density) that in turn affect nutrient cycling processes.

10.6 NITROUS OXIDE INHIBITION USING BANANA WASTE BIOCHAR

Biochar has the sorption capacity, empirically, to trap inorganic nitrogen (N) (Lan et al. 2017), organic compounds, and micro- and macro-elements (Wang et al. 2015; Danmaliki & Saleh 2017). Banana peel consists of 30–40% of its total fruit weight, which contains about 60–65% cellulose, 6–8% hemicellulose, and 5–10% lignin (Pokharel et al. 2018).

Based on banana peel biochar, Tanveer et al. (2019) applied five treatments of amendments: no amendment (control), banana peel 1% (P1), banana peel 2% (P2), biochar 1%, and biochar 2%. They found that biochar amendment significantly decreased cumulative nitrous oxide (N$_2$O) emissions (37.1–54.8%), whereas banana peels (non-biochar) amendment did not significantly decrease cumulative N$_2$O emissions (1.3–5.3%) as compared to control.

Therefore, with a high waste-to-fruit ratio of 4:1 (Pokharel et al. 2018), and annual banana fruit harvests worldwide being 120 million tons (FAO 2023, such a large amount of banana waste (480 million tons) promises the avoidance of substantial amounts of methane (via anaerobic decomposition) in the atmosphere.

PROBLEM 10.6.1
Signs of nitrous oxide decrements

How effective is banana biochar as a sequester of nitrous oxide?

SOLUTION 10.6.1

Using banana waste as a base, Sial et al. (2019) found that biochar application decreased the soil ammonium nitrogen (NH_4^+-N) and nitrate nitrogen (NO_3^--N) with an increasing rate. Cumulative carbon dioxide (CO_2) emissions for B1 and B2 treatments decreased by 20.0% and 24.0% in comparison to the banana peel amendment, respectively. Cumulative methane (CH_4) emissions were higher in peel waste than in biochar amendment.

Biochar amendment significantly increased soil enzyme activities (urease, invertase, and alkaline phosphatase). In contrast, banana peel amendment increased soil ammonium nitrogen, soil microbial biomass carbon (MBC) and microbial biomass nitrogen (MBN), β-glucosidase, and urease activities. Therefore, the relative lack of nitrogen in the former case limits the availability of the raw material for any conversion to N_2O in the soil.

PROBLEM 10.6.2
Mechanism of nitrous oxide diminution

How does biochar decrease nitrous oxide?

SOLUTION 10.6.2

Because of the fixed carbon, slow decomposition, and longtime stability in soil (Danmaliki & Saleh 2017), biochar application into soil can sequestrate carbon, reduce CO_2, and decrease the microbial activity (Zhang et al. 2017; Pokharel et al. 2018). Biochar application can potentially reduce N_2O emission production from the soil due to adsorption of ammonium (NH_4^+) and nitrate (NO_3^-) on the biochar surfaces and pore space (Lei et al. 2022), thereby decreasing the total nitrogen during the denitrification process in the soil (Pokharel et al. 2018).

REFERENCES

1st Products (2022) How to reduce your fertilizer cost? https://1stproducts.com/blog/how-to-decrease-your-fertilizer-cost/

Ameloot N, Maenhout P, De Neve S, Sleutel S (2016) Biochar-induced N_2O emission reductions after field incorporation in a loam soil. Geoderma 267, 10–16. https://doi.org/10.1016/j.geoderma.2015.12.016

Australian Government (2022) Feeding nitrates to beef cattle method. https://www.cleanen ergyregulator.gov.au/Choosing-a-project-type/Opportunities-for-the-land-sector/ Agricultural-methods/Reducing-greenhouse-gas-emissions-by-Feeding-Nitrates-to- Beef-Cattle

Bateman, EJ, Baggs, EM (2005) Contributions of nitrification and denitrification to N_2O emissions from soils at different water-filled pore space. *Biol Fertil Soils 41*, 379–388. https://doi.org/10.1007/s00374-005-0858-3

Butenhoff CL, Khalil MAK (2007) Global methane emissions from terrestrial plants. Environ Sci Technol 41(11), 4032–4037.

CAFÉ (2015, June 22) Plant nutrients—Major & minor. Center for Agriculture, Food, and the Environment (CAFE). Retrieved November 19, 2022, from https://ag.umass.edu/fruit/ ne-small-fruit-management-guide/general-information/plant-nutrients-major-minor

Cao Y, Sanchez NP, Jiang W, Griffin RJ, Xie F, Hughes LC, Zah CE, Tittel FK (2015) Simultaneous atmospheric nitrous oxide, methane and water vapor detection with a single continuous wave quantum cascade laser. Opt Express. Feb 9;23(3):2121–32. doi: 10.1364/OE.23.002121. PMID: 25836083.

Coleman BSL, Easton ZM, Bock EM (2019) Biochar fails to enhance nutrient removal in woodchip bioreactor columns following saturation. J Environ Manage 232, 490–498. https://doi.org/10.1016/j.jenvman.2018.11.074

Danmaliki GI, Saleh TA (2017) Effects of bimetallic Ce/Fe nanoparticles on the desulfur- ization of thiophenes using activated carbon. Chem Eng J 307, 914–927. https://doi. org/10.1016/j.cej.2016.08.143

Denman KL, Brasseur G, Chidthaisong A, Ciais P, Cox PM et al. (2007) Couplings between changes in the climate system and biogeochemistry. In Climate Change 2007: The Physical Science Basis. Contribution of Working Group I to the Fourth Assessment Report of the Intergovernmental Panel on Climate Change. Cambridge University Press, Cambridge, United Kingdom and New York, NY, USA.

DOE (2022) Coal. Retrieved October 13, 2022, from https://www.netl.doe.gov/research/Coal/ energy-systems/gasification/gasifipedia/nitrogen-oxides

Dong W, Walkiewicz A, Bieganowski A, Oenema O, Nosalewicz M, He C, Zhang Y, Hu C (2020) Biochar promotes the reduction of N_2O to N_2 and concurrently suppresses the production of N_2O in calcareous soil. Geoderma 362, 114091. https://doi.org/10.1016/j. geoderma.2019.114091

FAO (2023) Markets and Trade. Bananas. https://www.fao.org/markets-and-trade/commodities/ bananas/en/

Flatley K (2022) How is nitrous oxide produced naturally? TimesMojo. https://www.times- mojo.com/how-is-nitrous-oxide-produced-naturally/ https://www.sciencedirect.com/ science/article/abs/pii/0360128592900383

Gillbard E (2021) How to calculate nitrogen use efficiency on-farm? Farmers Weekly. https:// www.fwi.co.uk/arable/crop-management/nutrition-and-fertiliser/how-to-calculate- nitrogen-use-efficiency-on-farm

Iowa State University (2019) Greenhouse gas lifetimes in the atmosphere.

IPCC (2007) Summary for Policymakers. In Climate Change 2007: The Physical Science Basis. Contribution of Working Group I to the Fourth Assessment Report of the Intergovernmental Panel on Climate Change. Cambridge University Press, Cambridge, United Kingdom and New York, NY, USA.

Kroeze C, Mosier AR (2000). New Estimates for Emissions of Nitrous Oxide. https://www. semanticscholar.org/paper/New-Estimates-for-Emissions-of-Nitrous-Oxide-Kroeze- Mosier/1e3614d97c4a4e83b657ab5d0f3595a830cec1a0

Lan ZM, Chen CR, Rezaei RM, Yang H, Zhang DK (2017) Stoichiometric ratio of dissolved organic carbon to nitrate regulates nitrous oxide emission from the biochar-amended soils. Sci Total Environ 576, 559–571. https://doi.org/10.1016/j.scitotenv.2016.10.119

Lunga DD, Brye KR, Slayden JM, Henry CG., Wood LS (2021) Relationships among soil factors and greenhouse gas emissions from furrow-irrigated rice in the mid-southern, USA. Geoderma Reg 24, e00365. https://doi.org/10.1016/j.geodrs.2021.e00365

National Geographic (2022) Permafrost. National Geographic Resource Library Encyclopedia. Retrieved November 4, 2022, from https://education.nationalgeographic.org/resource/permafrost

OMAFRA (2016) Nutrient management. Ontario Ministry of Agriculture, Food and Rural Affairs. Retrieved November19, 2022, from https://www.nutrientmanagement.ca/about/what-is-nutrient-management/

Petter FA, de Lima LB, Júnior BHM, de Morais LA, Marimon BS (2016) Impact of biochar on nitrous oxide emissions from upland rice, Journal of Environmental Management, 169, pp. 27–33, ISSN 0301-4797. https://doi.org/10.1016/j.jenvman.2015.12.020.

Pokharel P, Kwak J-H, Ok YS, Chang SX (2018) Pine sawdust biochar reduces GHG emission by decreasing microbial and enzyme activities in forest and grassland soils in a laboratory experiment. Sci Total Environ 625, 1247–1256. https://doi.org/10.1016/j.scitotenv.2017.12.343

Ravishankara AR, Daniel JS, Portmann RW (2009) Nitrous oxide (N_2O): The dominant ozone-depleting substance emitted in the 21st century. Science 326(5949), 123–125.

Schilt A, Baumgartner M, Blunier T, Schwander J, Spahni R, Fischer H, Stocker TF (2010) Glacial–interglacial and millennial-scale variations in the atmospheric nitrous oxide concentration during the last 800,000 years. Quater Sci Rev 29(1–2), 182–192.

Schultheis L, Whitney T, Lesoing G, Gross P, Cates A et al. (2020) What is the carbon-to-nitrogen ratio within soil? Soul Health Nexus. https://soilhealthnexus.org/resources/soil-properties/soil-chemical-properties/carbon-to-nitrogen-ratio-cn/

Sial TA, Khan MN, Lan Z, Kumbhar F, Ying Z, Zhang J, Sun D, Li X (2019) Contrasting effects of banana peels waste and its biochar on greenhouse gas emissions and soil biochemical properties, Process Safety and Environmental Protection, 122, 366–377, ISSN 0957-5820 https://doi.org/10.1016/j.psep.2018.10.030.

Skalska K, Miller JS, Ledakowicz S (2010) Kinetics of nitric oxide oxidation. Chemical Papers 64(2), 269–272.

Strauss J, Biasi C, Sanders T, Abbott BW, von Deimling TS (2022) A globally relevant stock of soil nitrogen in the Yedoma permafrost domain. Nat Commun 13(1), 6074. https://pubmed.ncbi.nlm.nih.gov/36241637/

Strauss J, Schirrmeister L, Grosse G, Fortier D, Hugelius G, Knoblauch C, Romanovsky V et al. (2017) Deep Yedoma permafrost: A synthesis of depositional characteristics and carbon energy systems vulnerability. Earth Sci Rev 172, 75–86.

Wang X, Song D, Liang G, Zhang Q, Ai C (2015) Maize biochar addition rate influences soil enzyme activity and microbial community composition in a fluvo-aquic soil. Appl Soil Ecol 96, 265–272. https://doi.org/10.1016/j.apsoil.2015.08.018

Zhang A, Cheng G, Hussain Q, Zhang M, Feng H (2017) Contrasting effects of straw and straw–derived biochar application on net global warming potential in the loess plateau of China. Field Crops Res 205, 45–54. https://doi.org/10.1016/j.fcr.2017.02.006

Zhang G, Yang C, Serhan M, Koivu G, Yang Z, et al. (2018) Chapter Three – Characterization of Nitrogen-Containing Polycyclic Aromatic Heterocycles in Crude Oils and Refined Petroleum Products, Editor(s): Bing Chen, Baiyu (Helen) Zhang, Zhiwen (Joy) Zhu, Kenneth Lee, Advances in Marine Biology, Academic Press, Volume 81, 2018, pp. 59–96, ISSN 0065-2881, ISBN 9780128151051, https://doi.org/10.1016/bs.amb.2018.09.006.

Zhong L, Li G, Qing J, Li J, Xue J, Yan B, Chen G, Kang X, Rui Y (2022) Biochar can reduce N_2O production potential from rhizosphere of fertilized agricultural soils by suppressing bacterial denitrification, European Journal of Soil Biology, 109, 103391, ISSN 1164-5563 https://doi.org/10.1016/j.ejsobi.2022.103391.

Zimov SA, Schuur EAG, Chapin III SF (2006) Permafrost and the global carbon budget. Science 312(16), 1612–1613.

Section III

Atmospheric Emissions

11 Sulfur Dioxide

SULFUR DIOXIDE: MAIN POINTS

- 22 million tons of SO_2 released into the atmosphere by human activities
- Sulfur in burnt coal is the main cause of acid rain
- Irritating, choking odor; produced naturally in volcanoes and geysers
- Added to petroleum products to remove O_2 that would cause rusting, e.g., of pipes
- Added to soya beans and corn to destroy molds and preserve the products from decay
- Prevents oxidation in wine and dried fruits (brown discoloration of oxidation products)

11.1 PROS AND CONS OF SO_2

In addition to its health effects on humans, SO_2 has important consequences on the environment and physical environment (Figure 11.1). Chronic exposure of plants to sulfur dioxide (SO_2) causes chlorosis, a bleaching or yellowing of the normally green portions of the leaf. Plant injury increases with increasing relative humidity. Plants incur most injury from sulfur dioxide when their stomata (small openings in plant surface tissue that allow interchange of gases with the atmosphere) are open. For most plants, the stomata are open during daylight hours, and most damage from sulfur dioxide occurs then. Long-term, low-level exposure to sulfur dioxide can reduce the yields of grain crops such as wheat or barley. Sulfur dioxide in the atmosphere is converted to sulfuric acid. So in areas with high levels of sulfur dioxide pollution, plants may be damaged by sulfuric acid aerosols. Such damage appears as small spots where sulfuric acid droplets have impinged on leaves.

For *Manihot esculenta* (cassava) processing, almost all large-scale (industrial cassava) factories use sulfur dioxide as a bleaching and antimicrobial agent – in Thailand, only one of the 59 factories does not use sulfur dioxide (Sriroth et al. 1999). Sulfur dioxide is discharged mainly into the atmosphere. If used, 1.6 kg of sulfur is burned to sulfur dioxide for each ton of starch. For the production of 200 tons of starch, resulting in 2323 tons of wastewater, about 38 and 237 kg of sulfur dioxide were found in the starch and wastewater, respectively. The rest is released into the atmosphere. Release of sulfur dioxide to the atmosphere by a cassava processor is assumed to be safe, but no systematic study has been undertaken to substantiate this assumption (FAO 2002). Sulfur dioxide is a common air pollutant and an intermediary in the production of sulfuric acid, according to the following reactions:

$$S + O_2 \rightarrow SO_2 \tag{11.1.1}$$

$$2SO_2 (g) + O_2 (g) \rightarrow 2SO_3 (g) \tag{11.1.2}$$

$$SO_3 (g) + H_2O (aq) \rightarrow H_2SO_4 (aq) \tag{11.1.3}$$

DOI: 10.1201/9781003341826-15

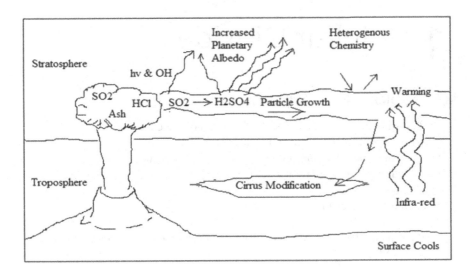

FIGURE 11.1   Some atmospheric processes during a volcanic eruption.

Source: Adapted from NASA.

Sulfuric acid fumes and mists are severely corrosive poisons, being irritants to the skin, eyes, mucus membranes, and respiratory system (the liquid form readily penetrating the skin to reach subcutaneous tissue), and are largely responsible for acid rain, while industrial exposure has caused tooth erosion in factory workers (Manahan 2005).

11.2   ENVIROMENTAL EFFECTS OF SO_2

The dangers of smog, epitomized in 1952 in London with aerosols of sulfur dioxide, beginning on December 5, affecting other areas of the southeast with icy roads causing several road accidents and spreading as far as northern France, Belgium, and the Netherlands, forced a large ship to anchor off the French coast, unable to get into port due to poor visibility (National Geographic 1952).

11.3   CALCULATING IMPURITY LEVELS

Natural releases of sulfur dioxide together with traces of sulfur trioxide follow volcanic action, whilst "unnatural" release results from the burning of fossil fuels. Coal contains 0.5–4.0% sulfur, some of which is in the form of extraneous minerals such as iron pyrites (FeS_2) and inorganic sulfates. Removal of extraneous sulfur-bearing minerals from coal is an effective way of reducing this pollution problem. Fuel oils also contain sulfur at similar levels, the highest concentration being in residual oils from the distillation process. Human beings are affected by quite low levels of sulfur dioxide in the air: 1–5 ppm causes discomfort whilst exposure to 10 ppm for 1 h usually results in severe distress. The California State Department of Health (CDH 2020) defines 5 ppm for 1 h as a "serious" level of pollution. Low concentrations of the gas cause the fine tubules of the lungs to contract, thus making breathing difficult.

PROBLEM 11.3.1
Calculating impurity levels

A certain grade of coal contains 2.3% sulfur. Assume that the burning of the sulfur compounds can be represented by the following equation:

$$S + O_2 \rightarrow SO_2 \qquad (11.3.1)$$

What is the mass of the pollutant sulfur dioxide in the gases emitted to the atmosphere per long ton of coal burnt if the coal-burning process is 84% efficient?

SOLUTION 11.3.1

As the given equation is balanced, it shows that 1 mole of sulfur produces 1 mole of sulfur dioxide. Converting 2.3% to a decimal:

$$\text{Moles of sulfur} = (0.023 \times 2240 \text{ lb / ton}) \times (454 \text{ g / lb} \times 1 \text{ mole S / } 32 \text{ g})$$

$$= (0.023 \times 2240 \text{ lb. / ton}) \times (454 \text{ g / lb. } \times 1 \text{ mole S / } 32 \text{ g})$$

$$= 730.94 \text{ moles S per ton}$$

$$= 730.94 \text{ moles SO}_2 \text{ (assuming 100\% yield)}$$

But as the process is only 84% efficient,

$$\text{Moles of SO}_2 = 730.94 \times 0.84$$

$$= 613.99 \text{ (84\% yield)}$$

$$\text{Mass of SO}_2 = 613.99 \text{ moles} \times 64 \text{ g / 1 moles}$$

$$\text{Answer} = 39295.33 \text{ g} = 3.9 \times 10^4 \text{ g} = 86.6 \text{ lbs.}$$

PROBLEM 11.3.2
SO₂ released from anhydrous gypsum

Calculate the sulfur dioxide released when anhydrous gypsum produces lime (CaO).

SOLUTION 11.3.2

Using anhydrous gypsum ($CaSO_4$) to produce CaO releases no carbon dioxide except that which is released from the combustion of the fossil fuel used to heat the gypsum:

$$2CaSO_4 = 2CaO + 2SO_2 + O_2 \qquad (11.3.2)$$

136 tons of $CaSO_4$ produce 64 tons of SO_2.
Nevertheless, the SO_2 released can lead to acid rain.

PROBLEM 11.3.3
SO_2 scrubbing

If a ton of SO_2 is released, what mass of NaOH is required to scrub it from exhaust gases?

SOLUTION 11.3.3

Sulfur dioxide mixed with other waste gases often emerges in exhaust waste during the burning of coal and oil at power plants. Sodium hydroxide can remove SO_2 by scrubbing it from the exhaust gas mixture.

$$2NaOH + SO_2 \rightarrow H_2O + Na_2SO_3 \tag{11.3.3}$$

$$2(23+16+1) + (32+32) \rightarrow H_2O + Na_2SO_3$$

If 64 kg of SO_2 is scrubbed by 80 kg of NaOH, then 1 kg of SO_2 is scrubbed by 80/64 kg of NaOH

Therefore, 1 ton of SO_2 will be scrubbed by 80/64 × 1000 kg = 1.250 ton.

SO_2 can be removed by $CaCO_3$ in an acid-base reaction, though with the undesirable production of CO_2.

PROBLEM 11.3.4
Neutralizing SO_2

What mass of calcium carbonate is required to react with the sulfur dioxide that is produced by burning 1 ton of coal that contains 4% sulfur by mass?

SOLUTION 11.3.4

4% of 1000 kg = 40 kg of sulfur

$$CaCO_3 + SO_2 \rightarrow CaSO_3 + CO_2 \tag{11.3.4}$$

Ratio of $S:O_2$ = 32:32 = 1:1
32 g of S reacts with 100 g of $CaCO_3$.
Hence, 1 g of S reacts with 100/32 g of $CaCO_3$.
Therefore, 50 kg of S reacts with 100/32 × 50 kg = 156.25 kg of $CaCO_3$.

PROBLEM 11.3.5
Effects of atmospheric conditions on SO_2 concentration

A 1 m^3 sample of air contained 95 µg · m^{-3} of SO_2. The air temperature and pressure concurrently taken were 27°C and 104.152 kPa, respectively. What was the concentration of SO_2 in ppm?

SOLUTION 11.3.5

First, the M_W (molar mass) of SO_2 = 64.06 g $mole^{-1}$.
Second, converting 27°C to K = 300 K.

Concentration $= [(95\,\mu g \cdot m^{-3} / 64.06\,g \cdot mole^{-1}) \times 22.414\,L \cdot mole^{-1} \times (300 / 273\,K) \times$
$101.325\,kPa / 104.152] \cdot m^{-3} \times 10^3\,L \cdot m^{-3}$

Dividing 95 μg by 10^6 for grams and "cutting" 1 m^3 into 1 million parts for ppm cancel out each other.

Therefore, concentration = 0.035 ppm of SO_2.

11.4 CYANIDE REMOVAL BY SO₂

The maximum allowable concentration of sulfur dioxide for prolonged exposure is 5 ppm. SO_2 and HCN have immediately dangerous to life or health concentrations (IDLH). The chosen IDLH of SO_2 is based on the statement by the American Industrial Hygiene Association that 50–100 ppm is considered the maximum concentration for exposures of 0.5-h (Hickey) 1943). Short-term exposures to high levels of sulfur dioxide can be life-threatening. Exposure to 100 parts of sulfur dioxide per million parts of air (ppm) is considered immediately dangerous to life and health.

With a half-life (the time needed for half of the material to be removed) of hydrogen cyanide in the atmosphere of about 1–3 years, gaseous hydrogen cyanide is not easily removed from the air by settling, rain, or snow. People who breathed 546 ppm of hydrogen cyanide died after a 10-min exposure; 110 ppm of hydrogen cyanide was life-threatening after a 1-h exposure (ATSDR 2011). Yet, combining the gas with atmospheric SO_2 promises detoxification of both (Harris 2019).

REFERENCES

ATSDR (2011) Public health statement for cyanide. https://wwwn.cdc.gov/TSP/PHS/PHS. aspx?phsid=70&toxid=19

CDH (2020) CID Outbreak Definition and Reporting Guidance – California Department of Health California. www.cdph.ca.gov

FAO (2002) Impact of cassava processing on the environment. Retrieved December 1, 2018, from www.fao.org/docrep/007/y2413e/y2413e0d.htm

Harris MA (2019) Confronting Global Climate Change: Experiments and Applications in the Tropics. CRC-Taylor & Francis Publications. London, New York, Boca Raton. DOIhttps://doi.org/10.1201/9780429284847.

Hickey FC (1943). Noxious Gases and the Principles of Respiration Influencing Their Action. *J. Chem. Educ.* 1943, 20, 5, 259 Publication Date: May 1, 1943 https://doi.org/10.1021/ed020p259.2

IPCC (2005) Carbon dioxide Capture and Storage – IPCC https://www.ipcc.ch/report/carbon-dioxide-capture-and-storage/

Manahan S (2005) Environmental Chemistry. Taylor and Francis, Boca Raton, Fla. Retrieved June 11, 2014, from https://www.iucn.org/our-union/commissions/group/iucn-ssc-invasive-species-specialist-group

National Geographic (1952) https://education.nationalgeographic.org/resource/great-smog-1952

Sriroth, K, Chollakup R, Chotineeranat S, Piyachomkwan K, Oates CG (1999) Processing of cassava waste for improved biomass utilization. *Bioresour Technol* 71(1), 63–69.

12 Cement, Pozzolans, and Ashes

CONCRETE: MAIN CONCERNS

- Concrete: a surprising contributor to climate change
- Ten percent of global CO_2 emissions comes from the production of concrete.
- Producing 1 ton of cement generates 1 ton of CO_2
- Around 4 billion tons of concrete cement per year worldwide

12.1 SOURCES OF CEMENT, FLY ASH

Burning coal produces coal combustion residuals (CCR), or by-products, which includes fly ash (FA) (Figure 12.1), bottom ash, boiler slag, flue-gas desulfurization residues, and fluidized bed combustion ash (Bhatta et al. 2019). Over 70% of waste coal ash is categorized as fly ash, fine particulates captured by particulate control equipment, ranging in size from 0.5 to 300 μm (Koukouzas et al. 2006; Yao et al. 2015).

Cement manufacture, the third largest producer of anthropogenic (human-made) CO_2 in the world at 4–5% globally, produces 780 kg of CO_2 for every ton of cement made. These figures can be greatly reduced by the addition of fly ash as a pozzolan additive during the making of concrete. Pozzolans are a broad class of siliceous and aluminous materials (also occurring naturally as volcanic ash; non-crystalline silica glass in the case of fly ash) which, in themselves, possess little or no cementitious value but which will, in finely divided form and the presence of water, react chemically with calcium hydroxide.

$Ca(OH)_2$ at ordinary temperature forms compounds possessing cementitious properties (Hemalathaa & Ramaswamy 2017). The quantification of the capacity of a pozzolan to react with calcium hydroxide and water is given by measuring its pozzolanic activity (Koukouzas et al. 2006).

Experiments have shown that iron rusts rapidly when it is in contact with air containing traces of sulfur dioxide (Kilincceker et al. 2011), ash (Oze et al. 2014), and smoke (Hirschler and Smith 1989). Removal of any one of these three factors causes a sharp decline in the rate of rusting. An increase in metal rusting induces more metal smelting as replacements, which increases atmospheric CO_2. Other than when used in concrete, several other repurposed uses of fly ash, such as a soil stabilizer, soil additive, or embankment construction (Yao et al. 2015), can be environmentally detrimental.

PROBLEM 12.1.1
Calculating ash volumes

A coal-fired plant converts 31% of the coal's energy into electricity. (It is given that the air pollution control system collects fly ash as heavier ash particles fall to the floor of the combustion chamber. Together, these comprise 99% of the ash produced.)

DOI: 10.1201/9781003341826-16

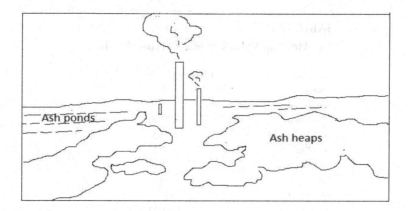

FIGURE 12.1   Coal ash surrounds a power plant.

Source: epa.gov

   To produce 680 MW electrical output, what is the approximate volume of ash produced in a year if the bituminous coal has a net heating value (NHV) of 28.5 MJ · kg^{-1}
(Table 12.1) and an ash content of 7.3% and the bulk density of the ash is 618 kg · m^{-3}?

SOLUTION 12.1.1 CALCULATING ASH VOLUMES

As indicated in the schematic diagram, transformations produce the masses and
energy of by-products.

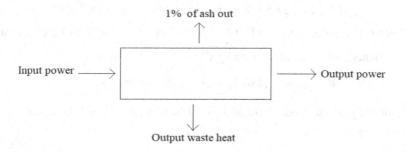

Based on a definition of efficiency,

$$\eta = \text{output power / input power}$$

Hence, converting the equation, the input energy for the power plant is

$$\text{Input power} = \text{Output power} / \eta = 680 \text{ MW}_e / 0.31 = 2194 \text{ MW}_t$$

Here, MW$_e$ refers to electrical power, while MW$_t$ refers to thermal power.

TABLE 12.1

Net Heating Values of Fuels (Typical Values)

Material	Net Heating Value (MJ · kg⁻¹)
Charcoal	26.3
Anthracite	25.8
Bituminous	28.5
Fuel oil	45.5
Gasoline	48.1
Natural gas	53.0
Peat	10.4
Wood (oak)	13.3–19.3
Wood (pine)	14.9–22.3

Source: EIA (1999)

Converting power to energy, $1 \text{ W} = 1 \text{ J} \cdot \text{s}^{-1}$:

$$(2194 \text{ MW}_t) (1 \text{ MJ} \cdot \text{s}^{-1} / \text{MW}_t) = 2194 \text{ MJ} \cdot \text{s}^{-1}$$

The amount of ash is found by balancing the masses involved in coal burning:

Coal ash in = Coal ash in chimney-stack gas + Coal ash in bottom of boiler

$$\text{NHV (net heating value, given)} = 28.5 \text{ MJ} \cdot \text{kg}^{-1}$$

28 MJ is produced by 1 kg of coal; therefore:

$$1 \text{ MJ (i.e., 28.5 / 28.5) is produced by} = 1 \text{ kg} / 28.5 \text{ MJ}$$

$$2194 \text{ MJ is produced by} = (1 \text{ kg} / 28.5 \text{ MJ}) \times 2194 \text{ MJ s}^{-1} = 76.98 \text{ kg s}^{-1} \text{of coal}$$

$$\text{Hence, ash going in} = (76.98 \text{ kg s}^{-1})(0.073) = 5.62 \text{ kg s}^{-1}$$

$$\text{But a 99\% ash capture} = (5.62 \text{ kg s}^{-1}) \times (0.99) = 5.56 \text{ kg s}^{-1}$$

$$\text{In one year, this would be} (5.56 \text{ kg s}^{-1})(86{,}400 \text{ s day}^{-1})(365 \text{ days year}^{-1})$$
$$= 1.75 \times 10^8 \text{ kg year}^{-1}$$

The volume of ash would be $= 1.75 \times 10^8 \text{ kg year}^{-1} / 618 \text{ kg m}^{-3}$
$$= 2.83171 \times 10^5 \text{ m}^3 \text{year}^{-1}$$

This large volume of ash would cover 70 football (or soccer) fields (90 m × 45 m) to a depth of 1 m.

PROBLEM 12.1.2
Heating of limestone and climate change

Calcining, a thermal treatment process bringing about a thermal decomposition below the melting point of the product, releases CO_2 from limestone to produce calcium oxide, which reacts with pozzolans during the manufacture of Portland cement.

a. Calculate the mass in metric tons of CO_2 released per metric ton of limestone which is used in this process (calcining) of converting limestone to quicklime.
b. Determine the mass of carbon added to the air for each gram of carbon dioxide that enters the atmosphere.

SOLUTION 12.1.2

a. Carbon dioxide (CO_2) is released thus:

$$CaCO_3 (s) \rightarrow CaO(s) + CO_2 (g)$$

$$40 + 12 + 48 = 100\,g \rightarrow (40 + 16) + (12 + 32 = 44\,g)$$

$$100\,g \times 10^6 = 1 \text{ metric ton of limestone } (CaCO_3)$$

$$44\,g \times 10^6 = 440 \text{ kg } CO_2 \text{ added to atmosphere for each calcined metric ton}$$
$$\text{of } CaCO_3$$

b. In CO_2: C = 12 g; CO_2 = 44 g.

Therefore, 1 g of carbon adds 44/12 g of CO_2 = 3.666 g of CO_2.

PROBLEM 12.1.3
Calculating CO_2 volumes

a. Calculate the mass in metric tons of CO_2 released per metric ton of limestone which is used in this process of converting limestone to quicklime.
b. Determine the mass of carbon added to the air for each gram of carbon dioxide that enters the atmosphere.

SOLUTION 12.1.3

a. Carbon dioxide is released into the atmosphere when limestone (calcium carbonate) is heated to produce quicklime (calcium oxide), as is the case in the manufacture of cement:

$$CaCO_3 (s) \rightarrow CaO(s) + CO_2 (g)$$

$$40 + 12 + 48 = 100\,g \rightarrow (40 + 16) + (12 + 32 = 44\,g)$$

$$100\,g \times 10^6 = 1 \text{ metric ton of limestone} (CaCO_3)$$

$$44\,g \times 10^6 = 440 \text{ kg of } CO_2 \text{ added to the atmosphere}$$

b. In CO_2: C = 12 g; CO_2 = 44 g.

Therefore, 1 g of carbon adds 44/12 g of CO_2 = 3.666 g.

12.2 ADVANTAGES OF TIRE-DERIVED FUEL

Tires can be used as fuel either in shredded form – known as tire-derived fuel (TDF) – or whole, depending on the type of combustion device, being typically used as a supplement to traditional fuels such as coal or wood (EPA 2016). Generally, tires need to be reduced in size to fit in most combustion units. Besides size reduction, the use of TDF may require additional physical processing, such as de-wiring.

There are several advantages to using tires as fuel (EPA 2016):

- Tires produce the same amount of energy as oil and 25% more energy than coal.
- The ash residues from TDF may contain a lower heavy metal content than some coals.
- Results in lower NO_x emissions when compared to many US coals, particularly the high-sulfur coals.

According to EPA (2016), the advantage of utilizing whole tires as kiln fuel is that there is no cost to create tire chips, and the removal of the steel is unnecessary since cement kilns have a need for iron in their processes, but tire chips may also be utilized because there is very little manual labor involved in handling chips versus whole tires. Nevertheless, producing chips from whole tires increases costs.

TDF is an appealing alternative fuel in this market due to its steady British thermal unit (Btu) value, low moisture content, and affordability when compared to other supplemental fuels (EPA 2016).

PROBLEM 12.2.1
TDF efficiency as a fuel

How many pounds of TDF would one need for a 24-h workday of production in a paper mill?

SOLUTION 12.2.1

In a pulp and paper mill, the workday carries on for 24 h with two 12-h shifts (Rākau 2022). Within this time the pulp goes through a lengthy process, a few steps of which involve steam. Between 1.1 and 2.2 kg of steam from water is used, with the amount of steam varying based on the grade of paper. The total steam consumed by the process is 13.9 kg · s⁻¹ (Edelmann et al. 1996). Considering TDF can produce 15,500 Btu · lb⁻¹ and it takes 1 Btu to raise 1 lb of water by 1°F, we can determine how much TDF is required in a single workday.

Btu means the amount of heat needed to raise 1 lb of water at the maximum density through 1°F, equivalent to 1.055×10^3 J.

First, it is necessary to find the total steam needed per day:

$$\text{Total steam consumed (kg} \cdot \text{s}^{-1}) \times 1 \text{ day (s)}$$

$$\text{Seconds in a day: } 24 \times 60 \times 60 = 86{,}400 \text{ s}$$

$$\text{Total steam consumed: } 13.9 \text{ kg} \cdot \text{s}^{-1}$$

$$13.9 \text{ kg} \cdot \text{s}^{-1} \times 86{,}400 \text{ s} = 1{,}200{,}960$$

Total steam needed per day: 1,200,960 kg (of steam)

Next, we can find the total water required to produce the amount of steam needed per day:

Total steam needed per day (kg) / total of steam produced from 1 kg of water (kg)

According to Edelmann et al. (1996), the average total of steam produced from 1 kg of water can be determined: $(1.1 + 2.2) / 2 = 1.65$ kg

Total steam needed per day: 1,200,960 kg

$$1{,}200{,}960 \text{ kg} / 1.65 \text{ kg} = 727{,}854.55 \text{ kg}$$

Water required to produce steam needed per day: 727,854.55 kg (or 1,604,644.60 lbs).

Assuming the water begins the process at room temperature (68°F) and has a boiling point of 212°F, the total energy per day is the total energy required to move 727,854.55 kg of water from 68°F to 212°F. Hence,

Total energy needed: $(212 - 68) \text{ °F} \times 1{,}604{,}644.60 \text{ lb} \times 1 \text{ Btu} \cdot \text{°F}^{-1} \cdot \text{lb}^{-1}$

$$= 144 \text{ °F} \times 1{,}604{,}644.60 \text{ lbs}$$

$$= 231{,}068{,}822.40 \text{ Btu}$$

Since TDF produces $15{,}500 \text{ Btu} \cdot \text{lb}^{-1}$ and the process requires 231,068,822.40 Btu, we can find how many pounds of TDF we would need to provide enough energy for the day:

Total energy needed (Btu) $\cdot \text{Btu}^{-1}$ produced per lb of TDF (Btu $\cdot \text{lb}^{-1}$)

Total energy needed: 231,068,822.40 Btu

Btu $\cdot \text{lb}^{-1}$ of TDF: $15{,}500 \text{ Btu} \cdot \text{lb}^{-1}$

$$231{,}068{,}822.40 \text{ Btu} / 15{,}500 \text{ Btu} \cdot \text{lb}^{-1}$$

TDF needed for energy for 1 day = 14,907.67 lbs

Therefore, in a single workday at the pulp and paper mill we would require 14,907.67 lbs of tire-derived fuel to produce enough steam for the paper making process. Moreover, at $15{,}500 \text{ Btu} \cdot \text{lb}^{-1}$, TDF is 25% more efficient than the highest quality coal, as 1 lb coal (anthracite)releases 12,700 Btu). TDFs are cheaper than coal, produce cleaner emissions, and have low moisture content (USEPA 2016). The US EPA testing has shown that (a) tire ash residues contain lesser heavy metals than some coals and (b) tire combustion results in less NO_x emissions compared to burning high-sulfur coal (USEPA 2016).

12.3 SCRUBBING CO_2

Owing to the increasing annual rate of CO_2 in the atmosphere, "scrubbing" carbon from the atmosphere gains importance. Though it has primarily been used as a soil amendment up to this point, biochar shows great promise for accomplishing this.

The use of biochar as a concrete addition to absorb carbon has been widely discussed, but no practical research has been conducted (USBI 2021).

PRACTICE PROBLEM 12.3.1

Coal-based power plants contribute heavily to greenhouse emissions. An older anthracite-fired power plant recently retrofitted with an air pollution control system collects fly ash while heavier ash particles fall to the floor of the combustion chamber. The plant now converts 37% of coal's energy into electricity. What is the approximate volume of ash produced in a year if the coal has an ash content of 7.3% and the bulk density of the ash is 598 kg m^{-3} (the plant produces 780 MW electrical output, and the fly ash collected and the bottom ash comprise 99.5% of the ash produced)? Table 12.1 shows net heating values.

SOLUTION 12.3.1

(Hint: the exact answer should not be very different from that of Problem 12.1.1.)

PRACTICE PROBLEM 12.3.2

List three benefits of pozzolan use in cement and concrete.

SOLUTION 12.3.2

PRACTICE PROBLEM 12.3.3

Replacing a proportion of Portland cement with ash decreases atmospheric CO_2. Find out and tabulate the maximum proportions of ash required for concrete of varying compressional and tensile strengths.

SOLUTION 12.3.3

Proportions in concrete	Concrete strength
Comments	

12.4   DRAWBACKS: RED MUDS AND FLY ASH

The main drawback of red mud waste is the high pH levels. Having developed a chemical equilibrium model for the fly ash/red mud system, Khaitan et al. (2007) used it to investigate the reactions governing the acid-base chemistry of the system. Despite having obtained estimates of the pH that can be achieved for various doses of fly ash (modeled as silica) added to red mud, overall, the results of their study (Khaitan et al. 2009) indicate that large doses of acidic fly ash are required to neutralize red mud, and the rate of neutralization is slow. Moreover, trace amounts of heavy metals in fly ash includes lead, zinc, copper, chromium, and cadmium, and Leelarungroj et al. (2018), after burying fly ash in soils, found that the dissolution of chromium and zinc from fly ash was amphoteric and controlled by oxide minerals at high and low pH.

PROBLEM 12.4.1
Fly ash disposal methods

The use of fly ash in concrete dates to the late 20th century and its advantages and disadvantages have been widely researched, with up to 35% by mass replacing cement (Hemalathaa & Ramaswamy 2017). In 2015, fly ash utilization rates were 70% for China, 43% for India, and 53% for the United States (Yao et al. 2015). This leaves a large potential for increased utilization, thereby decreasing disposal volumes and costs and replacing non-renewable and climate-change exacerbating processes.

What three other processes promise environmentally sustainable repurposing of fly ash?

SOLUTION 12.4.1

PRACTICE PROBLEM 12.4.2
Fly ash disposal ponds (group or individual work)

Fly ash from fossil fuels such as burnt coal contains toxic and heavy metals. The ponds where fly ash is usually dumped are poorly managed. Fly ash becomes dry as temperature increases and gets airborne. Thus, it becomes one of the major sources of air and water pollution. A systematic sampling procedure is one in which values are selected in a regular way. Line and point transect sampling are the primary distance methods. For this activity, you will test for the concentration of heavy metals outward from a power plant near you.

1. Obtain a Google map surrounding the nearest coal-fired power plant to your location.
2. On Google map, draw a 2-km circle around the power plant.
3. Do eight equally spaced transects radially outward from the power plant.
4. Plan sampling points at equally spaced distances along each transect.
5. Sample dust on tree leaves (equal surface areas) and/or that on vertical surfaces.
6. Make water solutions by shaking the dust in test tubes.
7. Test the filtered, clear solutions for heavy metals using "ICP".

8. Analyze the variance using ANOVA to determine whether there is greater variation between points within transects or between transects (see Appendix A of this book for ANOVA details).
9. Discuss your findings in at least 300 words.

PRACTICE PROBLEM 12.4.3
Fly ash disposal methods

Coal ash storage dams can fail as the EPA (2015) reports:

> On December 22, 2008, at approximately 1:00 a.m., a failure of the northwest side of a dike used to contain coal ash occurred at the dewatering area of the Tennessee Valley Authority (TVA) Kingston Fossil Plant (Figure 12.4.1), located at 714 Swan Pond Road in Harriman, Roane County, Tennessee. After the dike failure, approximately 5.4 million cubic yards (CYs) of coal ash was released into Swan Pond Embayment and three adjacent sloughs, eventually spilling into the main Emory River channel. The release extended approximately 300 acres outside of the fly ash dewatering and storage areas of the plant.

1. On the assumption that the dam failure spread ash evenly around the dam in the above example, calculate the area of ground contaminated in the above example._____
2. What processes can managers adopt to increase the viscosity of the ash, thereby decreasing the likelihood of failure and restricting contamination of the surrounding area in case of a dam failure?
   _____
3. Compare and contrast the physical properties, chemistry, and environmental effects of volcanic ash (Figure 12.2) with those of coal ash.

FIGURE 12.2 This pile of volcanic ash appears similar in appearance, texture, and grain size to the ash piles found adjacent to many coal-powered power plants.

Source: usgs.gov

PRACTICE PROBLEM 12.5.4
Dam failures

Dam failures, avalanches, and landslides often have a "most probable cause."

Make a table of dam failures (coal ash, red mud, etc.) occurring in major coal-burning countries during the last four decades. Classify these failures based on location, geological setting, topography, dam materials, configuration, length per square meter of surface, and volume. Using Appendix A (statistical tests) at the back of this book, do a Chi-square (or ANOVA) test for any relationships among the variables.

REFERENCES

Bhatt Arpita, Priyadarshini Sharon, Aiswarya Acharath Mohanakrishnan, Arash Abri, Sattler Melanie, Techapaphawit Sorakrich (2019) Physical, chemical, and geotechnical properties of coal fly ash: A global review. Case Stud Constr Mater 11, e00263. https://doi.org/10.1016/j.cscm.2019.e00263

Edelmann K, Kaijaluoto S, Timofeev O, Saarenko T, Kiiskinen H, & Karlsson M (1996) The impact of new paper drying technologies on energy consumption. https://www.research gate.net/publication/266026021_The_impact_of_new_paper_drying_technologies_on_energy_consumption#pf2

EIA (1999) U.S. Energy Information Administration (EIA). https://www.eia.gov

EPA (2015) EPA response to Kingston TVA coal ash spill. https://www.epa.gov/tn/epa-response-kingston-tva-coal-ash-spill

EPA (2016) Tire-derived fuel. https://archive.epa.gov/epawaste/conserve/materials/tires/web/html/tdf.html

Hemalatha T, and Ramaswamy A (2017) A review on fly ash characteristics – Towards promoting high volume utilization in developing sustainable concrete J Clean Prod 147, 546–559. https://doi.org/10.1016/j.jclepro.2017.01.114

Hirschler M, Smith GF (1989) Corrosive effects of smoke om metal surfaces. Fire Saf J 15(1), pp. 57–93, ISSN 0379-7112, https://doi.org/10.1016/0379-7112(89)90048-9.

Khaitan, S, Dzombak, D, Lowry G (2009). Chemistry of the acid neutralization capacity of bauxite residue. Environ Eng Sci 26(5), 873–881.

Khaitan Sameer, Dzombak David A, Lowry Gregory V (2007) Neutralization of bauxite residue with acidic fly ash. Published Online:3 Feb 2009 Environ Eng Sci 26(2) https://doi.org/10.1089/ees.2007.0232

Kilincceker G, Taze N, Galip H, Yazici B (2011) The effect of sulfur dioxide on iron, copper, and brass. Anti-Corros Metals Mater 58(1) pp 4–12.

Koukouzas NK, Zeng R, Perdikatsis V, Xu W, Kakaras EK (2006) Mineralogy and geochemistry of Greek and Chinese coal fly ash. Fuel 85(16), 2301–2309.

Leelarungroj K, Likitlersuang S, Chompoorat T, Janjaroen D (2018) Leaching mechanisms of heavy metals from fly ash stabilized soils. Waste Manag Res 36(7), 616–623.

Oze C, Cole J, Scott A, Wilson T, Wilson G et al. (2014) Corrosion of metal roof materials related to volcanic ash interactions. Nat Hazards 71, 785–802.

Rākau, KMP (2022) Pulp and paper mill operator. https://www.careers.govt.nz/jobs-database/manufacturing/manufacturing/pulp-and-paper-mill-operator/about-the-job

USBI (2021) Biochar research. US Biochar Initiative. https://biochar-us.org/biochar-research

USEPA (2016) Tire-derived fuel. https://archive.epa.gov/epawaste/conserve/materials/tires/web/html/tdf.html

Yao ZT, Ji XS, Sarker PK, Tang JH, Ge LQ, Xia MS, Xi YQ (2015) A comprehensive review on the applications of coal fly ash. Earth. Rev 141, 105–121.

13 Automobile Emissions

MAIN AUTOMOBILE POLLUTANTS:

- Gases
- Particulates
- Aerosols (small droplets of liquid suspended in gas aerosols)
- Primary pollutants – emitted directly from the point source (e.g., CO, NO_2, SO_2)
- Secondary pollutants – resulting from the interaction of primary pollutants PAN, smog, ozone, etc.

13.1 ROLE OF TURBULENCE

Turbulence in the atmosphere is a principal mechanism for preventing undesirable levels of contaminants from building up. For a given wind pattern, turbulence will be greater over steep hills or tall buildings than over a flat plain.

Carbon monoxide (CO), relatively stable in the atmosphere and with a half-life of about 0.2 years, forms mainly from incomplete combustion (Figure 13.1) of carbon in gasoline-fueled engines where gasoline normally burns at high temperature and pressure. It is converted into CO_2 by interaction with the hydroxide radical (OH^-) in the tropopause. Each year, over 290 million tons of CO are released into the world's atmosphere, mostly in the northern hemisphere. Concentration also depends on weather patterns.

CO, a colorless, odorless, potent gas, highly poisonous to living beings because of its ability to block the delivery of oxygen to the organs and tissues and a significant air pollutant in areas of consistently heavy traffic, affects individuals with respiratory diseases and anemia even at low concentrations.

PROBLEM 13.1.1
Avoiding exhaust gases

From a citizen's standpoint, little can be done to reduce the CO in exhaust from vehicles. Moreover, though CO is less dense than air and normally disperses quite rapidly after emission, it can remain undispersed, mainly in enclosed spaces such as garages and tunnels. Moreover, though ground level temperature inversions are a predominantly winter phenomenon, they regularly occur to varying degrees on cold mornings in every season. How, then, can residents in such areas decrease their intake of such high concentrations of atmospheric CO?

SOLUTION 13.1.1

Based on the premise "dilution is the answer to pollution," Figure 13.2 shows techniques for dispersion of ground-level CO in living areas.

DOI: 10.1201/9781003341826-17

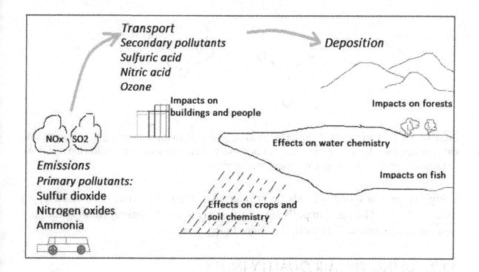

FIGURE 13.1 The fate and effects of automobile emissions.

By reducing the size of the aperture, the constriction at the left side in Figure 13.2 increases the velocity within the constricted area but preserves the same kinetic energy prior to the constriction based on the "Law of Conservation of Energy" (Equation 13.1).

$$KE = (1/2) \times \rho \times v^2 \tag{13.1.1}$$

where

ρ = mass density of the fluid

v = flow velocity

This increase in speed, accompanied by a compensatory drop in pressure (to conserve the original energy), produces a vacuum at the end of the aperture. But the

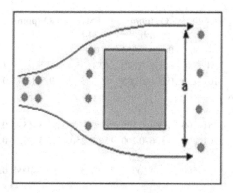

FIGURE 13.2 Dilution of incoming pollutants by the venturi effect.

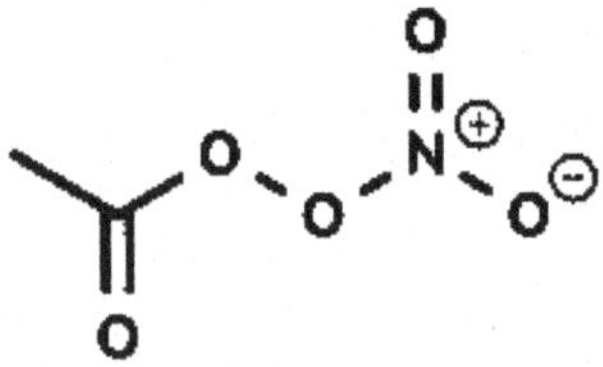

FIGURE 13.3 Peroxyacetyl nitrate, a secondary air pollutant produced by the photochemical epoxidation of hydrocarbons often in smog. Note the two reactive sites, allowing reactions in the atmosphere with chemicals from other species.

energy change in speed is adiabatic or internal to the system (i.e., not caused by an external force). The vacuum pulls air continuously through, thereby diluting the gaseous or particulate pollutants in the wider living area.

13.2   USING THE AIR QUALITY INDEX

In organic chemistry, peroxyacetyl nitrates (Figure 13.3) (also known as acyl peroxy nitrates, APN, or PANs) are powerful respiratory and eye irritants (i.e., a lachrymatory gas) present in photochemical smog. They are nitrates produced in the thermal equilibrium between organic peroxy radicals by the gas-phase oxidation of a variety of volatile organic compounds (VOCs), or by aldehydes and other oxygenated VOCs oxidizing in the presence of NO_2, coming from the high energy conditions of internal combustion engines. Being thermally unstable, peroxyacetyl nitrates decompose into phenoxyethanol radicals and nitrogen dioxide gas.

According to the US EPA Air Quality Index (AQI) (US EPA 2022), one may think of the AQI as a yardstick that runs from 0 to 500, where the higher the AQI value, the greater the level of air pollution and the greater the health concern (Table 13.1). For

TABLE 13.1

Daily Air Quality Index Basics

Color	Category	AQI	O_3 (ppm) 8-h	$PM_{2.5}$ 24-h	CO (ppm) 8-h	SO_2 (ppm) 1-h	NO_2 (ppm) 1-h
GREEN	Good	0–50	0.000–0.054	0.0–12.0	0.0–4.4	0–0.035	0–0.053
YELLOW	Moderate	51–100	0.055–0.070	12.1–35.4	4.5–9.4	0.30–0.075	0.054–0.100
ORANGE	Unhealthy for Sensitive Groups	101–150	0.071–0.085	35.5–55.4	9.5–12.4	0.076–0.185	0.101–0.360
RED	Unhealthy	151–200	0.086–0.105	55.5–150.4	12.5–15.4	0.186–0.304	0.361–0.649
PURPLE	Very Unhealthy	201–200	0.106–0.200	150.5–250.4	15.5–30.4	0.305–0.604	0.605–1.249
DK. RED	Hazardous	301–500	>0.200	250.5–500.4	30.5–1004	0.005–1.004	1.250–2.049

Source: US EPA (2022)

example, an AQI value of 50 or below represents good air quality, while an AQI value over 300 represents hazardous air quality. They continue:

> For each pollutant, an AQI value of 100 generally corresponds to an ambient air concentration that equals the level of the short-term national ambient air quality standard for protection of public health. AQI values at or below 100 are generally thought of as satisfactory. When AQI values are above 100, air quality is unhealthy: at first for certain sensitive groups of people, then for everyone as AQI values get higher.
>
> The AQI is divided into six categories. Each category corresponds to a different level of health concern. Each category also has a specific color. The color makes it easy for people to quickly determine whether air quality is reaching unhealthy levels in their communities.

13.3   FIVE MAJOR POLLUTANTS

The EPA establishes an AQI for five major air pollutants regulated by the Clean Air Act, with each of these pollutants having a national air quality standard set by EPA to protect public health (US EPA 2022). The list comprises the following:

- Ground-level ozone
- Particle pollution (also known as particulate matter, including $PM_{2.5}$ and PM_{10})
- Carbon monoxide
- Sulfur dioxide
- Nitrogen dioxide

THE PROCESS

- Air quality monitors measure $PM_{2.5}$ and PM_{10} concentrations in g m^{-3}.
- Local, regional, and national governments decide how to disseminate and monitor measurements to the public.
- Preferred way to communicate is via a color-coded AQI that is easy for the public to understand.

CALCULATING AQI VALUES

Breakpoints for the AQI can be observed in Table 13.2. The AQI is the highest value calculated for each pollutant as follows:

a. Identify the highest concentration among all of the monitors within each reporting area and truncate as follows:
   Ozone (ppm) – truncate to three decimal places
   $PM_{2.5}$ ($\mu g \cdot m^{-3}$) – truncate to one decimal place
   PM_{10} ($\mu g \cdot m^{-3}$) – truncate to integer
   CO (ppm) – truncate to one decimal place
   SO_2 (ppb) – truncate to integer NO_2 (ppb) – truncate to integer
b. Using Table 13.2, find the two breakpoints that contain the concentration.
c. Using Equation 13.2, calculate the index.
d. Round the index to the nearest integer.

TABLE 13.2
Breakpoints for the AQI

Color	AQI Values	8-h Avg O2 (ppm)	24-h Avg PM$_{10}$ ($\mu g \cdot m^{-3}$)	24-h Avg PM$_{2.5}$ ($\mu g \cdot m^{-3}$)
GREEN	0–50	0–0.059	0–54	0–15.4
YELLOW	51–100	0.06–0.075	55–154	15.5–35.4
ORANGE	101–150	0.076–0.095	155–254	35.5–65.4
RED	151–200	0.096–0.115	255–354	65.5–150.4
PURPLE	201–300	0.116–0.374	355–424	150.5–250.4
DK RED	301–500	Not defined	425–604	250.500.4

Source: US EPA (2022)

We calculate the AQI using pollutant concentration data using the following equation:

[Breakpntdiff AQI / Breakpnt diff pollutant] × (Conc pollutant – Breakpnt low) – AQI low

It means:

$$Ip = [(I_{High} - I_{Low}) / (BP_{High} - BP_{Low})] \times (Cp - BP_{low}) + I_{Low}$$

$$Ip = [(I_{High} - I_{Low}) / (BP_{High} - BP_{Low})] \times (Cp - BP_{Low}) + I_{Low} \qquad (13.3.1)$$

where

Ip = the index for pollutant p
Cp = the truncated concentration of pollutant p
BP_{High} = the concentration breakpoint that is greater than or equal to Cp
BP_{Low} = the concentration breakpoint that is less than or equal to Cp
I_{High} = the AQI value corresponding to BP_{Hi}
I_{Low} = the AQI value corresponding to BP_{Low}

PROBLEM 13.3.1
Identifying air quality index

A worker reported an 8-h ozone value of 0.08654333. What is the corresponding air quality index?

SOLUTION 13.3.1

First, truncate the value to 0.086. Then refer to the 8-hour ozone in Table 13.2 for the values that fall above and below our value (0.071–0.085). In this case, the 0.078 value falls within the index values of 101–150. Now we have all the numbers needed to use the equation.

$$Ip = [(I_{Hi} - I_{Lo}) / (BP_{Hi} - BP_{Lo})] \times (Cp - BP_{Lo}) + I_{Lo}$$

$$Ip = [(200 - 151) / (0.105 - 0.086)] \times (0.086 - 0.086) + 151$$

$$Ip = 151$$

So, an 8-h value of 0.08654333 corresponds to an index value of 151.

PROBLEM 13.3.2
More than one pollutant

In addition to reporting an 8-h ozone value of 0.086 ppm, the same worker in the above-mentioned case recorded a $PM_{2.5}$ value of 37.8 μg · m^{-3} and a CO value of 10.2 ppm. What is the AQI value?

SOLUTION 13.3.2

The AQI value of the major constituent comprises the AQI value.
For three pollutants, we apply the Equation 13.2 three times:

$$O_3 = [(200 - 151) / (0.105 - 0.086)] \times (0.086 - 0.086) + 151 = 151$$
$$PM_{2.5} = [(150 - 101) / (55.4 - 35.5)] \times (37.8 - 35.5) + 101 = 103.3$$
$$CO = [(150 - 101) / (12.4 - 9.5)] \times (10.2 - 9.5) + 101 = 112.8$$

The AQI is 151, with ozone as the responsible pollutant.

PROBLEM 13.3.3
Multiple readings

How do we use both ozone 1-h and 8-h values?

SOLUTION 13.3.3

We must calculate the 8-h values, and we may also calculate the 1-h values. If we calculate both, we must report the higher AQI value. Suppose we had a 1-h value of 0.162 ppm and an 8-h value of 0.078 ppm. Then we apply the equation twice. In this case, the index is 148 (the maximum of 148 and 126) and the responsible pollutant is ozone.
1 150 101 164 125 162 125 101 148 8 150 101 085 071 078 071 101 126 - - - - + =
- - - - + = h: () (. .) (. .) : () (. .) (. .) O PM CO 3 150 101 085 071 078 071 101 126 150
101 55 4 35 5 35.9 35 5 101 =102 100 51 9 4 4 5 8 4 4 5 51 90: () (. .) (. .) : () (. .) () :
() (. .) (. .) - - - + = - - - + - - - + = () (. .) (. .). . 150 101 085 071 078 071 101 49 014
007 101 125.5 126

PROBLEM 13.3.4
$PM_{2.5}$

A reading for $PM_{2.5}$ (Table 13.2) in Greensville showed a concentration of 20 μg · m^{-3}. If there are no other significant concentrations of pollutants in the air, what is the air quality index?

SOLUTION 13.3.4

US EPA Air Quality Index:

$$I = [I_{high} - I_{low} / C_{high} - C_{low}](C - C_{low}) + I_{low}$$

where

I = air quality index
C = pollutant concentration
C_{low} = conc breakpoint < C
C_{high} = conc breakpoint > C
I_{low} = index breakpoint for C_{low}
I_{high} = index breakpoint for C_{high}

From the breakpoint table for $PM_{2.5}$:

$$I = [100 - 51 / 35.4 - 12.1](20 - 12.1) + 51$$
$$= 123.8.$$

At this stage, the health risk is moderate.

PRACTICE PROBLEM 13.3.1
Using the air quality index

On three blank maps, plot the real-time AQI for O_3, $PM_{2.5}$, PM_{10}, CO, and NO_2 over major industrial cities in China, India, Canada, and your immediate location. Account for the disparities.

PRACTICE PROBLEM 13.3.2
Using the air quality index

What is the AQI for NO_2, O_3, and $PM_{2.5}$ in Mississauga, Canada, right now?
http://www.env.gov.bc.ca/epd/bcairquality/readings/map/station.html#E229797

PRACTICE PROBLEM 13.3.3
Using the air quality index

Multi-pollution index (MPI) – multi-pollutant air pollution (i.e., several pollutants reaching very high concentrations simultaneously) frequently occurs in several regions of some rapidly industrializing countries (Hu et al. 2015).

$$MPI = (1 / n)\left[\Sigma\{(ACi - GCi) / GCi\}\right]$$

where

ACi = actual concentration for pollutant i
GCi = guideline value for pollutant i
n = number of pollutants used in the index

What is the MPI for the following 3-h concentrations?
$O_3 = 0.087$ ppmv, $SO_2 = 0.226$ ppmv, $PM_{2.5} = 53.7 \ \mu g \cdot m^{-3}$

$$MPI = (1/3)\{(0.87-0.064)/0.064\}+\{(0.226-0.034)/0.034\}+\{(53.7-15.4)\}/ 15.4$$

$$= 1/3 \times (12.59+5.647+2.487) = 6.95$$

PROBLEM 13.3.5
Upper air pollution

To determine the upper air concentration of automobile-induced atmospheric pollutants over a large city, a 7000-L balloon will transport the required instruments. What mass of hydrogen gas is required to fill the balloon at 1 atm and 25°C?

SOLUTION 13.3.5

$n = PV / RT$ (as n = number of moles)

$$= (1 \ atm)(7000 \ L)/(298 \ K)^{-1}(0.082 \ L \cdot atm \cdot mole^{-1} \cdot K^{-1})^{-1}$$

$$= 286 \ mole$$

NB: Hydrogen is reactive and flammable. An innocent floating hydrogen-filled balloon exposed to something like fire would explosively react.
As hydrogen weighs $1 \ g \cdot mole^{-1}$,
$(286 \ mole)(1.0 \ g \cdot mole^{-1}) = 286 \ g$

PROBLEM 13.3.6
Air pollution and substandard automobile tires

In developing countries of Africa, South America, and surprisingly Australia, millions of cheap, low-quality automobile tires currently dominate the markets. What are some of the drawbacks of such tires?

SOLUTION 13.3.6

Drawbacks of these tires include those listed below (Meister 2022; US EPA 2022). Also, from several decades of experience with this kind of tire, the author of this book observed, and corroborates, the shortcomings listed here.

- Cheap tires typically have weaker sidewalls, which means premature replacement due to sidewall damage.
- Shorter lifespan, which means you have a more frequent replacement.
- Risking one's life on the road due to tire failures.
- Burst in extreme heat.
- Wear out unevenly and quickly.
- Gets punctured too easily.

- Okay performance on a dry road at low speeds of a maximum 50 mph.
- Notorious for short lifespan; very short lifespan, 10,000 miles at most.
- Easily damaged by road bumps and potholes.
- The products of these tire companies lack basic safety measures and quality control.
- Many have resulted in individual injury and wrongful death.

As shown above, these tires fail after less than a year of use and require replacing four to five times as often as the traditional tires. Their widespread usage can therefore exacerbate the effects of global climate change (and pollution of air, land, and water). As shown above, consumers reducing or refusing such products can significantly reduce the carbon footprint of many.

13.4   AGRICULTURE AS A CARBON SINK

Agriculture is a source of greenhouse gas emissions; however, it has the potential to become a large carbon sink. The introduction of perennial crops to orchards can increase carbon sequestration, produce more economic yield, and help mitigate climate change. Thyme among almond trees mitigates climate change and increases the land's productivity. The crop selected greatly affects the concentration of carbon dioxide sequestered.

PROBLEM 13.4.1
Agriculture and reducing emissions

How can perennial intercropping counteract automobile emissions?

SOLUTION 13.4.1

Values used:

Car engine = 2.3 kg of CO_2 produced per liter of gasoline consumed

Travel distance = 10,000 km annually

Mileage = 1 L of gasoline provides 14k

Total organic carbon of soil taken from results of a hypothetical report:
Tree details = 9 m (29.5 ft) tall, 20 cm (7.8 inches) diameter, and 15 years old
Step 1: Calculate the CO_2 emissions of subject car annually

10,000 km traveled annually
1 L of gasoline provides 14 km
Thus, 714.2 L used annually
1 L of gasoline consumed produces 2.3 kg of CO_2 emissions
Thus, 1642.6 kg of CO_2 is emitted annually and 16,426 kg over 10 years.

Step 2: Calculate the increase in CO_2 sequestration provided by perennial intercropping
According to the hypothetical report,

Total organic carbon increased at a depth of 10 cm from 3.85 kg $\cdot$ m^2 to 4.62 $\cdot$ m^2
4.62 kg – 3.85 kg = 0.77 kg

0.77 kg / 3.85 kg = 0.22 × 100 = 20
Thus, an increase of TOC at 10 cm depth from 3.85 kg shows a 20% increase in TOC and carbon sequestered.

Step 3: Calculate the total soil carbon for a small orchard with intercropping

If the orchard contains 50 trees with the above-mentioned characteristics,
Carbon sequestered = Carbon content weight × 3.67
Carbon content weight = 390.3 lbs (dry weight) × 0.5 = 195.1 lbs
195 × 3.67 = 716 lbs
Therefore, 716 × 50 = 35,800 lbs sequestered by trees over 10 years
Thus, with intercropping, 35,8000 increased by 20%
Therefore, 35,800 lbs + 20% = 42,960 lbs over 10 years and 4296 lbs annually

Step 4: Compare emissions of CO_2 and CO_2 sequestered

16,426 kg / 36,213 lbs CO_2 emitted over 10 years by engine
16,238 kg / 35,800 lbs of CO_2 sequestered by a 50-tree orchard without perennial intercropping
19,486 kg / 42,960 lbs of CO_2 sequestered by a 50-tree orchard with perennial intercropping

CONCLUSION

Perennial intercropping can noticeably increase the sequestration of land, allowing them to become great carbon sinks that can more significantly counter higher CO_2 emissions.

REFERENCES

Hu J, Ying Q, Wang Y, Zhang H (2015) Characterizing multi-pollutant air pollution in China: Comparison of three air quality indices. Environ Int 84, 17–25. https://pubmed.ncbi.nlm.nih.gov/26197060/

Meister P (2022) Top 10 Worst Tire Brands to AVOID Purchasing in 2022. https://carfromjapan.com/article/industry-knowledge/tire-brands-to-avoid-purchasing/

US EPA (2022) Air Quality Index. www.airnow.gov

14 Internal Combustion Engines Operating at Suboptimum Temperatures, Climate Change, and Health

ENGINE THERMOSTAT – MAIN USES:

- Rectifies engine temperature
- Saves fuel
- Extends engine life
- Reduces atmospheric pollutants

14.1 INGREDIENTS OF VEHICULAR EMISSIONS

Vehicular emissions contributing to air pollution are a major ingredient in the creation of climate change (Seinfeld & Pandis 1998) and smog in some large cities (USEPA 2013a. The escape of a large amount of unburnt hydrocarbons into the air (Figure 14.1), plus nitrogen oxides and carbon monoxide (Omave 2002), diminishes only as fossil-fueled internal combustion engines such as those in motor vehicles reach operating temperature (The Queensland Government 2016). As polycyclic aromatic hydrocarbons (PAHs) emanate from insufficiently burnt fuel, vehicles in which the engines have not warmed up are among the major contributors to the PAH level in cities (Baird & Cann 2013). Fine atmospheric particles – smaller than one-thirtieth of the diameter of a human hair – were identified more than 20 years ago as the most lethal of the widely dispersed air pollutants in the United States (Cohen et al. 2005). Although PAHs constitute only about 0.1% of airborne particulate matter, many of them are carcinogenic, at least in test animals (Baird & Cann 2013).

Preliminary results from a statistical study of children listed in the California Cancer Registry born between 1998 and 2007 suggest a 5–15% increase in the likelihood of some cancers due to traffic pollution (Reinberg 2013). These pollutants include secondary pollutants such as peroxyacetyl nitrates formed according to Equation 14.1.1:

$$\text{Hydrocarbons} + O_2 + NO_2 + \text{Light energy} \rightarrow CH_3COOONO_2 \qquad (14.1.1)$$

A WHO study found that diesel fumes cause an increase in lung cancer (BBC 2012). It concluded that the exhausts were a cause of lung cancer and may also cause

DOI: 10.1201/9781003341826-18

FIGURE 14.1 Unburnt hydrocarbons on a used spark plug.

tumors in the bladder. It based the findings on research on high-risk workers such as miners, railway workers, and truck drivers. Separate MIT studies have shown that traffic fumes account for 53,000 deaths per year in the United States and 5000 in the United Kingdom (Caiazzo et al. 2013). Small particles can penetrate deeply into sensitive lung tissue and damage it, worsening respiratory diseases such as emphysema and bronchitis and causing premature death in extreme cases (USEPA 2013b). It may also aggravate existing heart disease (USEPA 2013a), with such emissions being a major ingredient in smog which often affects large cities (USEPA 2013b).

PROBLEM 14.1.1
Shortcomings of running a cold engine

(a) Why is an engine thermostat necessary? (b) Why should it remain in place?

SOLUTION 14.1.1

(a)
- Because a cold engine cannot vaporize fuel completely, it releases unburnt hydrocarbons.
- Second, catalytic converters, the very devices meant to break down unburnt gases, are very inefficient below their ideal internal operating temperature (Farrauto and Heck 1999), and such a temperature occurs only when sufficient time elapses after ignition.
- Despite having markedly reduced that lag time by closer positioning of catalytic converters to the source of engine heat, and other successful innovations such as extra insulation, computer-controlled fuel injection, shorter intake lengths, and pre-heating of fuel and/or inducted air, even more so placing a small, yet rapid-response converter directly at the exhaust manifold, whereby this small converter handles the start-up

emissions, which allows enough time for the larger main converter to heat up (Pulkrabek 2004), some lag time still exists (USEPA 2013b). Exacerbation of this lag time can occur if vehicle operators fail to observe manufacturers' stipulations.

- The thermostat allows the engine to rapidly attain an ideal operating temperature at which fuel burns most efficiently.

(b)

- Removing a thermostat allows water to circulate and cool the engine block at a time when cooling is not required, thereby delaying operating temperature and releasing more unburnt emissions into the atmosphere.
- Fuel economy consequently decreases.
- Carbon buildup increases, shortening the life of the engine.
- Additionally, removing the thermostat can cause engine overheating (which increases global warming) after the operating temperature is achieved by facilitating a higher frequency of coolant circulation through the radiator whereby sufficient contact time between hot engine and coolant is not achieved. This shorter contact time between engine block and coolant concentrates heat in the atmosphere while ruining engines, thus causing the unnecessary production of replacement engines which entails increased burning of fossil fuels.
- Moreover, Mazzi and Dowlatabadi (2007) revealed an increase in PAHs from diesel vehicles after gaseous emissions from large automobiles were reduced by tax disincentives in the UK. Such environmental changes exhibit the potency of well-meaning policies (and, by extension, traditional activities of a population of persons) to exacerbate atmospheric pollution levels.

PROBLEM 14.1.2
The extent of the problem

How widely is thermostat removal practiced in the tropics?

SOLUTION 14.1.2

The extent to which thermostat removal is routinely practiced in several tropical locations was not studied. But while shopping for a car, the author of this book observed the phenomenon (Figure 14.2) in the tropics (Jamaica, W.I.).

Nevertheless, Harris (2019) found that missing thermostats in automobiles due to deliberate removal, in some cases under the quoted claim of "acclimatizing the car," widely occurred in several tropical countries. On the other hand, missing thermostats were almost non-existent in South Australia, despite the latter location experiencing higher temperatures than any tropical country during the summer season. A positive correlation between the education levels of many tropical mechanics and the extent of thermostat removal practiced was presented (Harris 2019) as a possible explanation for the phenomenon of widespread missing thermostats in the tropics.

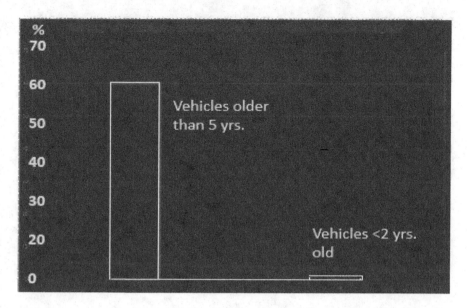

FIGURE 14.2   Thermostat removal, widely advocated among the less informed auto mechanics in Jamaica, and anecdotally in the wider tropics, wastes fuel & exacerbates climate change. Of the Suzuki SUVs observed in Kingston, Jamaica, the thermostat was missing in 60% of the >5-yr-old vehicles, but rarely missing in <2-yr-old vehicles.

REFERENCES

Baird C, Cann M (2013) Environmental Chemistry (5th ed.). W. H. Freeman. New York.

BBC (2012) Diesel exhausts do cause cancer, says WHO. www.bbc.com/news/health-18415532

Caiazzo F, Ashok Akshay, Waitz Ian A, Yim Steve HL, Barrett Steven RH (November 2013) Air pollution and early deaths in the United States. Part I: Quantifying the impact of major sectors in 2005. Atmos Environ 79, 198–208. https://search.yahoo.com/search/?p=Caiazzo%20F%2C%20Ashok%20Akshay%2C%20Waitz%20Ian%20A%2C%20Yim%20Steve%20HL%2C%20Barrett%20Steven%20RH%20(November%202013)%20Air%20pollution%20and%20early%20deaths%20in%20the%20United%20States.%20Part%20I%3A%20Quantifying%20the%20impact%20of%20major%20sectors%20in%202005.%20Atmos%20Environ%2079%2C%20198%E2%80%9320 8.

Cohen AJ, Ross Anderson H, Ostro Bart, Pandey KD, Krzyzanowski M et al. (2005) The global burden of disease due to outdoor air pollution. J Toxicol Environ Health 68 (13–14), 1301–1307. doi:10.1080/15287390590936166.

Farrauto Robert J, Heck Robert M (1999) Catalytic converters: state of the art and perspectives, Catalysis Today, 51(3–4), 1999, pp. 351–360 ISSN 0920-5861, https://doi.org/10.1016/S0920-5861(99)00024-3.

Harris M (2019) Confronting Global Climate Change Experiments & Applications in the Tropics. CRC Press, Boca Raton, Fla.

Mazzi E, Dowlatabadi H (2007) Air quality impacts of climate mitigation: UK policy and passenger vehicle choice. Environ Sci Technol 41(2), 387. doi:10.1021/es060517w.

Omave ST (2002) Metabolic modulation of carbon monoxide toxicity. Toxicology 180(2), 139–150. doi:10.1016/S0300-483X(02)00387-6. https://pubmed.ncbi.nlm.nih.gov/12324190/

Pulkrabek WW (2004) Engineering Fundamentals of the Internal Combustion Engine. Pearson Prentice Hall, New Jersey. https://search.yahoo.com/search/?p=Pulkrabek%20 WW%20(2004)%20Engineering%20Fundamentals%20of%20the%20Internal%20 Combustion%20Engine.%20Pearson%20Prentice%20Hall%2C%20New%20Jersey

Reinberg S (2013) Smog exposure during pregnancy might raise child's cancer risk: Study. US News, April 9.

Seinfeld John and Pandis Spyros (1998) Atmospheric Chemistry and Physics: From Air Pollution to Climate Change (2nd ed.). Hoboken, New Jersey: John Wiley & Sons, Inc. p. 97. ISBN 0-471-17816-0.

USEPA (United States Environmental Protection Agency) (2013a) State and county emission summaries: Nitrogen oxides. Air emission sources. United States. https://www.epa.gov/

USEPA (United States Environmental Protection Agency) (2013b) State and county emission summaries: Volatile organic compounds. Air emission sources. https://www.epa.gov/

15 Indoor Emissions

INDOOR POLLUTION – MAIN POINTS AND CHEMICAL SPECIES:

- Composition and exchange with outdoor air
- Indoor materials (construction, consumer goods)
- Indoor activities (cooking, heating, smoking)
- CO, CO_2, NO_x, VOCs, PM, Rn
- Specific VOCs of concern: H_2CO, BTEX, halocarbons
- PAHs, PBDEs, PCBs, phthalates

15.1 COMPOSITION AND EXCHANGE OF INDOOR AIR

Polluted air, originating from the discharge into the atmosphere of foreign gases, vapors, droplets, and suspended particles, can devastatingly exacerbate underlying morbidities and claim lives (WHO 2021). Never in all of Earth's history has civilization experienced atmospheric CO_2 due to fossil fuels as they do now.

Davis (2007) states that Americans spend approximately 90% of their time indoors, in homes, daycare agencies, schools, and workplaces. This percentage climbs even higher for sub-populations such as infants and young children, the elderly, and people living with disabilities (Davis 2007). Restricted air movements also exacerbate biological morbidities. The body metabolizes toxic substances to detoxify them. Benzene was once widely used as a solvent and reagent in organic chemistry laboratories, but it is now discontinued because of its potential to cause blood abnormalities and possibly leukemia. It is metabolized by cytochrome P-450 enzymes to products such as phenol and trans-muconic acid that are eliminated through urine. Unfortunately, benzene oxepin and benzene oxide (Figure 15.1) are reactive intermediates that react with biomolecules in the body to produce the toxic effects of benzene (Manahan 2005).

Viruses can remain infectious for long periods of time if airborne in closed spaces. The open air appears to have a bactericidal as well as a viricidal activity, associated by some authors with the air pollution complex of oxidants and ozone. Measurements of the viricidal activity of open air show that the sterilizing power of air increases with decreasing size of the particles containing the viruses (Benbough & Hood 1971).

PROBLEM 15.1.1
Factory air monitoring

Obtain photographs of a factory and its surroundings in your area (after requesting permission from the manager). Using "Fixed station modeling and sampling," determine the existing levels of indoor pollutants in the building. Using explanatory schematic diagrams as above, modify the atmospheric design of your factory and/or its environs to:

a. improve the comfort of workers and
b. reduce atmospheric pollutants which may flow into the factory.

DOI: 10.1201/9781003341826-19

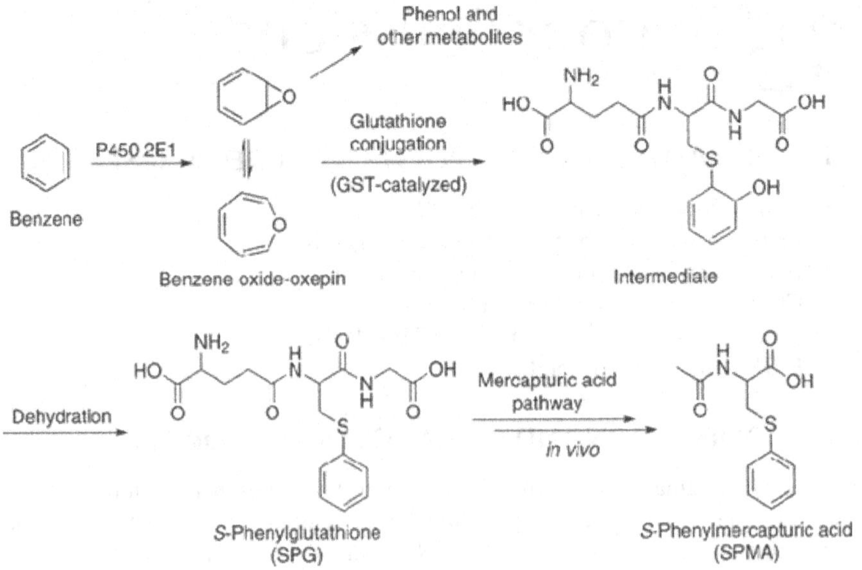

FIGURE 15.1   The benzene ring and ensuing metabolites.

Source: tools.niehs.nih.gov.

SOLUTION 15.1.1

The NIH (2022) reports that the effects of equipment heat and noise, as well as occupant inconvenience, sampling, and monitoring equipment, should (and usually can) be in remote locations outside the building being evaluated. Hence, it is common practice to locate the instruments outside the building space and draw the air samples to them.

Discuss your proposal with the factory's directors and promise to give them a copy of the report including any recommendations for remediation. Follow a procedural step outlined by the NIH (2022) by going to the following site:

https://www.niehs.nih.gov/research/supported/translational/peph/newsletter/2022/09/index.cfm

15.2   INDOOR FOSSIL FUEL-BASED MATERIALS (CONSTRUCTION, CONSUMER GOODS)

Indoor activities include cooking, heating, painting, and spraying.

PRACTICE PROBLEM 15.2.1
Residential monitoring

Sources of chemical fumes include paints and solvents, cleaning products, outdoor air pollutants, combustion gases, carbon monoxide, gases, and radon seeping in from foundations. Some health-threatening fossil fuel-based or wood-burning gases found

TABLE 15.1
Some Indoor Pollutants

Gas Symbol	Name	Sources (Minimum 3)	Threats or Possible Threat
NO_x	Nitrogen oxides	Unvented gas stoves, kerosene heaters, wood stoves	Irritated lungs, children's colds, headaches
	Carbon monoxide		
	1,1,1-Trichloroethane		
	Paradichlorobenzene		
$CHCl_3$			
CH_2O			
	Styrene		
	Tetrachloroethylene		
	Benzo-α-pyrene		
CH_2Cl_2			
	Paint		
	Hydrogen sulfide		
	Lacquer thinner		
	Fingernail polish		

indoors are given in Table 15.1. As shown in the first row, fill in the blanks, including indoor sources.

PROBLEM 15.2.1
Air changes per hour (ACH)

Having an adequate amount of external air entering a dwelling or workplace avoids the buildup of pollutants and helps control humidity levels, temperature, odor, and various other elements that can affect the comfort and health of those living inside.

Air changes per hour (ACH) tells us how many times a ventilation device can fill up the full volume of a room with air (EPA 2021). This is especially useful when comparing different air purifiers or air conditioners.

ACH for older homes and buildings: ACH ~ 1 h^{-1}

Newer homes and buildings ACH ~ two to five air changes per hour (h^{-1})

$$ACH = CFM \times 60 \, / \, Area \times Height \ of \ ceiling \qquad (15.2.1)$$

An air purifier with 220 CFM airflow is placed in a 200 sq ft room with a standard-height ceiling (8 ft). How many air changes per hour does the unit make?

SOLUTION 15.2.1

220 CFM = 220 cubic ft per minute.

In 1 h (60 min), air moved = $60 \times 220 = 13,200$ cubic ft per hour.

Volume of the room = 200 sq ft $\times$ 8 ft = 1600 cubic ft.

ACH = $220 \times 60 \, / \, 200$ sq ft $\times$ 8 ft

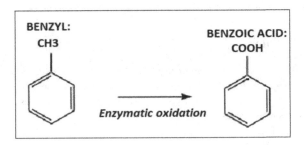

FIGURE 15.2   Toluene oxidation to benzoic acid.

Rate that air purifier can change the volumetric air in the room = 8.25 times · h⁻¹. *Why are PAHs of concern? It is because of their following characteristics:*

- Genotoxic
- Mutagenic
- Carcinogenic
- Of ubiquitous, constant input (limited or no control of non-point sources)

15.3   REMEDIATION BY SUPERSEDING BENZENE WITH TOLUENE

Benzene has been replaced in the laboratory with toluene, which has solvent and chemical properties that are largely like those of benzene. However, the $-CH_3$ side group on toluene is readily oxidized by body enzymes to benzoic acid (Figure 15.2), a harmless, common food metabolite. Benzoic acid is conjugated with one of the body's natural amino acids, glycine, to produce hippuric acid, which is eliminated through urine (Manahan 2005).

REFERENCES

Benbough JE, Hood AM (1971) Viricidal activity of open air. https://www.ncbi.nlm.nih.gov/pubmed/4944178

Davis A (2007, May 31) Home environmental health risks. *Online J Issues in Nurs* 12(2). https://pubmed.ncbi.nlm.nih.gov/21848351/

EPA (2021) Carbon monoxide's impact on indoor air quality. https://www.epa.gov/indoor-air-quality-iaq/carbon-monoxides-impact-indoor-air-quality#:~:text=%5BACGIH%20TLV%5D%20The%20American%20Conference,15%5D

Manahan Stanley (2005) Environmental Chemistry. Taylor and Francis, Boca Raton, Fla. https://www.iucn.org/our-union/commissions/group/iucn-ssc-invasive-species-specialist-group. Accessed June 11, 2014.

NIH (2022) Monitoring and modeling of indoor air pollution. https://www.ncbi.nlm.nih.gov/books/NBK234059/

WHO (2021, September 22) Household air pollution and health. https://www.who.int/news-room/fact-sheets/detail/household-air-pollution-and-health

16 Air Pollution Meteorology

AIR POLLUTION – MAIN FACTORS:

- Atmospheric particulates
- Atmospheric stability
- Chimney height
- Height of inversion layer
- Storms, hurricanes, and El Nino
- Humidity
- Radiation and agriculture
- Wind speed and evaporative water loss
- Wind speed
- Topography
- Mixing depth
- Remediation of aerosols

16.1 ATMOSPHERIC PARTICULATES

Complex chemistry involving hundreds of reactions, where photochemical responses occur when photons of ultraviolet radiation split molecules apart to create reactive fragments with unpaired electrons known as free radicals, dominates atmospheric interactions. Dust, smoke, and directly emitted gases (primary pollutants) moved into the air or resulting from the transformation of gaseous pollutants consisting of particulate matter (PM) are one of the major air pollutants in urbanized regions (Figures 16.1 and 16.2), the effects on human and environmental systems having been discussed by many of scientists (Özdemir 2015). Particulates with aerodynamic diameters <10 μm (PM_{10}) and <2.5 μm ($PM_{2.5}$) cause lung cancer, asthma, morbidity, and mortality (Akyüz & Cabuk 2009; Brandt et al. 2011). So, air pollution control and monitoring are very important for preserving human health.

Sources of VOCs:

1. Natural sources: forest fires, volcanoes
2. Mobile sources: volatile organic compounds (VOCs), NO_x, PM (mainly from motorized vehicles)
3. Stationary sources: NO_x, SO_2, PM (from power plants and factories)
4. Area sources: dry cleaners, petrol stations

VOCs' particles come from natural sources (e.g., volcanic eruptions) and human activities such as burning fossil fuels, incinerating wastes, and smelting metals. PM is one of the six EPA "criteria pollutants" that have been determined to be harmful

DOI: 10.1201/9781003341826-20

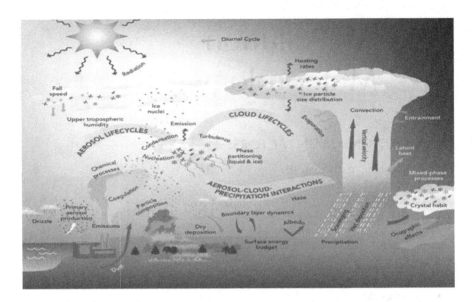

FIGURE 16.1   Interactions among anthropogenic substances and natural atmospheric systems.

Source: www.asr.science.energy.gov

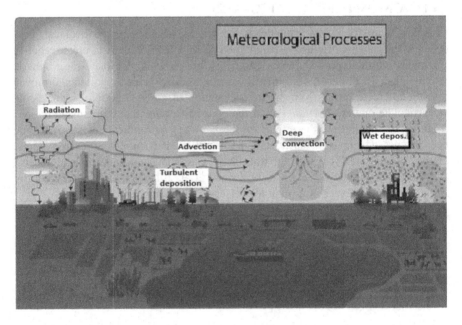

FIGURE 16.2   Meteorological processes showing turbulence, convection, deep convection, radiation, and wet deposition.

Source: https://www.epa.gov/cmaq/meteorological-process-overview

to public health and the environment. (The other five are ozone, sulfur dioxide, nitrogen dioxide, carbon monoxide, and lead.)

16.2 ATMOSPHERIC AEROSOLS

Aerosols are tiny particles in the air that can be produced when we burn different types of fossil fuels – coal, petroleum, wood, and biofuels – in different ways. A significant man-made source of aerosols is pollution from cars and factories. These small liquid droplets or solid particles suspended in the atmosphere (0.01–$10\,\mu m$) (mist, fog, clouds, smog) (dust, smoke, pollen) are usually associated with "haze," causing incoherent scattering of visible light, thereby interfering with the optical transmission. Scattering occurs when aerosols are comparable to the wavelength of light, e.g., 400 nm (blue light) scattered by particles in the 0.04–$4\,\mu m$ range. Large particles ($>10\,\mu m$) settle out. Very small particles ($< 0.01\,\mu m$) coagulate to form larger particles. Classified by size, source, and type (pre-formed versus condensed), aerosols in the 0.01–$1\,\mu m$ range can remain suspended for months. Given that the earth's atmosphere is about 100 miles deep, it is sometimes compared to the oceans. The thickness and volume sometimes are suggested to be enough to dilute all the chemicals and particles thrown into it. However, 95% of this air mass is within 12 miles of the earth's surface and it is this 12-mile depth that contains the air humans breathe as well as the pollutants they emit (Beyond Discovery 2022). They say aerosol concentrations are controlled by atmospheric mixing, chemical transformation, emission, etc. This chapter evaluates the impact of factors such as atmospheric stability, chimney height, and height of the inversion layer on the distribution of such pollutants.

Natural sources:
- Wind-blown dust, silt, fine sand, etc.
- Volcanoes: dust, ash, H_2SO_4 (aq)
- Forest fires: smoke particles, soot
- Terpenes/isoprenes: naturally occurring VOCs and their breakdown products such as aldehydes
- Pollens
- Microbial aerosols: stable colloidal systems of microorganisms floating in the air as a single cell, a group of cells, or fused with dry or solid particles (Mirhoseini et al. 2021)
- When the microorganism is a virus, it is called a virus aerosol (Langmuir 1961).
- Droplets having a diameter of $>5\,\mu m$ are emitted when someone talks, coughs, or sneezes. Because atmospheric instability increases with temperature, a warming planet exacerbates the atmospheric travel distance of aerosols.

Anthropogenic sources:
- Industrial dust: e.g., cement, soot, fly ash (may be removed or reduced using scrubbers or precipitators)
- Agriculture: land clearing, tilling, etc.
- Transportation: roadway dust, diesel exhaust, PAHs, smog
- Oxidation of volatile gases, VOCs, aldehydes, acids, NO_x salts, and SO_x.

Effects:
- Health: large particles (>10 μm) trapped in the nose or upper respiratory tract. Small particles (<2.5 μm) are transported into the lower lung cavity, where they become immobilized and cause serious ailments and disease.
- Visibility, climate (affects radiation budget), soiling of materials
- Major participants in heterogeneous atmospheric reactions (e.g., ozone hole formation, acid rain production)

Aerosol processes

Aerosol concentrations are controlled by atmospheric mixing, chemical transformation, rates of emission, etc. (Choi et al. 2008). But Pateraki et al. (2012), who attest to interactions that affect the concentration of aerosols in the atmosphere as being "real," note that the understanding of the connection between meteorological factors (relative humidity, temperature, wind, rainfall, etc.) with particulate matter is not clear.

Movement of aerosols

The following processes control the existence of atmospheric aerosols:

- Diffusion
- Coagulation
- Condensation
- Chemical reactions
- Sedimentation

Sedimentation rate of aerosols

$$\text{Sedimentation rate } \alpha = \text{Size} \times (\text{Density}) \, / \, \text{Air viscosity} \qquad (16.2.1)$$

For example, 1 μm diameter droplet of H_2O (l) settles at about ~10^{-4} m · s^{-1}, whereas a 1 mm diameter droplet of H_2O (l) settles at 6.5 m · s^{-1}.

Settling velocity

At "settling velocity" (constant velocity of fall or terminal velocity), the upward and downward forces balance each other.

Stokes's law, which pertains to the settling of sediment in fresh water and to measurements of the viscosity of fluids (assuming spherical particles), depends on this principle. However, the validity is limited to conditions in which the particle motion does not produce turbulence in the fluid.

Here, the drag force F acting upward in resistance to the fall is equal to:

$$F = 6\pi r \eta v \qquad (16.2.2)$$

where
  r = radius of the sphere
  η = viscosity of the liquid ($\eta = 1.9 \times 10^{-2}$ g · m^{-1} · s^{-1} at T = 298 K)
  v = velocity of fall (m · s^{-1})

The downward force is equal to:

$$\frac{4}{3}\pi r^3 (d_1 - d_2)g \qquad (16.2.3)$$

where

$\frac{4}{3}\pi r^3$ = volume of a sphere
d_1 = density of the sphere $(g \cdot m^{-3})$
d_2 = density of the liquid $(g \cdot m^{-3})$
g = acceleration due to gravity

To solve for v, we can now equate the two expressions given above:

$$6\pi r \eta v = \frac{4}{3}\pi r^3 (d_1 - d_2)g \qquad (16.2.4)$$

$$6\eta v = \frac{4}{3} r^2 (d_1 - d_2)g$$

$$v = \frac{2}{9}(d_1 - d_2)gr^2 / \eta = \text{meters per second}$$

PROBLEM 16.2.1
Settling rate of aerosols

The rate of fall of aerosol pollutants can predict their longevity in the atmosphere. Two drops of the same radius are falling through the air with a steady velocity of 6 cm $\cdot$ s. If the two drops coalesce, what would be the terminal velocity?

SOLUTION 16.2.1

If R is the radius of the new bigger drop formed (coalesced), then:

$$\frac{4}{3}\pi R^3 = 2 \times \frac{4}{3}\pi r$$

Or

$$R = 2^{1/3} r$$

As v varies directly with r^2 (Equation 16.2.4),

$$v_0 / v_{01} = R^2 / r^2 = (2^{1/3})^2 / r^2 = 2^{2/3}$$

Transposing:

$$v_{01} = v_0 (2^{2/3}) = 6(4)^{1/3} \text{ cm/s}$$

Terminal velocity of coalesced drops = 9.6 cm $\cdot$ s^{-1}

PROBLEM 16.2.2

Particulates of fly ash from a power plant are falling through the atmosphere from a height of 200 m. About how long will it take for the particulates to settle if they are of the following average size: 5.0×10^{-8} mm?

SOLUTION 16.2.2

$$u = d^2 g (\rho_s - \rho_f) / 18\eta$$

where

 d = particle diameter
 ρ_s = particle density
 ρ_f = density of fluid
 $\eta = 1.0 \times 10^{-3}$ kg · m^{-1} · s^{-1}

(as the SI unit of viscosity is the pascal second (Pa · s) or kg · m^{-1} · s^{-1})
At 1013.25 hPa pressure (P = 1 atm) and at 15°C and at mean sea level (MSL):

$$\text{Air density} = \text{Approx. } 1.225 \text{ kg} \cdot \text{m}^{-3}$$

Average rock particle density $= 2.7 \times 10^6$ g · m^{-3} (and ash is mainly rock minerals)

For ash, u (converting mm to m):

$$= [5.0 \times 10^{-11} \text{m} \times (9.81 \text{ m} \cdot \text{s}^{-2}) (2.7 \times 10^6 \text{g} \cdot \text{m}^3 - 1.225 \times 10^3 \text{g} \cdot \text{m}^3)] / $$
$$18(1.0 \times 10^{-6} \text{g} \cdot \text{m}^{-1})\text{s}^{-1}$$
$$= 0.007354 \text{ m} \cdot \text{s}^{-1}$$

Falling through the 200 m would take 200 m / 0.007354 m · s^{-1} = 27,196 s = 453.26 min.
Therefore, time for particulates to settle = 7.55 h.

PRACTICE PROBLEM 16.2.1
Sahara Desert dust

Desertification is reported to have increased in recent decades. If fine Sahara sand grains, mainly of the mineral quartz, are slowly falling through the atmosphere from a height of 2800 m, how long will it take for the particulates to settle if they are of the following average sizes: 4.0×10^{-4} mm.

SOLUTION 16.2.1

Hint: See the solution to the preceding problem.

16.3 ATMOSPHERIC STABILITY

The adiabatic lapse rate occurs when the energy released from within a rising air mass under decreasing atmospheric pressure as altitude is gained cools the air mass (Figure 16.3). The environmental lapse rate (non-adiabatic) is considered positive when the temperature decreases with elevation, zero when the temperature is constant with elevation, and negative when the temperature increases with elevation (temperature inversion). There are four kinds of inversions: ground, turbulence, subsidence, and frontal.

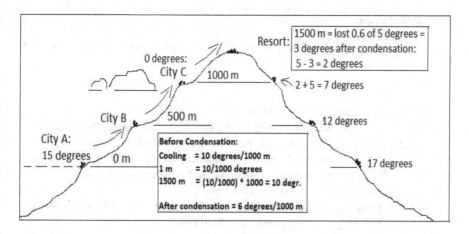

FIGURE 16.3 Effects of varying lapse rates and processes on temperature.

PROBLEM 16.3.1
Inversion characteristics

"Upside-down" lapse rates occur when air temperature increases with height, thereby trapping pollutants. What other attributes characterize an inversion layer under the headings: (a) stability, (b) time of day, (c) pressure systems (high or low), and (d) air quality?

SOLUTION 16.3.1

a. Stability? = Strongly stable
b. Does it occur during the day or at night? = Either
c. Is it associated with high or low air pressure systems? = High
d. Does it improve or deteriorate air quality? = Deteriorate

PROBLEM 16.3.2
Assumptions of the Gaussian model

Effects of vertical stability on mixing and concentrations at the ground determine the dispersion of atmospheric pollutants. The Gaussian model can be used to illustrate the resulting effect of wind fluctuations and speed on pollutant concentrations. In terms of the Gaussian model, what parameters determine pollution from a chimney?

SOLUTION 16.3.2

Gaussian dispersion modeling assumes the following conditions:

- Source pollutant emission rate = constant (steady state)
- Constant wind speed, wind direction, and atmospheric stability class
- Pollutant mass transfer is primarily due to bulk air

Gaussian dispersion equation:

$$C(x,y,z) = \frac{Q}{2}\pi\sigma y\ \sigma z\ u[\exp(-\frac{1}{2}\{y^2/\sigma^2y + (z-H)^2/\sigma^2z\}] \qquad (16.3.1)$$

where

- C = concentration of the pollutant
- Plume rise (Δh and z) is caused by the physical stack height (h), effective stack height (H or h_e), and wind speed (u[x], i.e., in the horizontal direction).
- σy and σz = f (downwind distance x and atmospheric stability; the greater the y, the greater the plume is spread out and greater the stability).
- σy and σz depend on the atmospheric conditions.
- Atmospheric stability classifications are defined in terms of surface wind speed, incoming solar radiation, and cloud cover.
- Q = pollutant emission rate ($g \cdot s^{-1}$)
- H = effective stack height (meters) = stack height + plume rise = $\sigma z + h_e$
- U = wind speed ($m \cdot s^{-1}$), which increases dispersion for x, y

Pollutant plume dispersion in the horizontal (y) and vertical (z) are produced by eddies and random shifts of wind direction (Figure 16.4, Equation 16.3.1).

Motion in the x-direction (assumptions):

- No pollutant chemical transformations occur
- Wind speeds are >1 $m \cdot s^{-1}$
- Limited to predicting concentrations >50 m downwind (to negate chimney effects)

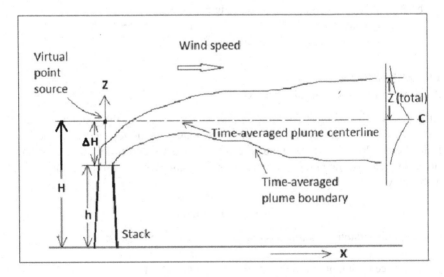

FIGURE 16.4 Gaussian plume model for atmospheric dispersion in local range (approximately <50 km).

Conclusion

Combining Equation 16.3.1 and Figure 16.4, and with the parameters being below all in the denominator of Equation 16.3.1, the following conclusions can be drawn:

- Concentration varies inversely with the horizontal spread of the pollutant, depicted here by the equation as a horizontal circle, i.e., $2\pi\sigma y\sigma zu$ ($=2\pi r$)
- The higher the plume height (h + z), the lower the concentration (C) of the pollutant
- The greater the speed (u), the lower the concentration

PROBLEM 16.3.3
Diluting air pollutants

Movement of air in the x, y, and z directions dilute air pollutants. What sequence of forces produces a convection cell of air movement?

SOLUTION 16.3.3

Horizontal motion seven-step development:
- Radiation absorbed by earth's surface – surface warms
- Air above ground warms by conduction – air expands, density decreases
- Air becomes unstable – convection sets in
- Air rises – pressure aloft increases (more molecules in motion)
- Air aloft pushed laterally away from the warm surface – surface pressure drops
- Surface pressure around warm area increases – air converges toward warm area
- Air subsides around the warm surface – complete convection cell established

16.4   ADIABATIC CONDITIONS

Pressure = Force per unit area. Pressure decreases exponentially with height.

$$10 \text{ mbar (millibar)} = 1 \text{ kPa (kilopascal)}$$

Any expanding system requires energy. Rising air expands but requires no external energy to expand. When a parcel exchanges no energy with its surroundings, we call this state adiabatic.

PROBLEM 16.4.1
Adiabatic lapse rate

Having supplied the required energy for its own expansion, an air mass system internally loses heat. How is the adiabatic lapse rate for dry air measured?

SOLUTION 16.4.1

The following equations quantify the adiabatic lapse rate:

$$0 = C_p \Delta T + g \Delta z \text{ (i.e., } 0 = \text{no energy lost, changed to movement?)}$$

where

   0 = no energy
   C_p = constant pressure specific heat (i.e., specific heat at same pressure)
   ΔT = change in temperature in K degrees
   g = gravitational acceleration
   Δz = change in height

$$c_p \, \Delta T = -g \, \Delta z$$

As $g = 9.81$ m · s² and the constant pressure specific heat for dry air is C_p = 1007 J · kg⁻¹ · K⁻¹, the adiabatic lapse rate for dry air is:

$$\frac{\Delta T}{\Delta z} = -\frac{g}{c_p} = -\frac{\left(9.81 \, \mathrm{m\,s^{-2}}\right)}{\left(1004 \, \mathrm{J\,K^{-1}\,kg^{-1}}\right)}$$

$$= 0.00977\,°C \cdot m^{-1}$$

$$\Gamma_d = \equiv -10\,°C \cdot km^{-1}$$

Here, Γ_d = dry air.

But when the temperature falls below the dew point, thereby saturating the air, moisture condenses, releasing latent heat. Then, the lapse rate decreases to approximately:

$$\Gamma_s \equiv -6\,°C \cdot km^{-1}$$

PRACTICE PROBLEM 16.4.1

Q1: If a rising parcel's temperature decreases at 10°C · km⁻¹ in an adiabatic process, does its temperature increase at 10°C · km⁻¹ as the parcel descends?

SOLUTION 16.4.1

Answer a) Yes or b) No

Q2: Why is the moist lapse rate lower than the dry lapse rate (which answer is correct)?

   a. Because condensation occurs in the moist adiabatic process
   b. Because evaporation occurs in the dry adiabatic process
   c. Because condensation occurs in the dry adiabatic process

PRACTICE PROBLEM 16.4.2

Topographical configurations can produce substantial contemporaneous temperature variations at the same altitude (Figure 16.5). Calculate the six unlisted temperatures in Figure 16.5.

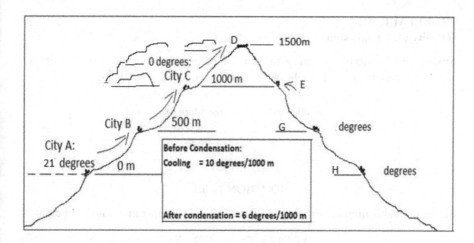

FIGURE 16.5   Effects of varying lapse rates and processes on temperature.

16.5   CHANGES TO ATMOSPHERIC STABILITY

16.5.1   Stability Activity

Most often, the air in the atmospheric environment cools more slowly (at the environmental lapse rate or ELR) than that of a rising parcel of air, thereby causing the parcel's continued ascent. Local heating causes air to rise. For instance, after being lifted, the air inside a "parcel" may have a different temperature than the surrounding air outside the parcel. A rising air parcel expands (using up its energy), and hence this loss of energy without any energy coming from outside the parcel will consequently cool the air in the parcel at the rate of 10°C per 1000 m of elevation. Outside the parcel, the ELR normally hovers around 6°C per 1000 m of elevation.

Whenever the air in the parcel acquires the same temperature as that of the surrounding air, it cannot, of its own volition, move in a vertical direction unless its water content differs from (i.e., is higher than) the surrounding air. If the air in the parcel continues to rise, the atmosphere is unstable. Yet, if that air, after rising, always comes back to its prior position, the atmosphere is stable.

But if condensation occurs, the parcel will release latent heat. This slows cooling, to approximately 6°C km, i.e., the saturated adiabatic lapse rate (SALR). The parcel is now unstable and rising. Had there been no condensation (i.e., the parcel remained "dry"), the parcel would have finally stopped rising at the elevation where its temperature equaled that of the environmental air. This elevation is lower than that of a saturated parcel.

PRACTICE PROBLEM 16.5.1
Kinds of stability

   a. What are the types of stability?
   b. What type of stability leads to calm air?
   c. A small lapse rate (ELR) leads to what type of air?

PROBLEM 16.5.1
Stability and dispersion

Atmospheric instability increases the dilution of air pollutants. What is the stability of the atmosphere, given the following temperature and elevation data?

Elevation (m)	Temperature (°C)
3.00	18.47
322.00	15.28

SOLUTION 16.5.1

Begin by establishing the existing lapse rate with T = temperature and Z = height.

$$\Delta T / \Delta Z = T2 - T1 / Z2 - Z1$$

$$= \frac{15.28 - 18.47}{319 - 3} = -\frac{3.19}{316} = -0.0100\,°C \cdot m^{-1} = -1.00\,°C\ 100 \cdot m^{-1}$$

As this equals the adiabatic lapse rate for "dry" air, the atmospheric stability is neutral.

Therefore, the answer is neutral stability.

However, had the figures been:

$$= 15.28 - 17.80 / 319 - 3 = -2.52 / 316 = -0.0079\,°C\ m^{-1} = -0.79\,°C\ 100 \cdot m^{-1}$$

the rising parcel of air cooling adiabatically at 1°C per 100 m would be colder and hence denser than the environmental air before it reached 100 m and so would cease to rise any further. This is unconditional stability.

PROBLEM 16.5.2
Effect of chimney height

The greater the effective height of a smoke plume, the greater the dilution and dispersion of its pollutants. A coal-burning plant emits sulfur dioxide at a rate of 1710.3 g · s⁻¹. The wind speed at the top of the 115-m-high chimney stack is 4.25 m · s⁻¹ on an overcast afternoon at 28.0°C. What is the height of the pollutant gas in the plume above the chimney stack?

SOLUTION 16.5.2

List the given information:

Stack attributes	Atmospheric conditions
Height = 115 m	Pressure = 94 kPa
Diameter = 1.3 m	Temperature = 24 °C
Exit velocity = 8.0 m · s⁻¹	
Temperature = 320 °C	

TABLE 16.1

Atmospheric Stability Classification

A	Very unstable
B	Unstable
C	Slightly unstable
D	Neutral
E	Slightly stable
F	Stable

The effective stack height (Figure 16.4) is the value of the gaseous plume height (ΔH) plus that of the physical stack (h):

$$H = h + \Delta H$$

ΔH can be calculated using a formula by Holland (1953) as follows:

$$\Delta H = vs\ d\ /\ u\left[1.5 + \left\{2.68 \times 10^{-2}\ (p)(Ts - Ta\ /\ Ts)d\right\}\right]$$

where

vs = stack velocity (in m s^{-1})
d = stack diameter (in m)
u = wind speed (in m s^{-1})
p = air pressure (in kPa)
Ts = stack temperature (K)
Ta = air temperature (K)

$$\Delta H = \left(8\ m\,s^{-1}\right)(1.3\ m)\ /\ 4.25\ m\,s^{-1}\left[1.5 + \{2.68 \times 10^{-2}\ (94\ kPa)(593 - 297\ K\ /\ 593\ K)\right.$$
$$\left.1.5\ m\}\right]$$
$$= 6.58\ m$$
$$H = 115 + 6.58\ m = 121.58\ m$$

The atmospheric stability class must then be selected. As it is overcast (given), upward air movement is restricted. Table 16.1 indicates that category D should be chosen.

16.6   EFFECTS ON MATERIALS

Atmospheric pollutants cause the following conditions.

- Corrosion of metal surfaces, fading
- SO_2 and water form H_2S – corrosion as well as disfigurement of statues made up of limestone or marble
- Air pollutants mix with rainwater and increase the acidity (acid rain) of water body and kill fish.
- Ozone causes cracking of rubber.

16.7 REMEDIATION OF AEROSOLS

In the United States, diesel vehicles are the major source of soot (the only aerosol having a large enough size to increase global warming), and

- Filters on exhaust pipes can help reduce the amount that they pump into the air (Unger 2009).
- In terms of sulfate aerosols, which are created by sulfur dioxide given off by power plants, the United States and Europe have very successfully used sulfur dioxide scrubbers in power plants to reduce these emissions over the past 20 years or so. But we can do more.
- According to Unger (2009), by reducing aerosol (soot) emissions, the earth's population can buy itself some climate time – about 5–10 years – while concurrently working on reducing emissions of greenhouse gases such as carbon dioxide. As CO_2 persists in the atmosphere for an extremely long time, from decades to centuries, even implementing cuts today will have no effect for years. She points out that aerosols, contrastingly, have much shorter lifetimes, such that working to reduce soot emissions now, which can enhance the global warming effect of CO_2 by 20–50%, would allow the climate impacts to be felt more rapidly.

16.8 REMEDIATING WATER LOSS BY CONTROLLING AIR HUMIDITY

While the energy supply from the sun and the surrounding air is the main driving force for the vaporization of water, the difference between the water vapor pressure at the evapotranspiration surface and the surrounding air is the determining factor for the vapor removal. Well-watered fields in hot dry arid regions consume large amounts of water due to the abundance of energy and the desiccating power of the atmosphere. In humid tropical regions, notwithstanding the high energy input, the high humidity of the air will reduce the evapotranspiration demand. In such an environment, the air is already close to saturation, so less additional water can be stored and hence the evapotranspiration rate is lower than in arid regions.

PROBLEM 16.8.1
Restricting water vapor diffusion

How can vapor condensation parameters be used to avoid current losses of immense quantities of water?

SOLUTION 16.8.1

- Cessation of net evaporation occurs when a steady state of dynamic equilibrium exists. Maximizing the ratio of the amount of water vapor in the air to the evaporating surface reduces evaporation, as:

$$\text{Relative humidity} = e \,/\, es\,(100) \tag{16.8.1}$$

where
Saturated: $e = es$ and relative humidity $= 100\%$
Unsaturated: $e < es$ and relative humidity $< 100\%$

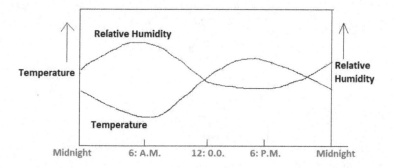

FIGURE 16.6 Relationship of daily variations in relative humidity with environmental temperatures. For an idealized diurnal cycle of temperature, the relative humidity will change, even if the amount of water vapor in the air remains the same.

- Saving water / reducing water loss under glasshouses occurs by reducing wind speeds and decreasing evapotranspiration. The high vapor pressures prevailing in glasshouses ensure low evaporation rates from vegetation. Growing crops in glasshouses (including transparent polythene "glasshouses") prevents the diffusion of water vapor from its source.
- Even orchards can be maintained in glasshouses by pruning the trees to attain a small height. By avoiding evapotranspiration losses, thereby reducing the need for irrigation, glasshouse agriculture would not only greatly conserve water worldwide but also produce millions of kilograms of toxicity-free air.
- Though solar radiation absorbed by the atmosphere and the heat emitted by the earth increase the air temperature, the sensible heat of the surrounding air transfers energy to the crop, thereby controlling the rate of evapotranspiration.
- Sunny, warm weather increases water loss by evapotranspiration over that which occurs in cloudy and cool weather, but the lack of a water vapor gradient inside a glasshouse permanently prevents such water losses. Similarly, for humid conditions, as is the case in glasshouses, the wind can only replace saturated air with slightly less saturated air and remove heat energy.
- Therefore, choosing the daytime periods of highest environmental humidity (Figure 16.6) for changing the air inside a glasshouse will minimize evaporative losses from the glasshouse.
- Consequently, the wind speed affects the evapotranspiration rate to a far lesser extent than under arid conditions where small variations in wind speed may result in larger variations in the evapotranspiration rate.
- Where glasshouses are not feasible, windbreaks can reduce wind speed and evapotranspiration.

16.9 AIR PRESSURE AND CHANGING ALTITUDES

Rising sea levels can force the relocation of many coastal airports to higher ground, where higher wind velocities increase the hazards of landing and take-off. As global warming continues, cities will become established at higher altitudes. With increasing altitude, wind speeds increase due to the loss of surface friction.

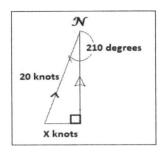

FIGURE 16.7 Calculating wind vectors. Rising sea levels may push populations to higher altitudes where wind speeds are higher, with ramifications for air transport.

Weather phenomena, such as crosswinds, may exacerbate headwinds and tailwinds on aircraft while landing or taking off will require calculations concerning these specific effects.

First, consider the orientation sign, where 0 degrees is at N and increases westward:

Therefore, 180 degrees occurs at the exact tail of the arrow above. But, as wind speeds incrementally increase between the sea level and the altitude of the geostrophic wind (i.e., at the altitude where wind speeds are greatest), the impact of crosswinds can greatly vary between two locations having the same geo-referenced coordinates but different altitudes.

PROBLEM 16.9.1
Increased wind speeds as altitudes increase

An aircraft landing with a wind of 210° degrees at 20 knots will experience a crosswind ("X"). Exactly what is the value of "X" (Figure 16.7)?

SOLUTION 16.9.1

1. The relevant trigonometric ratio for a right-angled triangle is:

Sine = Opposite/hypotenuse

2. Here, the crosswind component, "X," is opposite to the angle of 30 degrees.
3. Therefore, sine 30 degrees = X/20.
4. Hence, X = 20 (sine 30) = 20 (0.5) = 10 knots.

PROBLEM 16.9.2
Air pressure and temperature

The pressure of the helium gas inside a weather balloon with a fixed volume is 5.0 atm at 14°C. What is the pressure of the helium gas inside the weather balloon if the temperature changes to 259 K? Assume the amount of gas remains constant.

SOLUTION 16.9.2

At constant volume, volume constant is:

$$P1 / T1 = P2 / T2$$
$$5 / 287 \text{ K} = P2 / 259 \text{ K}$$
$$P2 = (5 / 287 \text{ K}) \times 259 \text{ K} = 4.51 \text{ atm}$$

PROBLEM 16.9.3
Air pressure and volume

A beach ball is inflated to a volume of 28 L of air at 17°C. During the afternoon, the volume increases by 2 L. What is the new temperature outside?

SOLUTION 16.9.3

$$V1 / T1 = V2 / T2$$
$$28 / 290 = 30 / T2$$
$$T2 = (30 / 28) \times 290 = 310.7 \text{ K}$$

PROBLEM 16.9.4
Air humidity and water conservation

Under what atmospheric conditions do glasshouse cropping systems conserve the most water?

PRACTICE PROBLEM 16.9.4
Wind speed and evaporation

Under what conditions are the largest conservations of evaporative water loss achieved?

SOLUTION 16.9.4

Water vapor removal depends largely on wind and air turbulence which both transfer large quantities of air over the evaporating surface. When vaporizing water, the air above the evaporating surface becomes gradually saturated with water vapor. If this air is not continuously replaced with drier air, that saturated air restricts water vapor removal, thereby decreasing the evapotranspiration rate. Generally, the greatest conservation of water occurs in glasshouses, but among glasshouses, those under hot dry, windy external atmospheric conditions conserve the most moisture.

PROBLEM 16.9.5
Elevation and air pressure

At the subtropical high-pressure belts, subsiding air restricts evaporation. In other words, evaporation generally varies directly with air pressure. What is the difference in atmospheric pressure between a location at mean sea level (MSL) and a location at 1000 m?

SOLUTION 16.9.5

Density and pressure decrease exponentially with height. For each 16 km in altitude, the pressure decreases by a factor of 10. As the atmospheric pressure P (the pressure exerted by the weight of the earth's atmosphere) decreases with altitude, evaporation at high altitudes is promoted due to low atmospheric pressure as expressed in the psychrometric constant. However, as the effect is small, an average reading is often used, being a simplification of the ideal condition:

$$P = P_{MSL} \times 10^{(-z/16\,km)}$$

where
$\quad$ Z = elevation in kilometers
$\quad$ P = pressure in mb at location
$\quad$ P_{MSL} = mean sea level pressure in mb

$$P = 1013\ mb \times 10^{(-1/16\,km)} = 877\ mb$$

Atmospheric pressure at 1000 m = 877 mb

PROBLEM 16.9.6
Using the psychrometric constant

What is the psychrometric constant and how is it used?

SOLUTION 16.9.6

The psychrometric constant, γ, is given by:

$$\gamma = C_p\ P\,/\,e\ \gamma = 0.665 \times 10^{-3}\ P$$

where
$\quad$ γ = psychrometric constant (kPa · °C^{-1})
$\quad$ P = atmospheric pressure (kPa)
$\quad$ C_p = specific heat at constant pressure, 1.013 10^{-3} (MJ kg^{-1}°C^{-1})
$\quad$ e = ratio molecular weight of water vapor/dry air = 0.622

The wet and dry bulb thermometer is called a psychrometer, i.e., an instrument that measures relative humidity. The specific heat at constant pressure is the amount of energy required to increase the temperature of a unit mass of air by one degree at constant pressure, and this value depends on the humidity, as the specific heat of water is higher than that of dry air. This specific heat value is the psychrometric constant and is also a function of altitude.

PROBLEM 16.9.7
Determination of atmospheric parameters

What is the atmospheric pressure and the psychrometric constant at an elevation of 1700 m?

SOLUTION 16.9.7

Given: z = 1700 m

From the air pressure equation $P = P_{MSL} \times 10^{(-z/16km)}$,

$$P = 1013 \text{ mb} \times 10^{(-1.7/16km)} = 793 \text{ mb}$$

From the psychrometric equation: $\gamma = C_p \, P \, / \, e \, \gamma = 0.665 \times 10^{-3} \, P$,

$$\gamma = 0.665 \times 10^{-3} \, (79.3) = 0.0527 \text{ kPa} \cdot {}^{\circ}C^{-1}$$

The average value of the atmospheric pressure is 79.3 kPa.
The psychrometric constant, γ, is 0.0527 kPa · °C⁻¹.

REFERENCES

Akyüz M, Cabuk H (2009) Meteorological variations of $PM_{2.5}/PM_{10}$ concentrations and particle-associated polycyclic aromatic hydrocarbons in the atmospheric environment of Zonguldak, Turkey. J Hazard Mater 170, 13–21. https://goo.gl/035UqZ

Beyond Discovery (2022) https://www.beyonddiscovery.org/.../meteorology-and-air-pollution.html

Brandt C, Kunde R, Dobmeier B, Schnelle-Kreis J, Orasche J, et al. (2011) Ambient PM10 concentrations from wood combustion-emission modeling and dispersion calculation for the city area of Augsburg, Germany. Atmos Environ 45, 3466–3474. https://goo.gl/ZTy1xn

Choi YS, Ho CH, Cheni D, Noh YH, Song CK (2008) Spectral analysis of weekly variation in PM10 mass concentration and meteorological conditions over China. Atmos Environ 42, 655–666. https://goo.gl/muZV4d

Holland JZ (1953) A Meteorological Survey of the Oakridge Area (U.S. Atomic Energy Commission Report No. ORO 99). U.S. Government Printing Office, Washington, DC, p. 540.

Langmuir (1961) Epidemiology of airborne infection. Bacterial Rev 25, 173–181.

Mirhoseini SH, Koolivand A, Bayani M, Sarlak H, Moradzadeh R, Ghamari F, Sheykhan A (2021) Quantitative and qualitative assessment of microbial aerosols in different indoor environments of a dental school clinic. Aerobiologia (Bologna) 37(2), 217–224. doi: 10.1007/s10453-020-09679-z.

Özdemir U (2016) Impacts of meteorological factors on particulate pollution: Design of optimization procedure. J Civil Eng Environ Sci. https://www.researchgate.net/publication/313960412_Impacts_of_Meteorological_Factors_on_Particulate_Pollution_Design_of_Optimization_Procedure

Pateraki S, Asimakopoulos DN, Flocas HA, Maggos T, Vasilakos C (2012) The role of meteorology on different sized aerosol fractions (PM10, PM2.5, PM2.5–10). Sci Total Environ 419, 124–135. https://goo.gl/s9J12K

Unger N (2009) Just 5 Questions: Aerosols. Interview by Amber Jenkins Global Climate Change/Jet Propulsion Laboratory. https://climate.nasa.gov/news/215/just-5-questions-aerosols/

17 Thermal Pollution and Remediation

OVERHEATING – ADVERSE EFFECTS:

- Power plants utilize only 1/3rd the energy they produce from fossil fuel; the rest is wasted as heat.
- Cold water taken from the water body for cooling is put back with an increase of 10–15°C.
- Metabolic activities increase at high temperatures, requiring more oxygen.
- O_2 penetration in water decreases as heat decreases O_2 dissolution.
- Heat kills fishes and other aquatic life. Spawning and migration are disturbed.
- The toxicity of pesticides and chemicals increases with temperature.
- The composition of flora and fauna changes. Temperature-tolerant species (often invasive) start developing.

17.1 HEAT DISSIPATION IN STREAMS

Elevated temperatures typically decrease the level of dissolved oxygen in water, as gases are less soluble in hotter liquids. Goel (2006) notes that adding to the fact of many aquatic species failing to reproduce at elevated temperatures, thermal pollution may also increase the metabolic rate of aquatic animals, such that these organisms consume more food in a shorter time than if their environment were not changed. Consequently, an increased metabolic rate may result in fewer resources such that the more adapted organisms moving in may have an advantage over organisms that are not used to the warmer temperature.

In addition to compromising food chains in the old and new environments, a thermal discharge can force some fish species to avoid stream segments or coastal areas adjacent to a thermal discharge, thereby decreasing biodiversity (Laws 2017; Kennish 1992). High temperatures limit oxygen solubility in water and, by extension, its dispersion into deeper waters, contributing to anaerobic conditions, which can lead to increased bacterial levels when there is ample food supply (Laws 2017).

Power plants require steam to drive turbines to produce electricity and for cooling machines; other industries requiring cooling of machines include pulp and paper mills, chemical plants, and steel mills; hence they all often occur near rivers and lakes. When the temperature of a natural body of water suddenly increases or decreases, thermal pollution occurs. This pollution is most often caused by industrial facilities releasing large quantities of heated wastewater. With climate change exacerbating increases in water temperature, thermal pollution is a growing concern.

Before entering water, such heat can be diverted. But the rate of dissipation of heat by materials to the surroundings affects the ongoing temperatures in the

 DOI: 10.1201/9781003341826-21

immediate environments. Dissipation of heat from a heat source varies directly with the enthalpy (heat of combustion) and mass of the heat source.

PROBLEM 17.1.1

What is the final temperature of 0.1 kg of ice at 0°C added to an insulated container filled with 4.5 kg of water at 20°C?

SOLUTION 17.1.1

NB: If energy goes into an object, the total energy of the object increases, and the values of heat (ΔT) are positive. If energy is being lost from an object, the values of heat and ΔT are negative.

$$\Delta H = m \times c_p \times \Delta T \tag{17.1.1}$$

where
ΔH = change in enthalpy (i.e., change in internal heat)
m = mass
c_p = specific heat at constant pressure
ΔT = change in temperature

For ice:

$$\Delta H = (0.1 \text{ kg})(2.10 \text{ kJ} \cdot \text{kg}^{-1} \cdot \text{K}^{-1})(T - 273.15)$$
$$= 0.21 \times (T - 273.15) = 0.21T - 57.36$$

For water:

$$\Delta H = (4.5 \text{ kg})(4.186 \text{ kJ} \cdot \text{kg}^{-1} \cdot \text{K}^{-1})(293.15 - T)$$
$$= 18.837 \times (T - 293.15) = 18.837T - 5522$$

Since equilibrium conditions mean the change in concrete, in absolute terms, balances the change in the water, the two equations are solved for temperature.

$$(\Delta H)_{ice} = (\Delta H)_{water}$$
$$0.21T - 57.36 = 5,522 - 18.837T$$
$$5579.36 = 19.047T$$
$$T = 292.92$$

T final (equilibrium) temperature = 292.92 K or 19.77°C

PROBLEM 17.1.2
Equilibrium temperature

In the United States, about 75–80% of thermal pollution is generated by power plants, causing heat transfer into fragile natural systems such as lakes and rivers. Thermal equilibrium refers to the condition under which two substances in physical contact with each other exchange no heat energy. Two substances in thermal equilibrium are said to be at the same temperature. A factory applies superheated water at 120°C to bare metal surfaces as a pre-treatment. The wastewater from this treatment must be cooled to 18°C before disposal. Forty-five cubic meters of this wastewater is placed for cooling into a concrete tank having a temperature of 22°C and a mass of 41,000 kg. On the assumption that there are no losses to the surroundings, what is the equilibrium temperature of the wastewater and the concrete tank? Is the water ready for disposal in a nearby stream?

SOLUTION 17.1.2

- When a process occurs at constant pressure, the heat evolved (either released or absorbed) is equal to the change in enthalpy.
- Enthalpy depends on the size of the system or on the amount of substance it contains. The SI unit of enthalpy is the joule (J). It is the energy contained within the system. It is the thermodynamic quantity equivalent to the total heat content of a system.
- Enthalpy is represented by the symbol H, and the change in enthalpy in a process is H2 – H1 or ΔH.

Hence, a non-phase change process occurring without a change in pressure requires a change in enthalpy (internal heat). That change is defined as:

$$\Delta H = M \cdot C_p \cdot \Delta T \tag{17.1.2}$$

where
  ΔH = change in enthalpy (i.e., change in internal heat)
  M = mass
  C_p = specific heat at constant pressure
  ΔT = change in temperature

Density of pure water = 1000 kg · m³
Specific heat of water = 4.186 kJ · kg⁻¹ · K⁻¹
Specific heat of concrete = 0.93 kJ · kg⁻¹ · K⁻¹
If the density of the water is 1000 kg · m⁻³, and the loss of enthalpy (loss of heat) of the superheated water is:

$$\Delta H = (1000 \text{ kg} \cdot \text{m}^{-3})(45 \text{ m}^3)(4.186 \text{ kJ} \cdot \text{kg}^{-1} \cdot \text{K}^{-1})(393.15 - T)$$
$$= 74{,}057{,}665 - 188{,}370 \, T$$
$$188{,}370 \times (393.15 - T) = 74{,}123{,}595 - 188{,}370 T$$

where the absolute temperature is 273.15 + 120 = 393.15 K
  The concrete tank gained enthalpy as follows:

$$\Delta H = (41{,}000 \text{ kg} \cdot \text{m}^{-3})(0.93 \text{ kJ} \cdot \text{kg}^{-1} \cdot \text{K}^{-1})(T - 295.15) = 38{,}130T - 11{,}253{,}306$$
$$38{,}130 \times (T - 295.15) = 38{,}130T - 11{,}254{,}069$$

Since equilibrium conditions mean the change in concrete, in absolute terms, balances the change in the water, the two equations are solved for temperature.

$$(\Delta H)_{water} = (\Delta H)_{concrete}$$

$$74,057,665 - 188,370 \, T = 38,130 \, T - 11,253,306$$

$$74,057,665 + 11,253,306 = 38,130 \, T + 188,370 \, T$$

$$85,377,664 = 226,500 \, T$$

$$T = 376.65 \text{ K or } 103.5°C$$

This water is still too hot for disposal into a small stream.

17.2 POLLUTANT RESIDUES FROM HEAT PRODUCTION

Increasing oxygenation to the surface area of a fuel improves its efficiency. Hence, the practice of converting coal into electricity, hydrogen, and other energy products by gasification, a thermo-chemical process that breaks down any carbon-based source, including coal, into its basic chemical constituents. The recalcitrant particles, primarily metal oxides, sulfates, and phosphates, remaining after burning coal, are called coal ash, principally consisting of fly ash and bottom ash which contain several toxic substances. People exposed to such particulates above a certain concentration may develop adverse health conditions.

PROBLEM 17.2.1
Heat energy and ash pollution

A coal-fired plant converts 31% of the coal's energy into electricity. To produce 780 MW electrical output, what is the approximate volume of ash produced in a year if the bituminous coal has a net heating value (NHV) of 28.5 MJ · kg⁻¹, an ash content of 7.3%, and the bulk density of the ash is 618 kg · m⁻³? (It is given that the air pollution control system collects fly ash as heavier ash particles fall to the floor of the combustion chamber. Together, these comprise 99% of the ash produced.)

SOLUTION 17.2.1

As indicated in Figure 17.1, transformations produce the energy and masses of by-products.

$$\text{Input power} = \text{Output power } 1 + \text{Output power } 2$$

Based on a definition of efficiency (η) for mass flow rate or concentration of a pollutant:

$$(\eta) = (\text{Mass in} - \text{Mass out} / \text{mass in}) \, (100\%)$$

Or $(\eta) = (\text{Concentration in} - \text{Concentration out} / \text{concentration in}) \, (100\%)$.

For power: $\eta = \text{Output power} / \text{input power} = 0.31 / 780 \text{ MW}_e$

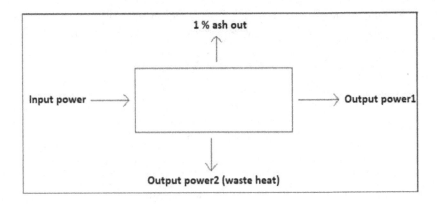

FIGURE 17.1   Mass and energy transformations.

Hence, converting the equation, the input energy for the power plant is

$$\text{Input power} = \text{Output power} / \eta = 780 \text{ MW}_e / 0.31 = 2194 \text{ MW}_t$$

where MW_e refers to electrical power, while MW_t refers to thermal power. Converting power to energy, $1 \text{ W} = 1 \text{ J} \cdot \text{s}^{-1}$ produces:

$$\left(2194 \text{ MW}_t \right) \left(1 \text{ MJ} \cdot \text{s}^{-1} / \text{MW} \right) = 2194 \text{ MJ} \cdot \text{s}^{-1}$$

Amount of ash

The amount of ash is found by balancing the masses involved in coal burning:

$$\text{Coal ash in} = \text{Coal ash in chimney-stack gas} + \text{Coal ash in bottom of boiler}$$

Combining the burning rate with the ash content of the coal determines the amount of coal ash entering the boiler.

NHV (net heating value) = $28.5 \text{ MJ} \cdot \text{kg}^{-1}$ (given)
1 MJ (i.e., 28.5 / 28.5) is produced by = 1 kg / 28.5 MJ
2194 MJ is produced by = (1 kg / 28.5 MJ) × 2194 MJ $\cdot$ s^{-1} = 76.98 kg $\cdot$ s^{-1} of coal
Hence, ash going in = (76.98 kg $\cdot$ s^{-1}) (0.073) = 5.62 kg $\cdot$ s^{-1}
But a 99% ash capture = (5.62 kg $\cdot$ s^{-1}) × (0.99) = 5.56 kg $\cdot$ s^{-1}
In 1 year, this would be (5.56 kg $\cdot$ s^{-1}) (86,400 s $\cdot$ day^{-1}) (365 days $\cdot$ year^{-1})
   = 1.75×10^8 kg $\cdot$ year^{-1}
The volume of ash would be = 1.75×10^8 kg $\cdot$ year^{-1} / 618 kg $\cdot$ m^{-3}
   = 2.83171×10^5 m^3 $\cdot$ year^{-1}

This large volume of ash would cover 28 football fields to a depth of 1 m. Could it be repurposed into concrete?

PROBLEM 17.2.2
Calculating required flow rate to waterway

If 17% of the waste heat from the power plant in the above question goes up the stack, and the remaining 83% must be removed by cooling water, what is the required flow rate of the cooling water if the temperature of the cooling water is limited to 12°C? Note: The stream flows at the rate of 59 $m^3 \cdot s^{-1}$, with a temperature of 17°C above the intake to the power plant.

SOLUTION 17.2.2

Figure 17.2 depicts an energy balance for the power plant.

At a steady state, the balance of energy is:

Input power = Power utilized + Waste heat (Figure 17.2)

MW_t = Megawatts thermal = Power input or output in a heat engine

MW_e = Power obtained as electricity

$1\ MW = 1\ MJ \cdot s^{-1} \times 1\ MJ \cdot s^{-1} = 1\ MJ \cdot s^{-1}$

From the above example, input power = 2194 MW_t and output power = 780 MW_e

So wasted power = 2194 – 780 MW_t = 1414 MW_t

But stack losses are:

(0.17) (1414 MW_t) = 240.38 MW_t, and the river water must sequester (given)

(0.83) (1414 MW_t) = 1173.6 MW_t (given)

Recalling the enthalpy equation:

$$\Delta H = M \cdot C_p \cdot \Delta T$$

$$dH / dt = C_p \cdot \Delta T \cdot dM / dt$$

where

$dH / dt = 1173.6\ MW = dH / dt = \Delta H$

C_p = specific heat of water = 4.186 $kJ \cdot kg^{-1} \cdot K^{-1}$

ΔT = the allowable rise in temperature of 12°C

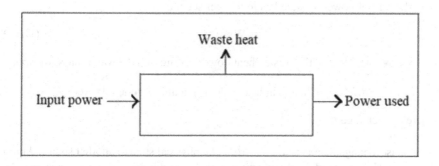

FIGURE 17.2 Energy balance of a power plant.

Converting megawatts to megajoules:

$$1173.62 \text{ MW}_t \ (1 \text{ MJ} \cdot \text{s}^{-1} \cdot \text{MW}^{-1}) = 1173.62 \text{ MJ} \cdot \text{s}^{-1}$$

As $\Delta T = 12°C = 12$ K, Equation 17.1.1 when transposed for dM / dt is:

$$dM / dt = (dH / dt) / Cp \cdot \Delta T$$

But mass did not change. Therefore,

$M = \Delta H / Cp \cdot \Delta T$
   $= (1173.62 \text{ MJ} \cdot \text{Mg}^{-1} \cdot \text{K}^{-1}) / (4.186 \text{ MJ} \cdot \text{Mg}^{-1} \cdot \text{K}^{-1}) (12 \text{ K}) = 23.36 \text{ Mg} \cdot \text{s}^{-1}$

Converting to volume $\cdot$ s^{-1}: density of water = 1000 kg $\cdot$ m^{-3} (or 1 Mg $\cdot$ m^{-3}):

Required volumetric flow rate = 23.36 Mg $\cdot$ s^{-1} / 1 Mg $\cdot$ m^{-3} = 23.36 m^3 $\cdot$ s^{-1}

To find the increase in the temperature of the heat-sequestering stream
Solving Equation 17.1.1 for ΔT:

$\Delta T = (dH / dt) / (C_p \cdot dt / dM)$
$\Delta T = \Delta H / (Cp \cdot M)$
$\Delta T = 1173 \text{ MJ} \cdot \text{s}^{-1} / (4.186 \text{ MJ} \cdot \text{Mg}^{-1} \cdot \text{K}^{-1}) (59 \text{ m}^3 \cdot \text{s}^{-1}) (1 \text{ Mg} \cdot \text{m}^{-3})$
      $= 4.75 \text{ K or } 4.75 °C$

As the upstream temperature was 17°C, the downstream temperature will now be:

$$17 °C + 4.75 °C = 21.75 °C.$$

This elevated temperature will be very harmful to aquatic life in the stream.

17.3   WHAT IS POWER?

A rubber catapult and a rifle may move similar masses through the same distance, but in accomplishing the task in a much shorter time, the rifle would have expended more energy. The rifle has greater power (with devastating effect). The watt, therefore, the unit of power, is stated as joules per second, or:

$$W = J \cdot s^{-1} \tag{17.3.1}$$

To show the effect of the waste "heat power" going into the water mass, we apply:

dH / dt = Change in heat energy in a unit of time = Power

where *d* = "change in…"

A phase-change means changing from one physical state to another such as liquid to gas or solid to liquid. The temperature of the medium is unchanged because the temperature required for that change is used up as energy required for the physical change.

PRACTICE PROBLEM 17.3.1
Fill in the blanks

Hence, a non-phase change process occurring without a change in pressure requires a change in enthalpy (internal heat). That change is defined as:

$$\Delta H = MC_p \cdot \Delta T \qquad (17.3.2)$$

where
  ΔH = change in enthalpy (i.e., change in internal heat)
  M = mass
  C_p = specific heat at constant pressure
  ΔT = change in temperature

Hence, a change in any one of the above components changes the enthalpy of the system.

Because $1\ W = 1\ J \cdot s^{-1}$, if $\Delta T = 10°C = 10°K$, transposition of the composite equation above yields
  $(C_p \Delta T) dM = dH(\cancel{dt} / \cancel{dt})$, which means (fill in the answer below in $Mg \cdot s^{-1}$):

$$dM / dt = (dH / dt) / (C_p \Delta T) = 1173.6\ MJ \cdot s^{-1} / (4.186\ MJ \cdot Mg^{-1} \cdot K^{-1})(10\ K)$$

$$= \underline{\hspace{2cm}} Mg \cdot s^{-1}$$

If the density of the water is $1000\ kg \cdot m^{-3}$, the required flow rate by volume is

$$\underline{\hspace{2cm}} Mg \cdot s^{-1} / 1\ Mg \cdot m^{-3} = \underline{\hspace{2cm}}, \text{ or } \underline{\hspace{2cm}} m^3 \cdot s^{-1}$$

PRACTICE PROBLEM 17.3.2
Calculating the increase in stream temperature

For the same power plant in the above example, if the river has a flow rate of $54\ m^3 \cdot s^{-1}$ and the temperature of the river water upstream of the power plant is $17°C$, what is the increase in stream temperature?

SOLUTION 17.3.2

Using Equation 17.3.2, $\Delta H = MC_p \Delta T$, we can solve for ΔT.

PRACTICE PROBLEM 17.3.3
As the upstream temperature was $17°C$, what would be the downstream temperature?

PROBLEM 17.3.1
Electrical energy calculated

For the above question, if the maximum discharge is $804\ m^3 \cdot s^{-1}$, what is the maximum amount of electrical energy that the generating plant can produce?

SOLUTION 17.3.1

$$\text{Power} = (9.81\ m \cdot s^{-2})(215\ m)(1000\ kg \cdot m^3)(804\ m^3 \cdot s^{-2}) = 1.69 \times 10^9\ J \cdot s^{-1}$$

$$= 1.69 \times 10^9\ W = 1690\ MW$$

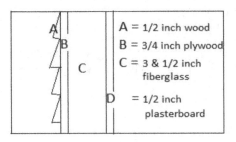

FIGURE 17.3 Wall layers in a residential building.

17.4 ROLE OF INSULATION

PROBLEM 17.4.1
Residential heat losses

A typical residential building outside of the tropics consisted of the layers depicted in Figure 17.3. If the inside temperature is to be maintained at 22° C and the outside temperature is 0° C, (a) what is the heat loss in that existing scheme and (b) with the addition of 25 cm of organic bonded glass filler insulation?

SOLUTION 17.4.1

According to NCC (2019), assuming 1-m^2 surface area, the total resistance for the original materials in $m^2 \cdot K \cdot W^{-1}$ is calculated as:

$$R = \text{Asphalt} + \text{Plyboard} + \text{Air} + \text{Insulation} + \text{Gypsum}$$
$$= 0.077 + 0.10 + (850/90)(0.4) + 3.32 + 0.56 = 7.834 \ m^2 \cdot K \cdot W^{-1}$$

Here, the ratio $850/90$ is the number of 90 mm air spaces in 850 mm of air in the attic. Using the following equation, heat loss (enthalpy) can be calculated as:

$$dH/dt = (1/R_T)(A)(\Delta T) \tag{17.4.1}$$

where
    A = surface area in m^2
    ΔT = difference in temperature (in K)
    R_T = resistance in $m^2 \cdot K \cdot W^{-1}$

$$dH/dt = (1/7.834 \ m^2 K \cdot W^{-1})(1-m^2)(22-0) = 2.80 \ W$$

The additional insulation will add resistance. This is in proportion to its thickness. Thus:

$$R = (27.7 \ m \cdot K \cdot W^{-1})(0.25 \ m) = 6.92 \ m^2 \cdot K \cdot W^{-1}$$

Hence, the new resistance is:

$$R_T = 7.83 + 6.92 = 14.75$$

And the new heat loss is:

$$dH \mathbin{/} dt = (1 / 14.75 \text{ m}^2 \cdot \text{K} \cdot \text{W}^{-1})(1 - \text{m}^2)(22 - 0) = 1.49 \text{ W}$$

This is 53% of the original heat loss.

PROBLEM 17.4.2
Utilization of natural insulators

Spathodea campanulata, a fast-growing tree, has invaded tropical, sub-tropical and extra-tropical regions. What are the insulating attributes of cured wood? What is the potential of the wood of *S. campanulata* as an indoor insulator for buildings?

SOLUTION 17.4.2

Some authors observed the following:

- The tree species *S. campanulata* is among the fastest growing tropical ever-greens, being often the first large tree to colonize wastelands (Pulipati 2013).
- It is rapidly invasive (ISSG 2017), especially in wet tropical fertile conditions as it propagates readily from seeds (wind-dispersed), suckers (from rapidly extending roots), or cuttings, the species having been nominated as among the 100 "World's Worst Invaders" (Pagad 2010) and being an invasive species in many tropical areas (Pulipati et al. 2013).
- *S. campanulata* is of low specific gravity (Bosch 2002), which suggests high porosity, and a high percentage of closed pores increases heat resistance of an insulator (Pederson & Helevang 2010).
- Its low efficiency as a firewood (Bosch 2002), being a difficult wood to burn (Pagad 2010), support the argument of its heat resistance.
- The author has repeatedly observed the relatively high ignition temperature of *S. campanulata*, thereby establishing value as a residential building material by the high fire resistance exhibited.

PROBLEM 17.4.3
High heat resistivity: reasons

What was the result of the above-mentioned test? How was thermal conductivity (k) calculated? What attributes account for the heat resistivity of *S. campanulata*?

SOLUTION 17.4.3

Thermal conductivity (k) was measured according to Fourier's law as follows:

$$Q / t = kA(T_{hot} - T_{cold}) / d \qquad (17.4.2)$$

where
  Q = heat (as watts)
  t = time (second)
  A = heat transfer area (m²)
  k = thermal conductivity of the material (W · m · K)
  T = temperature
  d = material (wood) thickness (m)

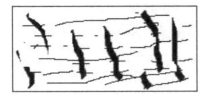 

FIGURE 17.4 *S. campanulata* (left): Note linkages across the grain. Cross-linkages bind linear cellulose polymers, thereby increasing stiffness and rupture point load. Structure of *P rigida* (right), a popular building-construction wood, shows no prominent cross-linkages, but a more streamlined internal structure.

From Equation 17.4.2, the heat transfer per unit area is calculated as:

$$1 / A(Q / t) = kA(T_{hot} - T_{cold}) / d$$

With three replicates each of 25 mm thickness and 300 mm square of each material, Harris (2019) showed that *S. campanulata* had the lowest heat conductivity of all woods tested, being 35% lower than that of *Pinus rigida* (pine wood, a major building timber). As heat resistivity reciprocates the k of thermal conductivity, *S. campanulata* had the highest heat resistance. Baxter (1998) noted that the connectedness of the solid phase also played a large role in the heat transfer rate, and it was shown that the cellulose grain pattern for *P. rigida* is more isotropic than that of *S. campulanata* (Figure 17.4) and that the cellulose fibers of *S. campanulata* are the most cross-linked of the three woods, thereby slowing down heat transfer. Baxter (1998) developed a model for the dependence of the average thermal conductivity on these structural properties, such that particles that connect at a single point or over a very small area typically conduct heat far worse than those that are connected over large fractions of their projected areas. Thus, in contrast to that of *P. rigida*, the cross-grained cellulose-lignin polymer network of *S campanulata* (Figure 17.4) increases connectedness at a single point, thereby enhancing tortuosity and delaying heat transfer.

17.5 ALBEDO

Human activities on the earth's surface may affect climate, most directly through the change of surface albedo, defined as the percentage of incident solar radiation reflected by a land or water surface (Manahan 2005). For example, if the sun radiates 100 units of energy per minute to the outer limits of the atmosphere, and the earth's surface receives 50 units per minute of the total, then reflects 25 units upward, the albedo is 50%. Some typical albedo values for different areas on the earth's surface are evergreen forests, 7–15%; dry, plowed fields, 10–15%; deserts, 25–35%; fresh snow, 85–90%; and asphalt, 8% (Manahan 2005).

PROBLEM 17.5.1
Mitigating environmental heat

Urban heat islands are one of the most pressing issues facing cities today – adversely affecting the health and comfort of residents. How can cities reduce such temperatures?

FIGURE 17.5   (a) Urban heat island, Atlanta. Note the correlation between temperature and intensity of landscapes – anthropogenic (aggravated) or natural (moderated).

Source: earthobservatory.nasa.gov www.globalchange.gov.

SOLUTION 17.5.1

By exploiting the heat-conversion to kinetic energy of evaporation effect of vegetation, as observed by MIT (2021), one of the most effective ways to fight the urban heat island effect is to reintroduce vegetation by expanding parkland, planting street trees, and installing "green roofs" designed to harbor plant life. Kurn et al. (1994) found that the presence of vegetation can lower nearby air temperatures by as much as around 4°F.

MIT (2021) presents another option: that of building cool roofs and pavements. Cool roofs feature bright coatings that reflect more sunlight and, therefore, absorb less heat. Cool pavements work similarly. They are made of brighter materials, like concrete and light-colored aggregates, or have been treated with reflective coatings. One model of concrete sustainability estimated that, if widely implemented, cool pavements could reduce the frequency of heat waves by 41% across all US urban areas (MIT 2021)

PRACTICE PROBLEM 17.5.1
Excess heat

In some heavily developed areas, anthropogenic (human-produced) heat (Figure 17.5a) release is comparable to solar input.

The anthropogenic energy release over the 60 km² of Manhattan Island averages about four times the solar energy falling on the area; over the 3500 km² of Los Angeles, the anthropogenic energy release is about 13% of the solar flux. Why is the anthropogenic heat in Manhattan different from that for Los Angeles?

17.6   HEAT REMEDIATION USING CONCRETE AS A REFRIGERATOR

Since concrete production exacerbates global warming, this section applies only to existing concrete buildings. Due to the large thermal mass of a concrete floor, which includes the ground (earth) to which it is physically connected, when the

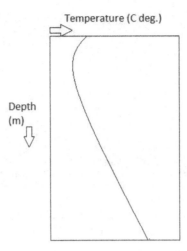

FIGURE 17.5 (b) Typical ground temperature profile during hot seasons, with a small decrease in the first 100 m below the surface.

aboveground surroundings are hot, a concrete floor will absorb heat and still retain its lower temperature. The equilibrium temperature between objects of small mass such as coins, in physical contact with a concrete floor, is effectively that of concrete. Hence, just as supermarkets store vegetables at just a few degrees lower than that of the surrounding air, concrete floors could be utilized as short-term refrigerators. Being a part of the ground, the effective mass of the concrete is almost limitless, and conduction of heat downgradient could be constant if the concrete is a fraction of a degree cooler than whatever is firmly connected to it (Figure 17.5b).

17.7 HEAT REMEDIATION BY NATURAL EVAPORATION

When a glass of water with a vacuum (no air) above it is left to stand, the molecules that are at the topmost layer of the water will evaporate when the water molecules gain enough kinetic energy (i.e., how fast they vibrate) to break the bonds that hold them to one another. Kinetic energy is dependent on temperature. So, the molecules vibrate faster, break their bonds, and enter the vacuum as a vapor. Some molecules will stay as a vapor in the vacuum, but others will re-enter the liquid. When the molecules enter the liquid as fast as they are leaving, then it is saturated. Hence, the rate of evaporation depends on the surface area of exposure, the strength of the molecule-to-molecule bonds, the temperature of the water, and the temperature of the surroundings. But evaporation is a cooling process. Large-scale evaporation causes drop in temperature which everyone feels when rain falls (if the prior humidity was not very high).

PROBLEM 17.7.1
Cooling the interior of buildings

How can cooling within buildings occur on hot days without using fossil fueled cooling fans or air conditioners, thereby reducing the CO_2 footprint?

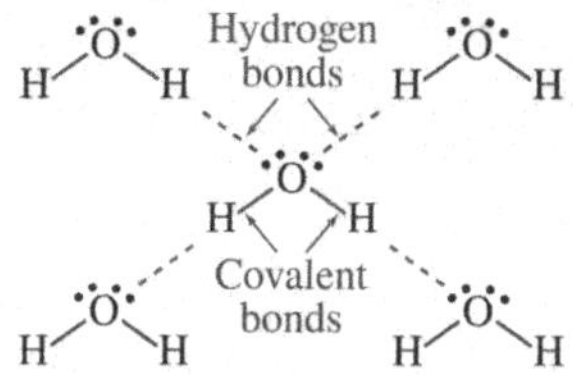

FIGURE 17.6 Hydrogen bonding among water molecules causing a high viscosity and a strong surface tension.

SOLUTION 17.7.1

If a liter of water spills out on a flat concrete floor, the layer of water so created will immediately begin to evaporate, even as every molecule in that liter remains bonded (directly or indirectly) to every other water molecule. With such a high collective bonding strength, evaporation from the surface is slow, and the ground could remain wet for many hours, even days, before complete evaporation dries the floor. However, if sheets of newspaper are spread on the wet floor, drying occurs much more rapidly. Why?

First, the millions of pores in the newspaper change one large mass of interconnected water molecules connected by hydrogen bonds (Figure 17.6) into millions of separated entities such that each water molecule now links with a far fewer number of other hydrogen-bonded molecules than before. The now weakly bonded molecules escape more rapidly, thereby increasing the intensity of evaporation. This increase in evaporation intensity extracts heat energy more efficiently from the surrounding air, producing greater cooling efficiency within a building. To avoid increasing the humidity of the room due to the rapid evaporation, an extraction fan operated at low speed can maintain comfort levels by preventing a buildup of excess water vapor.

In other words, by deliberately wetting a flat impermeable surface in a building and covering it with an absorbent material (preferably a recyclable one such as old newspaper), efficient cooling can be achieved while substantially reducing the carbon footprint. In fact, if the wet materials were hung out in layers (Figures 17.7–17.9) rather than on a floor, more sheets could be exposed, causing more rapid cooling.

PROBLEM 17.7.2
Cooling effect of trees

How do trees cool the environment?

SOLUTION 17.7.2

Shading and transpiration have been shown to decrease the temperature in buildings. Based on different tree characteristics on saving building energy use, Hsieh et al. (2018) developed four scenarios: Scenario 1 with treeless condition, Scenario 2 with existing trees condition, Scenario 3 with more tree shading condition, and Scenario 4 with more tree transpiration condition. The simulation results showed that compared to Scenario 1, Scenario 2 can reduce building energy use by 10.3% due to the benefits

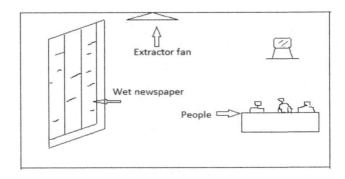

FIGURE 17.7 Evaporative adiabatic cooling occurs where a portable absorbent surface continuously converts thermal energy to kinetic energy in a closed system, thereby cooling hot air while minimizing the use of fossil fuels, with the aid of an extractor fan.

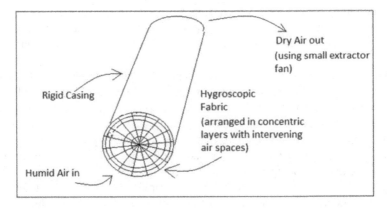

FIGURE 17.8 A water vapor trap for evaporative cooling. This device removes water vapor from the ambient air, thereby avoiding the counterproductive discomfort associated with increasing the humidity during evaporative cooling.

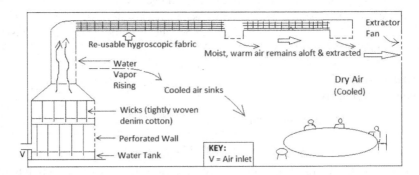

FIGURE 17.9 A self-contained evaporative cooling system. Note the air inlet at V, lower left, which allows fresh air in when needed. Continuous evaporation ensures continuing conversion of thermal to kinetic energy, thereby removing heat from the room. Note greater air space in second batch of hygroscopic fabric, to facilitate the pumping out of the humid air entrained.

from existing trees shading and transpiration. More shading in Scenario 3 and a higher transpiration rate in Scenario 4 reduced building energy use by 15.2% and 12.4%, respectively, in comparison to that in Scenario 1.

When the plant opens its stomata to let in carbon dioxide, water on the surface of the cells of the spongy mesophyll and palisade mesophyll evaporates and diffuses out of the leaf. This process is called transpiration. Water is drawn from the cells in the xylem to replace that which has been lost from the leaves. The hydrogen bonding based on the strong attraction between water molecules creates a strong cohesion between water molecules. Whenever the stomata open to admit CO_2, evaporation occurs continually atop the mechanism which pulls up this water column as the transpiration stream through the leaves. The kinetic energy for the evaporation comes largely from the surrounding air which consequently cools.

Moreover, trees improve comfort conditions outdoors within cities by blocking hot and dust-laden winds. Based on these findings, air entering a building can be cooled appreciably by vegetation prior to it being used in the evaporative cooling system previously proposed in Section 17.7. In this connection, cooling varies directly with (a) leaf size and (b) the number of leaves.

17.8   HEAT EXCHANGING AND HEAT PUMPING

When heat is being transferred, the temperature of one system of particles decreases, while the system of particles toward which the heat is transferring increases in temperature. Thus, heat exchangers located inside exhaust tail pipes or smokestacks capture waste heat as the hot exhaust gases drifting upward brush past aluminum fins with water flowing through them. The water carries the heat back into the plant, to warm cold gases fed into an engine or furnace, etc.

In an energy-efficient shower, heated water flowing down the plug runs through copper coils of a heat exchanger installed close by. Cold water feeding into the shower but not mixing with the dirty wastewater pumps up past the same coils, picking up some of the waste heat, thus reducing the heat needed to heat the shower.

Fluid used to cool down vehicle engines like those in buses is often passed through a heat exchanger, reclaiming heat which can warm cold air from outside that is then pumped up from the floor of the passenger compartment. That avoids additional electric heaters inside the bus while accomplishing three necessary tasks: cooling the water, heating the air, and maintaining the required engine operating temperature. Ceramic conduits used in high-temperature applications (over 1000°C or 2000°F) that would melt metals like copper, iron, and steel, also resist deterioration by corrosive and abrasive fluids at either high or low temperatures.

Air contains a high level of kinetic and thermal energy if it remains above 0 °K (or −273°C). Hence atmospheric heat is available in the cooler months. A heat pump gathers a large volume of low-temperature heat and concentrates it into a small volume of high-temperature heat. For example, an outside air temperature of 10°C can supply a heat pump with several extra degrees °C of heat from the air by heating heat a small volume of, say, 200 litres.

Hence, the greater the resulting temperature change due to the energy input (ΔE_t), the lower the specific heat of the substance. Water, for example, will not exhibit a large ΔT despite the input of a large increase in ΔE_t. Hence, its specific heat capacity changes very little in such a case; hence, heat pumps are effective on only small volumes of water.

PROBLEM 17.8.1
Thermal energy

Substance A has a mass of 10 kg, and its specific heat is 0.060 J · kg °C. The temperature difference before and after heat transfer is 25°C. What is the thermal energy of the substance?

SOLUTION 17.8.1

Given: m = 10 kg, c = 0.060 J · kg^{-1} °C, ΔT = 25°C.
   Since

$$Q = m \cdot c \cdot \Delta T$$

∴ Q = 10 × 0.060 × 25
∴ Q = 15.0 J

PROBLEM 17.8.2
Temperature increase due to specific heat

If we transfer into water an energy of 1900 J, how much does the temperature of 250 g of water rise? Water has a specific heat of 3482 J · kg^{-1} °C.

SOLUTION 17.8.2

Given: m = 250 g = 0.250 kg, c = 3482 J · kg^{-1} °C, Q = 1800 J
   Since

$$Q = mc\Delta T$$

∴ ΔT = Q / mc
∴ ΔT = (1900) / (0.250 × 3482)
∴ ΔT = 1900 / 870.5
∴ ΔT = 2.182°C

PROBLEM 17.8.3
Energy required for temperature increase

If the specific heat of copper is 453 J · kg^{-1} °C, how much energy is required to raise the temperature of 450 g of copper by 21°C?

SOLUTION 17.8.3

Given: m = 0.45 kg, c = 453 J · kg^{-1} °C, ΔT = 21°C

Since

$$Q = mc\Delta T$$

∴ Q = 0.45 × 453 × 21
∴ Q = 4280.85 J

PROBLEM 17.8.4
Determination of specific heat

A substance has a mass of 20 kg, and its thermal energy is 655 J. The difference in temperature recorded after an interaction is 27°C. Based on this information, what is the specific heat?

SOLUTION 17.8.4

Given: m = 20 kg, Q = 655 J, ΔT = 27°C
Since

$$Q = mc\Delta T$$

∴ c = Q / mΔ
∴ c = (655) / (20 × 27)
∴ c = 655 / 540
∴ c = 1.212 J · kg^{-1} °C

PRACTICE PROBLEM 17.8.1
Temperature change due to energy input

How high will the temperature of 27 kg of water rise after transmitting around 1800 J of energy, using the specific heat of water of 3820 J · kg^{-1} °C? _____

PRACTICE PROBLEM SOLUTION 17.8.1

Given: m = 27 kg, c = 3820 J · kg^{-1} °C, Q = 1800 J
Since

$$Q = mc\Delta T$$

∴ ΔT = _____
∴ ΔT = _____
∴ ΔT = _____
∴ ΔT = 0.02032°C

PRACTICE PROBLEM 17.8.2
Bond formation and temperature rise

Thermal pollution can also occur when high concentrations of some chemical wastes suddenly become effluents into small streams, with potentially adverse effects on aquatic life forms. When sodium hydroxide dissolves in water, the new bonds forming between the sodium and hydroxide ions with water expel 1.1 kJ of heat per gram of NaOH. If we dissolve 20 g of NaOH in 100 mL of water, how many degrees will the temperature of the water increase? _____

PRACTICE PROBLEM SOLUTION 17.8.2

1 g of NaOH expels 1.1 kJ of heat in water.
    ∴ 20 g will expel 20 × 1.1 kJ of heat, or 22,000 J during its dissolution.
    ∴ Q in the equation is 22,000.
    There are 100 mL of water, and the density of water is 1 g · mL^{-1}.
    So, there are 100 g of liquid water which has a heat capacity of 4.186 J · kg^{-1} · K^{-1} at 20°C.
    Solving for change in temperature,

$$\Delta T = Q / mc$$

∴ ΔT = _____

$$\Delta T = \underline{\hspace{2cm}} = \underline{\hspace{2cm}} = 45.26 \text{ K}$$

PROBLEM 17.8.5
Determination of the mass which produced the heat

The combustion of glucose releases a heat of 15.5 kJ · g^{-1}. Suppose I can do this reaction under nitrogen in a special container in which I can provide just enough oxygen for the reaction to reach completion. This way all the heat expelled goes into heating up the nitrogen. The container contains 10 g of nitrogen gas, and the reaction causes it to heat up to 5°C. How much glucose did I start with?

SOLUTION 17.8.5

Heat capacity of nitrogen = 1.040 J · g · K^{-1}
    Mass of nitrogen = 10 g
    Temperature raised by reaction = 5°C
    First, solve for Q using Q = mcΔT
    Q = 1.040J · (g · K)$^{-1}$ × 10 g × 5 K = 52 J (so, we used 52 J to raise 10 g of N by 5°C degrees)
    The glucose provided 15.5 kJ per gram or 15,500 J per gram.
    52 J / 15,500 J · g^{-1} = 0.0034 g or 3.4 mg of glucose.

PRACTICE PROBLEM 17.8.3
Determination of the mass which produced the heat

The combustion of glucose releases a heat of 15.5 kJ · g^{-1}. Suppose I can do this reaction under nitrogen in a special container in which I can provide just enough oxygen for the reaction to reach completion. This way all the heat expelled goes into heating up the nitrogen. The container contains 8 g of nitrogen gas, and the reaction causes it to heat up to 7 °C. How much glucose did I start with?

Hint:

Heat capacity of nitrogen = _____

Mass of nitrogen = _____

Temperature raised by reaction = _____

First, solve for Q using Q = mcΔT

Q = _____

The glucose provided. _____ J · g^{-1}

Ans. = _____ mg of glucose.

PROBLEM 17.8.6
Measuring evaporative cooling

The larger the evaporating surface, the greater the rate of evaporation. Therefore, to determine the quantitative requirements of evaporative cooling for a room, the quantity of water evaporated in time t is required thus:

$$\text{Cooling efficiency} = \Delta W / \Delta t$$

where

ΔW = mass of water evaporated (kg)

Δt = time (s)

How much energy is released when 1 kg of water is condensed?

SOLUTION 17.8.6

$$\text{Latent heat released} = \text{Mass of water} \times \text{Latent heat of vaporization}$$

$$= 1000\,g \times 2500\ Jg^{-1} = 1.0 \times 10^3\ g \times 2.5 \times 10^3\ Jg^{-1}$$

$$= 2.5 \times 10^6\ J$$

Considering that evaporation removes the same amount of heat from the environment as condensation produces (latent heat of vaporization = liquid-gas transition = 2500 J · g^{-1}), evaporative cooling is ineffective if the evaporated water vapor is allowed to condense in the same enclosed area. However, removing the evaporated water vapor avoids condensational heat.

PROBLEM 17.8.7
Heat of condensation

How fast would a typical compact car (1300 kg) move with the equivalent amount of energy required to condense 1 kg of water?

SOLUTION 17.8.7

$$E = \frac{1}{2}\left(mv^2\right)$$

where

E = energy

m = mass

v = velocity

$$\text{Latent heat released} = \text{Kinetic energy of car} = \left(\frac{1}{2}\right) \times \text{Mass of car} \times \text{Velocity}^2$$

$$2.5 \times 10^6 \text{ J} = \left(\frac{1}{2}\right) \times (1300 \text{ kg}) \times v^2$$

$$2.5 \times 10^6 \text{ kg} \cdot \text{m}^2 \cdot \text{s}^{-2} = \left(\frac{1}{2}\right) \times (1300 \text{ kg}) \times v^2$$

$$v = 62 \text{ m} \cdot \text{s}^{-1}$$

or about 139 mph

PROBLEM 17.8.8
Heat capacity of metals

By exploiting the varying thermal conductivity of metals and other materials, individuals can reduce their impact on global warming. Some applications exist in the choice of metal comprising cooking utensils. The specific heat capacity for copper is 385 J · kg · °C. Calculate the thermal energy change when the temperature of 2.00 kg of copper is changed by 10.0°C.

SOLUTION 17.8.8

$$\text{Change in thermal energy} = \text{Mass} \times \text{Specific heat capacity} \times \text{Change in temperature}$$

$$= \underline{\hspace{3cm}}$$

$$= 7700 \text{ J} (7.7 \text{ kJ})$$

By comparison, an iron cooking utensil of the same mass would require:

$$2.00 \times 921 \times 10.0 = 18,420 \text{ J, representing} >150\% \text{ of wasted heat.}$$

PRACTICE PROBLEM 17.8.4
Thermal capacity in water

The specific heat capacity of water is 4180 J · kg · °C. What is the thermal energy change when 0.200 kg of water cools from 100°C to 25.0°C? _____

PRACTICE PROBLEM SOLUTION 17.8.4

$$\text{Change in temperature} = (100 - 25) = 75.0°C$$

$$\text{Change in thermal energy} = \text{Mass} \times \underline{\hspace{2cm}} \times \underline{\hspace{2cm}}$$

$$= \underline{\hspace{2cm}} \times \underline{\hspace{2cm}} \times \underline{\hspace{2cm}}$$

$$= 62,700 \text{ J} (62.7 \text{ kJ})$$

An iron heater of the same mass and temperature of the water in winter would have released:

0.200 × 460.5 × 75.0 = 6,907.5 J of thermal energy = 10-fold less heat released from iron.

Hence, solar-heated "hot-water bottles" are very effective in reducing greenhouse gases.

REFERENCES

Baxter LL (1998) Influence of ash deposit chemistry and structure on physical and transport properties. Fuel Process Technol 56, 81–88.https://doi.org/10.1016/S0378-3820(97)00086-6.

Bosch CH (2002) *Spathodea campanulata* P. Beauv. [Internet] Record from protabase. In: Oyen LPA, Lemmens RHMJ (eds) PROTA (Plant Resources of Tropical Africa// Ressources végétales de l'Afrique tropicale). Wageningen. https://ethnobotanyjournal. org/index.php/era/article/download/4483/1683. Accessed June 14, 2023.

Goel PK (2006) Water Pollution – Causes, Effects and Control. New Age International, New Delhi, p. 179.

Harris MA (2019) Confronting Global Climate Change: Experiments & Applications in the Tropics. CRC Press, Boca Raton, FL.

Hsieh C-M, Li J-J, Zhang L, Schwegler B (2018) Effects of tree shading and transpiration on building cooling energy use. Energy Build 159, 382–397. https://doi.org/10.1016/j.enbuild.2017.10.045

ISSG (2017) Global Invasive Species Database (GISD) https://www.iucngisd.org/gisd/

Kennish MJ (1992) Ecology of Estuaries: Anthropogenic Effects. Marine Science Series. CRC Press, Boca Raton, FL.

Kurn DM, Bretz SE, Huang B, Akbari H (1994) The potential for reducing urban air temperatures and energy consumption through vegetative cooling. United States. https://doi.org/10.2172/10180633

Laws E (2017) Aquatic Pollution: An Introductory Text (4th ed.). John Wiley & Sons, Hoboken, NJ.

Manahan S (2005) Environmental Chemistry. Taylor and Francis, Boca Raton, FL. https://www.iucn.org/our-union/commissions/group/iucn-ssc-invasive-species-specialist-group. Accessed June 11, 2014.

MIT (2021) The MIT Concrete Sustainability Hub makes key impacts in three areas. https://climate.mit.edu/explainers/urban-heat-islands. Accessed September 5, 2023.

NCC (2019) Volume One Amendment 1. Specification J1.2 Material properties. National Construction Code. https://ncc.abcb.gov.au/editions/2019-a1/ncc-2019-volume-one-amendment-1/section-j-energy-efficiency/specification-j12

Pagad S (2010) Invasive species specialist group (ISSG) of the IUCN Species Survival Commission. https://www.iucn.org/our-union/commissions/group/iucn-ssc-invasive-species-specialist-group

Pederson C, Helevang K (2010) Determining Insulin and air infiltration levels using an infra-red thermometer. NDSU Extension Service, North Dakota State University, Fargo, ND.

Pulipati S, Hapsarana Parveen SK, Kiran Babu R, Vagdevi G, Srinivasa Babu P (2013) Pharmacognostical and physicochemical standardization of leaves of Spathodea campanulata. J Pharmacogn Phytochem 2(2), 189–192.

Section IV

Water

18 Water
Pesticides and Acidity

MAIN POINTS

- Most of the organic toxins are almost water insoluble and nonbiodegradable.
- They are therefore highly persistent toxins that are transferred from the lower trophic level to higher trophic level.
- Over time, they reach a level that causes serious metabolic/physiological damage.

18.1 PROBLEMS OF ANALYSIS

Analysis of water samples can be preceded by evaporation and crystallization, but more modern methods are also employed. Several decades ago, the ion micro-probe mass analyzer, a refinement of the mass spectrometer, detected impurity of as little as one part per billion in a substance.

PROBLEM 18.1.1
Interpreting results

A sample of lithium removed from a water reservoir was found on analysis to contain five atoms of mercury for every billion lithium atoms. Calculate the percentage by weight of mercury in the lithium sample.

SOLUTION 18.1.1

$$Hg = 200.6 \text{ amu (atomic mass units)} \times 5 = 1003 \text{ g}$$
$$Li = 28.09 \text{ amu} \times 10^9 = 2.809 \times 10^{10} \text{ g}$$
$$\% \ Hg = (1003 \text{ g} / 2.809 \times 10^{10} \text{ g}) \times 100$$
$$= 0.00000358\% \text{ Hg by weight}$$

Note: As the presence of such a low percentage of mercury attached to lithium is no indication of mercury contamination levels in the surrounding fluid medium, analysis of mercury in the water is required.

PROBLEM 18.1.2
Proportion of components

Analysis of 100 g of apparently pure crystals recovered after evaporation of a sample from municipal wastewater indicated a compound of xenon chloride containing 55.2% xenon and 44.8% chlorine by weight. What is the empirical formula?

DOI: 10.1201/9781003341826-23

SOLUTION 18.1.2

Ratio by Weight		Weight per Mole of Atoms		Ratio by Moles
Xenon	55.2 g	÷	131.3 g · mole^{-1}	=	0.420 mole
Chlorine	44.8 g	÷	35.5 g · mole^{-1}	=	1.262 mole

The ratio of moles, 0.420:1.262, is also the ratio of atoms (e.g., ratio of $US = ratio of $CA)

Here, the aim is to obtain the ratio of atoms as small whole numbers. We can quickly do this by dividing by the smallest (or smaller) number as follows:

Xenon	0.420 mole	÷	0.420	=	1
Chlorine	1.262 mole	÷	0.420	=	3.0

The above ratio, 1:3, indicates a compound having the formula $XeCl_3$ (xenon chloride).

PROBLEM 18.1.3

A petroleum-based pesticide sprayer got stuck while trying to ford a stream flowing at a rate of 136 L · s^{-1}. Pesticide leaked into the stream (river) for exactly 1 h and at a rate that contaminated the stream at a uniform 0.25 ppm of methoxychlor. How much pesticide leaked from the sprayer during this elapsed time?

SOLUTION 18.1.3

Amount of water flowed = 136 L × 3600 s = 489,600 L
   Contamination rate = 0.25 ppm = 0.25 / 1000,000
          = (0.25 / 1000,000) × 100 = 0.000025%
   Contamination time = 3600 s (1 h)
   Amount of pesticide lost = Amount which contaminated the water at 0.25 ppm
   As the amount of water that flowed is 489,600 L, contamination of water was 0.25 ppm of 489,600 L.
   0.25 ppm of 489,600 = 0.25 / 1,000,000 × 489,600
   Therefore, 0.0972 L leaked during the elapsed time.

PROBLEM 18.1.4
Toxicity thresholds

The NOEL for a pesticide is found to be 0.10 mg · kg^{-1} · day^{-1}. (a) What would the ADI or RfD (reference dose) value for adults be set at? (b) What mass of the chemical is the limit that a 60 kg woman should daily ingest?

SOLUTION 18.1.4 (A)

Persistent pesticides include organochlorides but not organophosphates (Figure 18.1) and are found in the fat tissue of humans. The *no observable effects level* (NOEL),

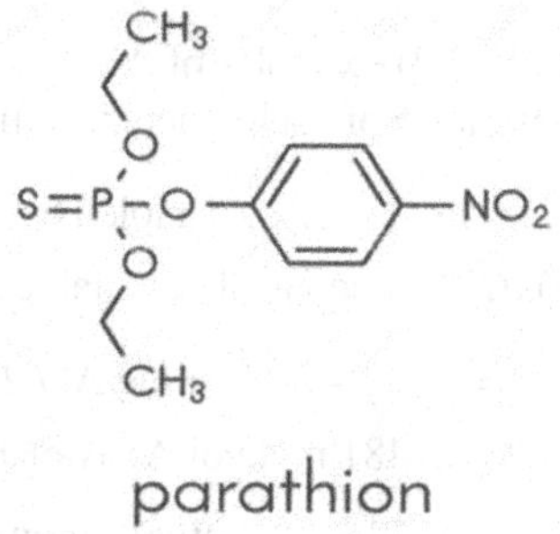

parathion

FIGURE 18.1   Parathion, an organophosphate, is faintly soluble in water ($25 \text{ mg} \cdot \text{L}^{-1}$).

normally expressed in milligrams of the substance per kilogram of body mass per day, is the level just below the highest dose at which no effects are seen. To protect the most sensitive individuals in a species, it is a common practice to divide the NOEL by a large number, which often is 100. The resulting figure is the acceptable daily intake (ADI), or, according to the USEPA, the *toxicity reference dose* (RfD).
    ADI or RfD = 0.01 / 100

SOLUTION 18.1.4 (B)

$0.00010 \text{ mg} \cdot \text{kg}^{-1} \cdot \text{day}^{-1} = \text{Safe dose for a 1 kg adult}$

$0.00010 \text{ mg} \cdot \text{kg}^{-1} \cdot \text{day}^{-1} \times 60 = \text{Safe dose for a 60 kg woman}$

$6.0 \times 10^{-3} \text{ or } 60 \times 1 / 10^{3} = 0.0060 \text{ mg}$, the maximum that should be ingested daily

18.2   OTHER INORGANIC WATER CONTAMINANTS

Several non-pesticidal wastewater effluents can contaminate water bodies.

PROBLEM 18.2.1
Mercury

If a solution that contains 21 g of mercury II chloride is placed in a glass container containing 4 g of powdered aluminum, which of the reactants is in excess? What is the mass of mercury formed? How much of the excess reactant is left if the reaction is complete?

SOLUTION 18.2.1

The above reaction occurs in the following mole ratio:

$$3HgCl_2 + 2Al \rightarrow 3Hg + 2AlCl_3$$

Hence, 3 moles of $HgCl_2$ react with 2 moles of Al and by simple proportion, 1 mole of $HgCl_2$ reacts with (2/3) moles of Al or

x moles of $HgCl_2$ react with $(2/3) \cdot$ x moles of Al.

To calculate the number of moles of each reactant initially available:

$$\text{moles of } HgCl_2 = 21.00 \text{ g } HgCl_2 \times 1 \text{ mole } HgCl_2 \, / \, 271 \text{g } HgCl_2$$
$$= 0.0774 \text{ mole } HgCl_2 \text{ available}$$

$$\text{Moles of Al} = 4.00 \text{g } Al \times 1 \text{ mole Al} \, / \, 27.0 \text{g } Al$$
$$= 0.1481 \text{ mole of Al available}$$

For each of the reactants available, the actual amount of each needed to react completely with the other can be calculated as follows:

Quantity of Al needed to react with 0.0774 mole $HgCl_2$
= 0.0774 mole $HgCl_2$ × 2 moles Al / 3 moles $HgCl_2$ (i.e., 2/3 the $HgCl_2$ molar quantity)
= 0.0516 mole Al (but as there are more moles of Al than is necessary for the complete reaction, i.e., 0.1481, the Al is in excess)

Amount of $HgCl_2$ needed to react with 0.0516 mole of Al
= 0.0516 × 3 moles $HgCl_2$ / 2
= 0.0774 mole $HgCl_2$ (but there is only 0.0774 of $HgCl_2$ available; hence, all the $HgCl_2$ will be completely used up and will limit the quantity of Al that can react).

If precise amounts necessary for a reaction to reach completion are not present, the reaction cannot continue to completion. The reactant which is in the deficit is the limiting reagent. Here, the limiting reagent is $HgCl_2$. Hence, the following calculations involving masses and products depend on the amount of $HgCl_2$ available:

$$\text{Moles Hg formed} = \text{Moles reacted}$$
$$\text{(The balanced equation above shows that they are equal)}$$
$$= 0.0744 \text{ moles of } HgCl_2 \text{ (and 1 mole Hg weighs 201 g)}$$

$$\text{Mass of Hg formed} = 0.0744 \text{ moles Hg} \times 201 \text{ g Hg} \, / \, 1 \text{ mole Hg}$$
$$= 14.95 \text{ g Hg}$$

$$\text{Moles of Al reacted} = 0.0516 \text{ (as shown above)}$$
$$\text{Moles of Al in excess} = 0.1481 \text{ (available initially)} - 0.0516$$
$$= 0.0965 \text{ mole of Al in excess}$$

0.0965 mole Al × 27.0 g Al / 1 mole Al = 2.605 g Al in excess
NB: Mercury atoms are 201 / 27 times as heavy as aluminum atoms.

18.3   REMEDIATION: ALTERNATIVES TO PERSISTENT PESTICIDES

PROBLEM 18.3.1
Organophosphates

What are the advantages of using organochloride insecticides compared to DDT?

SOLUTION 18.3.1

Though it is highly toxic, thereby requiring strict adherence to safety recommendations, parathion, an organophosphate (Figure 18.1), has the following advantages:

- It decays following spray applications on foliage with a half-life of only
  1 day, reaching low levels in a week or two (Howard 1989).
- In orange groves, the half-life of parathion is as long as one month, but usually it is closer to one or two weeks.
- In contrast, the biodegradation half-life of DDT, for example, in soil varies
  from 2 to 15 years, depending on conditions (CDC 2017).
- Parathion has little or no potential for groundwater contamination (Meister
  1992).
- In water, malathion, a far less toxic organophosphate pesticide than parathion, has a half-life between 2 and 18 days depending on conditions like
  temperature and pH.

These chemicals act by interfering with the activities of acetylcholinesterase,
an enzyme that is essential for the proper working of the nervous systems of both
humans and insects.

18.4   NEUTRALIZING ACID RAIN EFFECTS

According to the EPA (EPA 2022), the ecological effects of acid rain occur mainly in
aquatic environments, such as streams, lakes, and marshes where it can be harmful
to fish and other wildlife. They found that acidic rainwater can leach aluminum from
soil clay particles as it flows through soil and then flow into streams and lakes. The
more acid that is introduced to the ecosystem, the more aluminum is released. And
to the majority of life forms, aluminum is highly toxic.

PROBLEM 18.4.1
Neutralizing acid rain effects

Due to increased acidity, a fish-kill occurred in a small lake occupying a volume of
120,000 m^3. To provide a buffer of alkalinity, a 35 kg block of limestone ($CaCO_3$)
was added to the water. What mass (weight) of the limestone block remained after
the solution reached equilibrium with $CaCO_3$(s)?

SOLUTION 18.4.1

The reaction concerned is:

$$CaCO_3(s) = Ca^{2+} + CO_3^{2-} \quad K_s = 10^{-pK_s} = 10^{-8.48}$$

where

K_s = solubility product, i.e., the number of moles at saturation point
pK_s = negative log of solubility = 8.48

(Similarly, $pK_w = -\log [1.0 \times 10^{-14}] = 14$)

For any precipitation reaction, the K_s can be written in wing brackets, e.g.:

$$K_s = \{Ca^{2+}\}\{CO_3^{2-}\}$$

Assuming the solution is dilute, the reaction proceeds according to molar concentrations.

Hence, square brackets depicting molar concentration apply as such:

$$K_s = [Ca^{2+}][CO_3^{2-}]$$

For every mole of calcite that dissolves, one mole of Ca^{2+} and a mole of CO_3^{2-} are released into solution until, at equilibrium, both ions are equal, such that:

$$[Ca^{2+}] = [CO_3^{2-}] = s$$

Substituting s for each entity in the K_s expression:

$$10^{-8.48} = s^2$$

Solving for s (which is equivalent to Ca^{2+}), the concentration of calcium is:

$$Ca^{2+} = \sqrt{s^2} = 10^{-4.24} \text{ or } 1/10^{4.24} = 5.75 \times 10^{-5} \text{ M (i.e., moles per liter)}$$

Liters in 120 m^3 = 120,000
Moles of $CaCO_3$ in 120,000 L = 120,000 × 5.57 × 10^{-5} M = 668,400 × 10^{-5} M
Moles of $CaCO_3$ in the lake at equilibrium = 6.68 = 6.68 × (40 + 12 + 48) = 668 g
Amount of limestone left = 35 − 0.668 kg = 34.332 kg

PROBLEM 18.4.2
Acidity and turnover in lakes

If the lake water in the above problem has a turnover period (change of water) of 6 months, how long will it be before the limestone has been used up (assuming the water status remains unchanged)?

SOLUTION 18.4.2

Mass of $CaCO_3$(s) in lake = 34.332 kg ·

Years required to solubilize remaining amount = (34.332 / 0.668) × $\frac{1}{2}$

= 29.697 years

PROBLEM 18.4.3
Consider the above problem and the solution. If 35 g of calcite limestone is added to water and topped to make a 1.0 L solution containing 0.01 M NaCl, what would be the concentration of Ca^{2+} ions in the solution (on the assumption that the calcium in the solution is at equilibrium with calcite(s) and the temperature of the solution is 25°C)?

SOLUTION 18.4.3

The ionic strength formula calculates the sum of each ion's molar concentration multiplied by the valence squared.

The Davies equation can be used to calculate activity coefficients of electrolyte solutions at relatively high concentrations at 25°C. Such a case applies when the ionic strength is less than 0.50 M (Davis and Masten 2020), such that:

$$\text{Log } \gamma = -Az^2 \left(\left[\sqrt{I} / 1 + \sqrt{I} \right] - 0.2I \right)$$

where

A $\cong$ 0.5 for water at 25°C
z = charge of the ion
I = ionic strength of the solution = $\frac{1}{2}\Sigma c_i z_i^2$
c_i = molar concentration of each of the ith ion
z_i = charge on the ith ion

Because both an ion's concentration and its charge are important, we define the solution's ionic strength, I, as:

$$I = \frac{1}{2}(c_i^2 + z_i^2)$$

where

I = ionic strength
c_i and z_i = the concentration and charge of the ith ion, respectively

$$I = \frac{1}{2}\left\{ \left[Na^+\right] \times (+1)^2 + \left[Cl^-\right] \times (-1)^2 \right\}$$
$$\text{So}, I = \frac{1}{2}\left\{ (0.10) \times (+1)^2 + (0.10) \times (-1)^2 \right\} = 0.10 \text{ M}$$

For the two ions being investigated, calcium and carbonate, the absolute value (i.e., ±) of the charge on both is the same (i.e., two), where $\gamma Ca = \gamma CO_3$.

$$\text{Log } \gamma Ca = \gamma CO_3 = -(0.5)(2)^2 \left\{ \left[\sqrt{0.01} / 1 + \sqrt{0.01} \right] - 0.2(0.01) \right\} = -0.178$$
$$\gamma Ca = \gamma CO_3 = 10^{-0.178} = 0.664$$

Recalling {}, the effect of ionic strength on the solubility product is applied thus:

$$K_s = \left\{ Ca^{2+} \right\}\left\{ CO_3^{2-} \right\} = \gamma Ca[Ca^{2+}] = \gamma CO_3[CO_3^{2-}]$$
$$K_s / \gamma Ca \cdot \gamma CO_3 = [Ca^{2+}][CO_3^{2-}] = 10^{-8.48} / (0.664)(0.664) = 7.51 \times 10^{-9}$$

As was done in Problem 18.4.1,

$$\left[Ca^{2+}\right]\left[CO_3^{2-}\right] = s^2$$
$$\left[Ca^{2+}\right] = \left[CO_3^{2-}\right] = \sqrt{s^2} = \sqrt{(7.51 \times 10^{-9})}$$
$$= 8.66 \times 10^{-5} \text{ M}$$

PROBLEM 18.4.4
Regaining the pH in water bodies

If 85 mg of sulfuric acid (H_2SO_4) is added to a liter of pure water, bringing the final volume to 1.0 L, what is the final pH?

SOLUTION 18.4.4

Applying the molecular weight of H_2SO_4 to determine moles per liter,

(85 mg / 1 L H_2O) / (1 mole / 98 g) (1g / 1000 mg) = 8.67×10^{-4} · mole · L^{-1}

Ionization occurs in this reaction as:

$$H_2SO_4 \rightarrow 2H^+ + SO_4^{2-}$$

As sulfuric acid is a strong acid, the pH can be determined in the following manner:

As the concentration of sulfuric acid is 8.67×10^{-4} M, during the dissolution of the acid, $2(8.67 \times 10^{-4})$ M of H^+ occurs. As pH is the negative logarithm of this figure, pH is:

$$pH = -\log(8.67 \times 10^{-4})$$

Ans. pH = −log of 0.000867 = 3.06

PROBLEM 18.4.5
Concentration of an acid or a base

To neutralize accidental spills of alkaline wastes into a water source, remediation with acids can be applied. Sulfuric acid is commonly supplied commercially at concentrations of 78, 93, or 98 weight percent (wt%) solution, and sulfuric acid at 100% has a specific gravity of 1.839. Assuming that the temperature is 15°C, what is the concentration of a 78% solution in units of milligram per liter, molarity, and normality?

SOLUTION 18.4.5

Density = Mass per unit volume, and the density of a 100% sulfuric acid is:

$$(1000.0 g \cdot L^{-1})(1.839) = 1839 g \cdot L^{-1}$$

From the USGS (2018), at 15°C, 1.000 L of water weighs 99.103 g. Hence, a 78% solution of H_2SO_4 would have a density of:

$$(999.1026 g \cdot L^{-1})(0.22) + (1839 g \cdot L^{-1})(0.78) = 1654.2 g \cdot L^{-1} \text{ or } 1.7 \times 10^6 mg \cdot L^{-1}$$

The molecular weight of H_2SO_4 is found by multiplying the number of atoms by each atomic weight such that:

$$(H = 2 \times 1.008, \ S = 1 \times 32.06, \ O = 4 \times 15.9994 = 64) = 98.08 g \cdot mole^{-1}$$

To find the molarity (moles per liter), divide the concentration (grams per liter) by the molecular weight (grams per mole), i.e., to determine the fraction of a mole in the liter:

$$1654.2 \text{ g} \cdot \text{L}^{-1} / 98.08 \text{ g} \cdot \text{mole}^{-1} = 16.866 \text{ M}$$

Finding the normality (N) requires the following information:

Normality depends on the number of H ions involved in a precipitation reaction such as that of replacing the calcium in $CaCO_3$ to produce H_2CO_3, i.e., the number of equivalents per mole. As H_2SO_4 can donate two H ions, it has two gram equivalents.

$$N = nM$$

where n = the number of gram equivalents per mole.

Therefore, N = 16.866 mole $\cdot$ L^{-1} × (2 Eq $\cdot$ mole^{-1}) = 33.73 Eq $\cdot$ L^{-1} or 33.73 N.

PRACTICE PROBLEM 18.4.1
Concentration of solutions

Assuming that the temperature is 15°C, what is the concentration of a 60% solution of sodium hydroxide in units of milligrams per liter, molarity, and normality?

PROBLEM 18.4.6
Buffering capability of waters

The water in a lake has a measured pH value of 7.2. The concentration of bicarbonate was measured at 1.4×10^{-3} M. Assuming this system to be closed to the atmosphere, what are the concentrations of carbonate, carbonic acid, and total carbonate?

SOLUTION 18.4.6

Using the pK and molar concentration formulas:

$$pK_{a1} = \left[HCO_3^-\right]\left[H^+\right] / \left[H_2CO_3\right] = 10^{-6.3}$$
$$pK_{a2} = \left[CO_3^{2-}\right]\left[H^+\right] / \left[HCO_3^-\right] = 10^{-10.33}$$

To solve for [H_2CO_3]:

$$\left[H_2CO_3\right] = \left[HCO_3^-\right]\left[H^+\right] / pK_{a1} = \left(1.4 \times 10^{-3}\right)\left(10^{-7.2}\right) / 10^{-6.3} = 9.08 \times 10^{-5} \text{ M}$$

Solving for $\left[CO_3^{2-}\right]$, we obtain

$$\left[CO_3^{2-}\right] = pK_{a2}\left[HCO_3^-\right] / \left[H^+\right] = \left(10^{-10.33}\right)\left(1.4 \times 10^{-3}\right) / 10^{-7.2}$$
$$= \left(4.67 \times 10^{-11}\right) \times 0.0014 / 6.30 \times 10^{-8} = 1.03 \times 10^{-6} \text{ M}$$

$$C_T = [H_2CO_3] + \left[HCO_3^-\right] + \left[CO_3^{2-}\right] = 9.08 \times 10^{-5} + 1.4 \times 10^{-3} + 1.03 \times 10^{-6}$$

Total carbonate = 1.491×10^{-3} M $\cong 1.5 \times 10^{-3}$ M

PROBLEM 18.4.7
Carbonate buffers in solution

Most natural waterways contain a buffer system to resist sudden changes in pH, such as during acidic and alkaline spills. An acid that gets into the water increases the hydrogen ion concentration, thereby taking the system out of equilibrium. To regain equilibrium, the carbonate (CO_3^{2-}) combines with the free protons (H^+ ions) to achieve equilibrium. Bicarbonate already in the water also reacts to form carbonic acid (a weak acid), which subsequently dissociates to CO_2 and water. Hence, buffering capacity is the ability to resist acidity.

For example, consider water in a lake exhibiting a pH value of 7.65. On the assumption that the temperature = 25°C, Henry's constant for CO_2 is $10^{-1.47}$ M atm^{-1} at this temperature, and the partial pressure of carbon dioxide is $10^{-4.14}$ atm, what are the concentrations of carbonate, bicarbonate, carbonic acid, and total carbon (C_T)?

SOLUTION 18.4.7

As the partial pressure of carbon dioxide and Henry's law constant are given, the concentration of carbonic acid can be calculated using the following relationship:

$$[H_2CO_3] = K_H P_{CO2} = (10^{-1.47} \text{ M} \cdot \text{atm}^{-1})(10^{-4.14} \text{ atm}) = 10^{-5.61} \text{ M} (\text{i.e.}, CO_2 + H_2O$$
$$= H_2CO_3)$$

Or if the air were 100% CO_2 (1 atm of CO_2), the water below would have $10^{-1.47}$ M CO_2.

$$pK_{a1} = [HCO_3^-][H^+] / [H_2CO_3] = 10^{-6.3}$$

and

$$[CO_3^{2-}][H^+] / [HCO_3^-] = 10^{-10.33}$$

To solve for $[H_2CO_3^-]$, by transposing:

$$[HCO_3^-] = [H_2CO_3]pK_{a1} / [H^+] = (1 \times 10^{-5.61} \text{ M})(10^{-6.3}) / 10^{-7.65} = 10^{-4.26}$$
$$= 5.49 \times 10^{-5}$$

To solve for $[CO_3^{2-}]$:

$$[CO_3^{2-}] = pK_{a2}[HCO_3^-] / [H^+] = (10^{-10.33})(5.49 \times 10^{-5}) / 10^{-7.65} = 5.49 \times 10^{-7.68} \text{ M}$$

$$C_T = [H_2CO_3] + [HCO_3^-] + [CO_3^{2-}] = 10^{-5.61} + (5.49 \times 10^{-5}) + (5.49 \times 10^{-7.68})$$

$$= 0.00000245 + 0.0000549 + 0.0000001.47$$

$$= 0.00005749 \text{ or } 5.74 \times 10^{-5} \text{ M}$$

PROBLEM 18.4.8
Calculating the exact alkalinity

The difference between alkaline water and high alkalinity is like comparing, respectively, "flame" with "conflagration." Both alkaline water and high alkalinity are above a pH of 7, but water with high alkalinity has a high buffering capacity; i.e., there are many more alkaline ions available as the need arises. Because alkalinity is largely linked to calcium species, by convention, it is not expressed as moles per liter, but as milligrams of $CaCO_3$ per liter (or normality). Considering that $CaCO_3$ has two equivalent weights, converting concentrations of ions to milligrams per liter of $CaCO_3$ is done as:

$$Mg \cdot L^{-1} \text{ as } CaCO_3 = \left(Mg \cdot L^{-1} \text{ of species}\right)\left(EW \text{ of } CaCO_3 / EW \text{ of species}\right)$$

What is the exact alkalinity of a pond water which contained 105.0 mg $\cdot$ L^{-1} of CO_3^{2-} and 78.0 mg $\cdot$ L^{-1} of HCO_3^- at a pH of 10.5 and where $T = 25°C$?

SOLUTION 18.4.8

To equate units, first convert all species present to Mg $\cdot$ L^{-1} as $CaCO_3$.
To determine equivalent weights, list molecular weight (MW) and normality (n):

$$CO_3^{2-}: MW = 60, \ n = 2, \ EW = 30$$
$$H^+: MW = 61, \ n = 1, \ EW = 1$$
$$OH^-: MW = 17, \ n = 1, \ EW = 17$$
$$HCO_3^-: MW = 61, \ n = 1, \ EW = 61$$

Concentrations of H$^+$ and OH$^-$ are determined as follows: pH = 10.5; therefore,

$\left[H^+\right] = 10^{-10.5}$ mole $\cdot$ L^{-1}, and knowing that

$Mg \cdot L^{-1} = \text{Molarity} \times \text{Molecular weight} \times 10^3$,

$[H^+] = (10^{-10.5}$ mole $\cdot$ L$^{-1}) (1 \text{ g} \cdot \text{mole}^{-1}) (10^3 \text{ mg} \cdot \text{g}^{-1}) = 10^{-6.5}$ mg $\cdot$ L^{-1}

$\left[OH^-\right] = K_w / \left[H^+\right] = 10^{-14} / 10^{-10.5} = 10^{-3.5}$ mole $\cdot$ L^{-1}

Also,

$$\left[OH^-\right] = (10^{-3.5} \text{ mole} \cdot L^{-1})(17 \text{ g} \cdot \text{mole}^{-1})(10^3 \text{ mg} \cdot \text{g}^{-1})$$
$$= 0.0003 \times 17 \times 1000 = 5.1 \text{ mg} \cdot L^{-1}$$

Using the equivalent weight of $CaCO_3$ as 50, concentrations in mg $\cdot$ L^{-1} as $CaCO_3$ are

$$CO_3^{2-}: = 105.0 \times 50 / 30 = 175 \text{ mg} \cdot L^{-1} \text{ as } CaCO_3$$
$$H^+: = 10^{-6.5} \times 50 / 1 = 1.58 \times 10^{-5} \text{ mg} \cdot L^{-1} \text{ as } CaCO_3$$
$$OH^-: = 5.1 \times 50 / 17 = 15 \text{ mg} \cdot L^{-1} \text{ as } CaCO_3$$
$$HCO_3^-: = 78.0 \times 50 / 61 = 63.9 \text{ mg} \cdot L^{-1} \text{ as } CaCO_3$$

Therefore, the exact total alkalinity in milligrams per liter is:

$$175 + 15 + 63.9 - \left(1.58 \times 10^{-5}\right) g = 253.9 \text{ mg} \cdot L^{-1} \text{ as } CaCO_3$$

PROBLEM 18.4.9
Water treatment

Alum coagulation (or "flocculation") removes particulate matter, reduces the concentration of organic matter, and decreases the alkalinity of water according to the following equation:

$$(Al_2(SO_4)_3 \cdot 14\ H_2O) + 6HCO_3^- \rightleftharpoons 2Al(OH)_3\ (s) + 6CO_2 + 14H_2O + 3SO_4^{2-} \quad (18.4.1)$$

A water treatment plant with an average flow of $Q = 0.0514\ m^3 \cdot s^{-1}$ uses alum $(Al_2(SO_4)_3 \cdot 14\ H_2O)$ to treat its water at a dose of 30 mg $\cdot$ L^{-1}. If the organic matter concentration is reduced from 10 mg $\cdot$ L^{-1} to 4 mg $\cdot$ L^{-1}, calculate the total mass of alkalinity consumed and the total mass of dry solids removed each day.

SOLUTION 18.4.9

We first determine the total amount of alkalinity consumed by noting from Equation 18.4.1 that six moles of alkalinity, as bicarbonate, are removed. Then, using its molecular weight of 594.35 g $\cdot$ mole^{-1}, we convert the alum used into molar units:

$$(30\ \text{mg} \cdot L^{-1})(10^{-3}\ g \cdot mg^{-1}) / 594.35 g \cdot mole^{-1} = 5.04 \times 10^{-5}\ mole \cdot L^{-1} \text{ of alum}$$

Thus, the added 5.04×10^{-5} mole $\cdot$ L^{-1} of alum removed six times that amount of bicarbonate $(6HCO_3^-)$, or:

$$(6)(5.04 \times 10^{-5}) = 0.000302\ \text{mole} \cdot L^{-1} \text{ of } HCO_3^- \text{ or } 3.02 \times 10^{-4}\ eq \cdot L^{-1} \text{ of alkalinity}$$

To determine the alkalinity removed per day, we multiply the above result by the average flow of the plant:

$$(0.000302\ eq \cdot L^{-1})(0.0514\ m^3 \cdot s^{-1})(1000\ L \cdot m^{-3} \cdot s^{-1})(86,400\ s \cdot day^{-1})$$
$$= 1342.9\ eq \cdot day^{-1}$$

Converting to a mass basis, taking the equivalent weight of bicarbonate to be 61 g $\cdot$ mole^{-1}

$$= 1342.9\ eq \cdot day^{-1} \times 61 g \cdot eq^{-1} = 81916.9 g \cdot day^{-1} \text{ or } 81.9\ kg \cdot day^{-1}$$

Because every mole of alum added precipitates 2 moles of solid we have a hydroxide/alum ratio of 2:1,

$$(2\ \text{mole } Al(OH)_3 / \text{mole alum})(5.04 \times 10^{-5}\ mole \cdot L^{-1} \text{ of alum})$$
$$= 5.08 \times 10^{-4}\ mole \cdot L^{-1} \text{ of } Al(OH)_3$$

Calculating this amount of daily yields:

$$(5.08 \times 10^{-4}\ mole \cdot L^{-1})(0.0514\ m^3 \cdot s^{-1})(1000\ L \cdot m^{-3})(86,400\ s \cdot day^{-1})$$
$$= 2256\ mole \cdot day^{-1}$$

Converting to mass by using the molecular weight for aluminum hydroxide yields:

$$2256 \text{ mole} \cdot \text{day}^{-1} \times 78 \text{ g} \cdot \text{mole}^{-1} = 175968 \text{ g} \cdot \text{day}^{-1} \text{ or } 175.968 \text{ kg} \cdot \text{day}^{-1}$$

Note that the total solids removed include the settled organic matter, which was given as 10 mg $\cdot$ L^{-1} in and 4 mg $\cdot$ L^{-1} out (as effluent); hence, the total organic matter removed is $10 - 4 = 5$ mg $\cdot$ L^{-1}. Multiplying this number by the plant flow determines the total organic matter settled daily.

$$6 \text{ mg} \cdot \text{L}^{-1} \times (0.0514 \text{ m}^3 \cdot \text{s}^{-1})(1000 \text{ L} \cdot \text{m}^{-3})(86,400 \text{ s} \cdot \text{day}^{-1})$$
$$= 26,645,750 \text{ mg} \cdot \text{day}^{-1} \text{ or } 26.6 \text{ kg} \cdot \text{day}^{-1}$$

Therefore, to find the total amount of dry solids removed per day, we add the mass fluxes of aluminum hydroxide and organic settled organic matter sludges:

$$176 \text{ kg} \cdot \text{day}^{-1} + 26.6 \text{ kg} \cdot \text{day}^{-1} = 202.6 \text{ kg} \cdot \text{day}^{-1}$$

PROBLEM 18.4.10
Water treatment

The groundwater from an aquifer which contains 2.5×10^{-6} M CO_2 is being pumped at a rate of 220 L $\cdot$ s^{-1} as it supplies the residents of a small town. Calculate the mass in kilograms of hydrated lime that must be added each day to neutralize the carbon dioxide present in the water. Determine the mass in kilograms of calcium carbonate produced daily.

SOLUTION 18.4.10

The following equation shows that each mole of hydrated lime produces 1 mole of CO_2 in the water:

$$H_2CO_3 + Ca(OH)_2 \rightarrow 2CaCO_3 + 2H_2O \qquad (18.4.2)$$

Therefore, we need 2.5×10^{-5} M of hydrated lime. To determine the mass needed:

$$(2.5 \times 10^{-5} \text{ M})(220 \text{ L} \cdot \text{s}^{-1})(86,400 \text{ s} \cdot \text{day}^{-1})(74.096 \text{ g Ca(OH)}_2 \cdot \text{mole}^{-1})(10^{-3} \text{ kg} \cdot \text{g}^{-1})$$
$$= 35.2 \text{ kg} \cdot \text{day}^{-1}$$

Also, as shown in Equation 18.4.2, for every mole of carbon dioxide neutralized, 1 mole of calcium carbonate results. Hence, we have produced:

$$(2.5 \times 10^{-5} \text{ M})(220 \text{ L} \cdot \text{s}^{-1})(86,400 \text{ s} \cdot \text{day}^{-1})(100.09 \text{ g CaCO}_3 \cdot \text{mole}^{-1})(10^{-3} \text{ kg} \cdot \text{g}^{-1})$$
$$= 47.56 \text{ kg CaCO}_3 \cdot \text{day}^{-1}$$

PROBLEM 18.4.11
Water treatment

The above-mentioned groundwater contains not only 2.5×10^{-5} M but also 320 mg $\cdot$ L^{-1} as $CaCO_3$ of carbonate hardness due to calcium and 60 g $\cdot$ L^{-1} as $CaCO_3$ of carbonate hardness due to magnesium. It is to be treated at the same rates as in the previous problem above, but the worker is required to remove all the carbonate hardness due to calcium without removing the magnesium ions. On the assumption that the worker will remove all but 25 mg · L⁻¹ (as calcium ions), what mass of calcium sludge will be produced daily?

SOLUTION 18.4.11

According to the relevant equation below, each mole of calcium removed requires 1 mole of hydrated lime:

$$Ca^{2+} + 2HCO_3^- + Ca(OH)_2 \rightleftharpoons 2CaCO_3(s) + 2H_2O \qquad (18.4.3)$$

Additionally, the process produces 2 moles of calcium carbonate sludge. Because all but 25 mg of $CaCO_3$ is to be removed, only $320 - 25 = 295$ mg $CaCO_3$ is to be removed. Converting that amount to a molar mass yields:

$$(295 \text{ mg} \cdot L^{-1} \text{ as } CaCO_3) / 100.09 \text{ mg} \cdot \text{mmole}^{-1} CaCO_3$$
$$= 2.56 \text{ mmole} \cdot L^{-1} \text{ as } Ca^{2+} \text{ or } 2.56 \times 10^{-3} \text{ mole} \cdot L^{-1}$$

Therefore, based on the stoichiometry of Equation 18.4.3, the worker produces twice the amount of $CaCO_3$ sludge (5.12×10^{-3} mole · L⁻¹), which in kilograms per day is:

$$(5.12 \times 10^{-3} \text{ M})(220 \text{ L} \cdot s^{-1})(86,400 \text{ s} \cdot \text{day}^{-1})(109.09 \text{ g } CaCO_3 \cdot \text{mole}^{-1})(10^{-3} \text{ kg} \cdot g^{-1})$$
$$= 10616.7 \text{ kg} \cdot \text{day}^{-1} \text{ of } CaCO_3$$

In addition, 47.56 kg $CaCO_3$ · day⁻¹ of sludge resulted when the hydrated lime ($Ca(OH)_2$) neutralized the CO_2 of Equation 18.4.2. This adds up to a total mass of sludge produced daily of 10,664 kg.

PROBLEM 18.4.12
Henry's law

The concentration of carbon dioxide in water at 30°C is 1.00×10^{-6} M. The Henry's constant for CO_2 dissolution in water is $0.1.86 \times 10^{-4}$ M atm⁻¹ at 30°C. What is the partial pressure of CO_2 in the atmosphere above the water?

SOLUTION 18.4.12

$$C_{CO_2} = K'_H P_{CO_2}$$

Transposing:

$$P_{CO_2} = C_{CO_2} / K'_H$$

Hence,

$$P_{CO_2} = (1.00 \times 10^{-6} \text{ M}) / 1.86 \times 10^{-4} \text{ M atm}^{-1}$$
$$P_{CO_2} = 5.37 \times 10^{-4} \text{ atm}$$

PROBLEM 18.4.13
To prove that the sequestration by water of all greenhouse gases from the air increases as the water cools.

Consider the above problem. If the air and water were each only 10°C and not 30°C, and C_{CO2} was 1.00×10^{-6} M, what would be the partial pressure of CO_2 in the air?

SOLUTION 18.4.13

K'_H for CO_2 in air at $10°C = 0.104 \times 10^{-4}$ M atm^{-1}

$$P_{CO_2} = C_{CO_2} / K'_H = 1.00 \times 10^{-5} \text{ M} / 0.104 \times 10^{-4} \text{ M atm}^{-1}$$
$$= 9.61 \times 10^{-2} \text{ M atm}$$

REFERENCES

CDC (2017) Dichlorodiphenyltrichloroethane (DDT). https://www.cdc.gov/biomonitoring/DDT_BiomonitoringSummary.html

Davis and Masten (2020) Principles of Environmental Engineering & Science. 4th edition. ISBN10: 1259893545 I ISBN13: 9781259893544

EPA (2022) Effects of Acid Rain. The effects of Acid rain on Ecosystems. I US EPA https://www.epa.gov/acidrain/effects-acid-rain

Howard PH (1989) Handbook of Environmental Fate and Exposure Data for Organic Chemicals, Vol. III: Pesticides. Lewis Publishers, Chelsea, MI. ISBN 978087371151

Meister RT (ed.) 1992. Farm Chemicals Handbook '92. Meister Publishing Company, Willoughby, OH.

USGS (2018) Water density. https://www.usgs.gov/special-topics/water-science-school/science/water-density

19 Water
Pesticides in Fish

AQUATIC LIFE: MAIN CONCERNS

- Contamination
- Bioconcentration
- Biomagnification
- Toxicity threshold levels

19.1 CONTAMINATION

Rising air temperatures affect all aspects of insect life cycles and generally result in larger insect populations because these conditions will support earlier emergence and more generations per year. Many pests currently limited by cooler temperatures at higher latitudes will be able to expand their ranges into these areas as air temperatures become warmer (USDA 2019). Pesticides sprayed on agricultural fields and on urban landscaping can run off into nearby streams and rivers, thereby potentially harming aquatic life such as fish (Figure 19.1), algae, and invertebrates like aquatic insects.

PROBLEM 19.1.1

A pesticide sprayer got stuck while trying to ford a stream flowing at a rate of $136 \, L \cdot s^{-1}$. Pesticide leaked into the stream for exactly 1 h and at a rate that contaminated the stream at a uniform 0.25 ppm of methoxychlor, which is similar to DDT (Figures 19.2 and 19.3) but acts more rapidly, is less persistent, and does not accumulate in the fatty tissues of animals as DDT. How much pesticide was lost from the sprayer during this time?

SOLUTION 19.1.1

Amount of water flowed $= 136 \, L \times 3600 \, s = 489,600 \, L$

Contamination time $= 3600 \, s \, (1 \text{ hour})$

Amount of pesticide lost = Amount which contaminated the water at 0.25 ppm

As the amount of water that flowed is 489,600 L, contamination of water was 0.25 ppm in 489,600 L of water.

Total is 0.25 ppm of $489,600 = 0.25 / 1,000,000 \times 489,600 = 0.0972 \, L$ lost.

DOI: 10.1201/9781003341826-24

FIGURE 19.1 Fish kills occur where concentrations of toxic materials become too high in waterways.

Source: www.epa.gov.

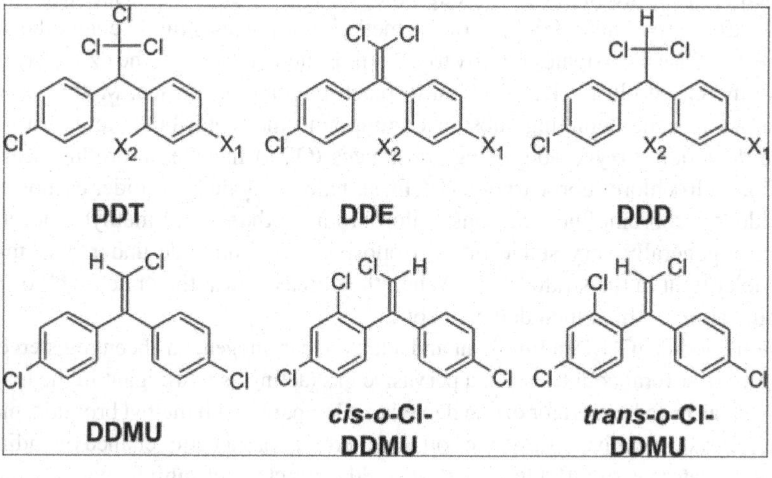

FIGURE 19.2 DDD is structurally like DDT, is a known metabolite of DDT, and is a probable human carcinogen.

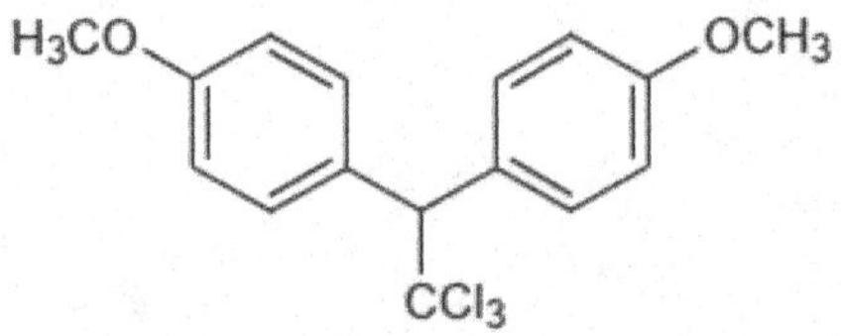

FIGURE 19.3 The structure of a major organochlorine insecticide methoxychlor, which is an acetylcholinesterase inhibitor.

19.2 BIOCONCENTRATION

Some compounds selectively diffuse into the bloodstream when water passes over fish gills and become concentrated over time. The bioconcentration factor (BCF) is the concentration of the substance in the organism compared to that which is in the surrounding war provided the diffusion mechanism is the only source of the compound in the fish.

As the BCF represents the equilibrium ratio of the concentration in the fish species compared to that in the surrounding water, several possible outcomes depending on body mass and concentration of chemicals in the surrounding water are addressed through worked examples and practice problems.

Soil fumigants can enhance greenhouse effects. If excess global warming increases the proliferation of crop pests, farmers may apply more pesticides. Fumigants are used to rid fields of pests and diseases before planting. The primary use of methyl bromide is as a fumigant in the soil to control fungi, nematodes, and weeds; in space fumigation of food commodities (e.g., grains); and in storage facilities (such as mills, warehouses, vaults, ships, and freight cars) to control insects and rodents. Agricultural growers inject methyl bromide about 2 ft into the ground to sterilize the soil before crops are planted. Although the soil is covered with plastic tarps immediately after a treatment, 50–95% of the methyl bromide eventually enters the atmosphere. Because it dissipates rapidly to the atmosphere, it depletes the ozone layer and allows increased ultraviolet radiation to reach the earth's surface. Methyl bromide is a Class I ozone-depleting substance compound that contributes to stratospheric ozone depletion. Oxygen-depleting substances (ODS) include chlorofluorocarbons (CFCs), hydrochlorofluorocarbons (HCFCs), halons, methyl bromide, carbon tetrachloride, hydrobromofluorocarbons, chlorobromomethane, and methyl chloroform. ODS are generally very stable in the troposphere and degrade under only intense ultraviolet light in the stratosphere. When they break down, they release chlorine or bromine atoms, which then deplete ozone.

Methyl iodide (CH_3I), a fumigant and a known carcinogen, has been registered as a fumigant in several countries. As a pervasive gas (acting as a fumigant in the holds of ships), it has been blamed for ozone depletion. Compared with methyl bromide, methyl iodide degrades relatively slowly in soil and therefore has a better chance of ending up in natural water via runoff. Once there, it could interact under sunlight with mercury to form methylmercury. Methylmercury is the most dangerous and toxic form of mercury and damages the brain and immune system. It can accumulate in fish.

PROBLEM 19.2.1

What is the mass, in milligrams, of mercury in a 2.5-kg lake carp fish that perfectly meets the North American standard of 0.50 ppm Hg? What mass of fish, at the 0–50 ppm Hg level, would a 60 g adult have to eat to ingest a total of 100 mg of mercury?

SOLUTION 19.2.1

a. 0.50 ppm Hg in 1.5 kg fish
   Mass of Hg is $0.5 / 10^6 \times (1.5 \times 10^6) = 0.75$ mg
b. 0.75 mg Hg is in 10^6 mg of fish
   Therefore, 1 mg Hg is in $10^6 / 0.75 = 1.3 \times 10^6$ mg

$$100 \text{ mg is in } 1.3 \times 10^8 \text{ mg}$$

Therefore 130 kg fish would have to be ingested.

PROBLEM 19.2.2
Insecticide uptake

For the insecticide toxaphene ($C_{10}H_8Cl_8$), log $K_{ow} = 5.3$, what would be the predicted concentration of toxaphene due to bioconcentration in the fat of fish that swim in waters containing 0.000010 ppm of the chemical?

SOLUTION 19.2.2

This requires a bioconcentration factor calculation:

Regarding the solubility of hydrophobic substances, 1-octanol, i.e., $CH_3(CH_2)_6CH_2OH$, has been experimentally found to be an approximate surrogate for the fatty tissue of fish. Octanol and water are almost immiscible. The n-octanol/water partition coefficient (K_{ow}) is defined as the ratio of the concentration of a chemical in n-octanol (a straight chain fatty alcohol) and water at equilibrium at a specified temperature. Therefore, a system that allows a substance to equilibrate between water and 1-octanol (n-octanol) will indicate the retention levels of the substance in the fat of aquatic organisms. Hence, the partition coefficient, K_{ow}, for a substance is defined as:

$$K_{ow} = [S]_{octanol} / [S]_{water}$$

where the bracketed items depict concentration in molarity or ppm units.

Based on the very low solubilities in water of many hydrocarbons, hence often very large values for the K_{ow}, a logarithmic scale is used. Further, substances with high log K_{ow} values tend to adsorb more readily to organic matter in soils or sediments because of their low affinity for water.

$$K_{ow} = [S]_{octanol} / [S]_{water}$$
$$\text{Log } K_{ow} = 5.3 = 199526.2315$$
$$\text{Log } K_{ow} \times [S]_{water} = [S]_{octanol}$$
$$199526.2315 \times 0.00001 = [S]_{octanol}$$

Therefore, concentration in fish fat = 1.99 ppm

19.3 CALCULATING BIOCONCENTRATION

Biomagnification is the process resulting in the accumulation of a chemical in the body of an organism at a higher concentration than occurs in its food. The chemical becomes more and more concentrated as it moves up the food chain. Bioconcentration and bioaccumulation happen within an organism, but biomagnification occurs across levels of the food chain. An example of bioconcentration is that of phytoplankton and other microscopic organisms, which take up methylmercury and then retain (concentrate) it in their tissues. Many pesticides, including organochlorides, are much more soluble in hydrocarbons than they are in water. Hence, they accumulate in the fatty tissue of fish.

Log K_{ow} serves as a measure of the relationship between lipophilicity (fat solubility) and hydrophilicity (water solubility) of a substance. The chemical, which is dissolved in the water, passes into the blood of fish (eventually mixing with the fish fat) as water moves over the gills of the fish. Experimentally, the chemical is allowed to equilibrate in a two-phase system between water and 1-octanol $CH_3(CH_2)_6CH_2OH$, a naturally occurring fatty alcohol in many plants, including mint, lavender, cannabis, hops, tea, oats, and ginger, and which has been found to be a suitable surrogate for the fatty portions of fish (Baird & Cann 2013).

Since the magnitude of concentration in fish is very large, for example, 100,000 (10^5) times that of the bioconcentration in the surrounding water, the log of that number is used; for example, the log of 100,000 is 5, so log K_{ow} = 5. The value is greater than the number "one" if a substance is more soluble in fat-like solvents such as n-octanol and less than one if it is more soluble in water. Log K_{ow} values are used, among others, to assess the environmental fate of persistent, nonionizable organic pollutants. Chemicals with high partition coefficients, for example, tend to accumulate in the fatty tissue of organisms (bioaccumulation). Under the Stockholm Convention, chemicals with a log K_{ow} greater than 5 are considered to bioaccumulate (Lipnick 1989).

PROBLEM 19.3.1
Bioconcentration factor

The overall BCF for a particular substance in a specified aquatic species (inclusive of fat and non-fat tissue) can be estimated as the value of the K_{ow} of that substance times the percentage of fat known for that species. Catfish taken from a particular lake were tested and 22 ppb parathion was found in their tissues. What is the concentration of the pesticide in the lake (the log K_{ow} is 3.8)?

SOLUTION 19.3.1

$$K_{ow} = [S]_{octanol} / [S]_{water} \qquad (19.3.1)$$

where the square brackets denote concentrations in molarity or ppm units at equilibrium.

Log K_{ow} is 3.8 = 6309 (i.e., the antilog, or 6309 times the concentration in water) ppb of 22 in fish tissues = 0.022 ppm

$$6309.53 = 0.022 / [S]_{water}$$

$$[S]_{water} = 0.022 / 6309.53 \times 0.05 \, (\text{as catfish fat} = 5\%)$$

Concentration in the water = 0.0000824 ppm or 0.08 ppb

19.4   TOXICITY THRESHOLD LEVELS

For some xenobiotics, there exists a level below which no response occurs. A dose slightly below that is referred to as the *no observable effects level* (NOEL). However, the NOEL can elicit observable effects in sensitive individuals in a population. For chronic exposures, the threshold is slightly below the NOEL in mg · kg^{-1} · day^{-1}. The acceptable daily intake (ADI), or toxicity reference dose (RfD), which the US EPA adopts, represents the maximum dose for sensitive individuals and is set at 100 times less than the NOEL.

If the lethal dose of the drug for 50% of a population (LD$_{50}$) is, for example, 427 mg · kg^{-1} of animal mass, the NOEL for an animal of average adult weight for that species (e.g., human = 70 kg) is calculated as:

$$\text{NOEL} = (\text{LD}_{50} \times 70 \text{ kg}) / 2000 \tag{19.4.1}$$

where

LD$_{50}$ = lethal dose

2000 = the accepted divisor constant of the LD$_{50}$

Hence, the above animal species has a NOEL of:

$$\text{NOEL} = 427 \text{ mg} \times 70 \text{ kg} / 2000$$
$$= 14.94 \text{ mg}$$

PROBLEM 19.4.1

The threshold NOEL level found for a particular soil fumigant from animal studies is 0.006 mg · kg^{-1} body weight · day^{-1}. The only source for the pesticide is freshwater fish, which occurs at an average level of 0.2 ppm. For a 50 kg child, what is the maximum average daily consumption of such fish that would keep the exposure level below the ADI or RfD for the compound?

SOLUTION 19.4.1

$$\text{NOEL} = 0.006 \text{ ppm}$$

ADI of RfD is NOEL / 100 = 0.00006 ppm or 6×10^{-5}

Contaminated fish has 0.2 mg · kg^{-1} (i.e., fumigant in a 1 kg fish) = 2 ppm

A 50 kg person must have 0.00006 mg · kg^{-1} body weight · day^{-1} = 50×0.00006
$$= 0.003 \text{ mg}$$

0.2 mg of fumigant is in 1 kg.

Therefore, 0.003 mg of fumigant is in $(0.003 / 0.2)$ of $1\ kg = 0.015\ kg$

Maximum ADI for a 50 kg child is approximately $= 15\,g$

19.5 REMEDIATION: DDT OR METHOXYCHLOR?

PROBLEM 19.5.1
Methoxychlor: an alternative to DDT?

Is there a safer alternative to DDT?

SOLUTION 19.5.1

Methoxychlor, a derivative of DDT and very similar, acts more rapidly, is less persistent, and does not accumulate in the fatty tissues of animals as DDT does. It is approximately just as effective and much less persistent in the environment than DDT. According to the EPA (2005), methoxychlor (2,2-bis(methoxyphenyl)-1,1,1-trichloroethane) and DDT (2,2-bis(chlorophenyl)-1,1,1-trichloroethane) undergo different hydrolytic degradation pathways in water at pHs common to the aquatic environment. For methoxychlor at common aquatic pHs, the reaction is pH independent, and at 27 °C the half-life is about 1 year. On the other hand, for DDT the reaction is pH dependent, and at 27 °C and pH 7, the half-life is about 8 years.

REFERENCES

Baird C & Cann M (2012) Environmental Chemistry (5th ed). W. H. Freeman. New York.
EPA (2005) Methoxychlor and DDT degradation in water: Rates and products. https://cfpub. epa.gov/si/si_public_record_report.cfm?Lab=NERL&dirEntryId=47526
Lipnick RL (1989) Narcosis, electrophile and proelectrophile toxicity mechanisms: Application of SAR and QSAR. Environ Toxicol Chem 8(1), 1–2.
USDA (2019) How could climate change affect insect pest populations? https://ask.usda. gov/s/article/How-could-climate-change-affect-insect-pest-populations

20 Rainwater
Collection, Pollution, and Remediation

RAINFALL: MAIN POINTS

- 90% produced by evaporation from water bodies
- 10% comes from transpiration from plants (USGS 2018)
- Earth's average annual rainfall ~39 inches
- Earth's average water vapor ~1 inch liquid over globe
- Evaporation from the surface returns as precipitation in ~9 days (Britannica 2020)

20.1 RAINFALL: SOURCES AND PATTERNS

The versatility of rainwater as a viable source of moisture depends on its storage against rapid losses to runoff and evaporation. Excess carbon dioxide in the atmosphere causing global heating could significantly alter rainfall patterns and turn currently productive areas of the earth into desert regions The interface between two masses of air that differ in temperature, density, and water content is called a front. A mass of cold air moving such that it displaces one of warm air is a cold front, and a mass of warm air displacing one of cold air is a warm front. Because cold air is more dense than warm air, the air in a cold air mass along a cold front pushes under warmer air, causing the warm, moist air to rise such that water condenses from it. The condensation of water releases energy, so the air rises further (Ahrens & Henson 2017). The net effect can be massive cloud formations (thunderheads) that may even penetrate the stratosphere, sometimes producing heavy rainfall, even hail, and occasionally violent storms with strong winds, including tornadoes (Ahrens & Henson 2017). Warm fronts, having gentler slopes, cause less spectacular effects as warm, moist (hence, less dense) air slides over colder air, where the front is usually much broader, therefore thinner and milder, typically resulting in widespread drizzle and less intense rain.

20.2 HENRY'S LAW FOR GAS SOLUBILITIES AND THE SOLUBILITY OF A GAS IN A LIQUID

Before it reaches a roof, rainwater can be polluted by dust, gases, and particles, particularly in large, highly industrialized areas, where it is largely unavoidable. But polluted (ground) runoff is one of the greatest threats to clean water (EPA 2022). Hence, soaking up rain before it reaches the ground reduces water pollution, thereby protecting valuable drinking water resources.

DOI: 10.1201/9781003341826-25

Unpolluted deposition (or rain), in balance with atmospheric carbon dioxide, has a pH of 5.6. In most parts of the world, the pH of rain is lower than this (Baird and Cann 2012). The main pollutants responsible for acid rain are sulfur dioxide (SO_2) and nitrogen oxides (NO_x). Nitrogen and sulfuric emissions come from natural and anthropogenic sources. Natural emissions include, for example, volcano emissions, lightning, and microbial processes. Power stations and industrial plants, like the mining and smelting of high-sulfur ores and the combustion of fossil fuels, emit the largest quantities of sulfur and nitrogen oxides and other acidic compounds. These compounds mix with water vapor, causing acid deposition with a pH of 4.2–4.7, i.e., 10 or more times the acidity of natural deposition. When the rate at which molecules of gas dissolve into the water equals the rate at which they come out of the solution into the air, the gas is in equilibrium with the aqueous (dissolved) form of the compound. In other words (for a nonreactive solution), Henry's law for gas solubilities states that the solubility of a gas in a liquid is linearly proportional to the partial pressure of that gas in contact with the liquid.

PROBLEM 20.2.1

The atmospheric concentration (gas phase) of dichloromethane (DCM) is measured at 0.003 mg · m^{-3}. What is the concentration in the rainwater when the temperature is 20 degrees Celsius?

SOLUTION 20.2.1

In very dilute solution, solute molecules are minimally restricted from escaping to the gaseous medium above; hence, the rate of their escape will be proportional to their concentration in the solution, and solute will accumulate in the gas until the return rate is equal to the rate of escape. This is the basis of Henry's law.

K'_H = Henry's Law (mol · L^{-1} · atm^{-1}) [(m · m^{-1} · water)$^{-1}$]
H = Henry's law constant (atm^{-1})
K'_H of DCM in water is 0.5 M · atm^{-1}
Concentration of DCM in air = 0.004 mg · atm^{-1}
K'_H = H / RT
H = (0.5 M · atm^{-1}) (0.082 L · atm^{-1} · mol^{-1}) (273 + 20 K) = 12.01
Solving for the aqueous concentration,
$Conc_{aq}$ of DCM = H ($Conc_{gas}$ · DCM)
= (12.01) (10^{-4} mg · m^{-3}) = 0.001201 or 1.2 × 10^{-3} mg · L^{-1}

20.3 RAINFALL POLLUTION SOURCES

PROBLEM 20.3.1
Rainfall analysis

A chemical analysis of a liter of rain falling on a city displayed the following results:

$$HCl = 0.349$$
$$HNO_3 = 1.42$$

$$H_2SO_4 = 3.25$$
$$NH_3 = 0.357$$

Is this acid rain?

SOLUTION 20.3.1

Acid rain occurs when pH values of rainfall fall below 5.6.

As all waters must be electroneutral, the ionic balance reflects electroneutrality. In this case, it is:

$$\left[NH_4^+\right]+\left[H^+\right]=\left[NO_3^-\right]+\left[SO_4^{2-}\right]+\left[Cl^-\right]+\left[OH^-\right]$$

OH^-, though not stated, is always present in water.

As molarity determines acidity, the preceding concentrations must be converted to moles per liter by dividing by the molar mass.

The molar mass of H_2SO_4, for example, is

$$1(2)+32+(16\times4)=98 \text{ g}\cdot\text{mole}^{-1}$$

Hence,

$$\left[(3.25 \text{ mg of } H_2SO_4L^{-1})(1 \text{ g}/1000 \text{ mg})\right]/98 \text{ g}\cdot\text{mole}^{-1} = 3.31\times10^{-5} \text{ M}$$

Table 20.1 shows results of the remaining calculations.

We can determine $[OH^-]$ only after calculating the $[H^+]$.

Total anions (presuming negligible OH^-) = $[NO_3^-] + [SO_4^{2-}] + [Cl^-] \gg [OH^-]$

Thus, while temporarily ignoring $[OH^-]$,

Total anions (as numbers, not mass)
$$=9.5\times10^{-6}+2.2\times10^{-5}+3.31\times10^{-5}+2.1\times10^{-5}=9.51\times10^{-5}$$

At equilibrium in water, total cations = total anions.

Therefore, $[NH_4^+] + [H^+]$, like the total anions present, = 9.51×10^{-5}.

(In an acidic medium, all the ammonia would be in the form of $[NH_4^+]$.)

$$[H^+] = \text{Total cations} - [NH_4^+]$$
$$=9.51\times10^{-5}-2.1\times10^{-5}=7.41\times10^{-5}$$

TABLE 20.1
Concentration of Pollutants

Chemical Species	Concentration (mg · L^{-1})	Molecular Weight (g · mole^{-1})	Concentration (M)
HCl	0.349	36.5	9.5×10^{-6}
HNO$_3$	1.42	63	2.25×10^{-5}
H$_2$SO$_4$	3.25	98	3.31×10^{-5}
NH$_3$	0.357	17	2.1×10^{-5}

We use the given concentration of NH_3.

$$pH = -\log(7.41 \times 10^{-5}) = 4.13$$
$$\left(As\ pH = -\log[H^+]\right)$$

Ans. $pH = 4.13 = $ Acid rain

The concentration of OH^- is $10^{-(14-4.13)}$ M $= 10^{-9.87}$ M $= 1.35 \times 10^{-10}$ M.

20.4   ACIDIFICATION AND AQUATIC LIFE

Burning of fossil fuels has produced acid rain, and the acidification of freshwater in an area is dependent on the quantity of calcium carbonate (limestone) in the soil. Limestone can buffer (neutralize) the acidification of freshwater. The effects of acid deposition are much greater on lakes with little buffering capacity.

Much of the damage to aquatic life in sensitive areas with this little buffering capacity is a result of "acid shock." This is caused by the sudden runoff of large amounts of highly acidic water and aluminum ions into lakes and streams, when snow melts in the spring or after unusually heavy rains.

20.4.1   ACID NEUTRALIZING CAPACITY (ANC)

This is a measurement of alkalinity. The difference between ANC and alkalinity is that ANC measures the net condition of the water. For example, an ANC below 0 means the water is acidic and has no buffering capacity. If the ANC is above 0, the water has some buffering ability.

20.5   EFFECTS ON AQUATIC LIFE

Most freshwater lakes, streams, and ponds have a natural pH in the range of 6–8. Acid deposition has many harmful ecological effects when the pH of most aquatic systems falls below 6 and especially below 5.

The EPA (2021) lists the following effects of increased acidity on aquatic systems:

- As the pH approaches 5, non-desirable species of plankton and mosses may begin to invade, and populations of fish such as smallmouth bass disappear.
- Below a pH of 5, fish populations begin to disappear, the bottom is covered with undecayed material, and mosses may dominate nearshore areas.
- Below a pH of 4.5, the water is essentially devoid of fish.
- Aluminum ions (Al^{3+}) attached to minerals in nearby soil can be released into lakes, where they can kill many kinds of fish by stimulating excessive mucus formation. This asphyxiates the fish by clogging their gills. It can also cause chronic stress that may not kill individual fish but leads to lower

body weight and smaller size and makes fish less able to compete for food and habitat.

- The most serious chronic effect of increased acidity in surface waters appears to be interference with the fish reproductive cycle. Calcium levels in the female fish may be lowered to the point where she cannot produce eggs or the eggs fail to pass from the ovaries or, if fertilized, the eggs and/or larvae develop abnormally.
- Extreme pH can kill adult fish and invertebrate life directly and can also damage developing juvenile fish. It will strip a fish of its slime coat and a high pH level "chaps" the skin of fish because of its alkalinity.

According to Lenntech (2022), a high pH of freshwater (e.g., 9.6) may kill fish, damage the outer surfaces like gills, eyes, and skin, and produce an inability to dispose of metabolic wastes. High pH may also increase the toxicity of other substances. For example, the toxicity of ammonia is ten times more severe at a pH of 8 than it is at pH 7. It is directly toxic to aquatic life when it appears in alkaline conditions.

PROBLEM 20.5.1
Neutralization measurements

To neutralize the extreme alkalinity in a freshwater lake produced by an accidental spill, a batch of concentrated commercial sulfuric acid is to be purchased as a 90% (by weight) solution prior to dumping it in the lake. Knowing that sulfuric acid at 100% has a specific gravity of 1.839, how can the true concentration of the purchased sulfuric acid be ascertained? If the 90% on the label is authentic, and the temperature of the solution is 17° C, what is the concentration of this solution in each of the following units?

A. milligrams per liter
B. molarity
C. normality

How much of this acid would be needed to neutralize the NaOH alkalinity of a 20,000 m³ lake having a pH of 9.5?

SOLUTION 20.5.1

A. The density of water at 17°C is 998.778 per liter (Table 20.2). Hence, a 90% solution of H_2SO_4 would have a density of:

$$\left(998.778 \text{ g} \cdot \text{L}^{-1}\right)(0.10) + \left(1839 \text{ g} \cdot \text{L}^{-1}\right)(0.90) = 1754.98 \text{ or } 1.7 \times 10^6 \text{ mg} \cdot \text{L}^{-1}$$

B. Molarity = Concentration of H_2SO_4 divided by the molar weight of 98.08 g L⁻¹:

$$1754.98 \text{ g} \cdot \text{L}^{-1} / 98.08 \text{ g} \cdot \text{mole}^{-1} = 17.89 \text{ M}$$

TABLE 20.2

Density of Water at Various Temperatures at 1 Atm

Temperature (°F/°C)	Density (g · cm⁻³)
32/0	0.99987
39.2/4.0	1.00000
40/4.4	0.99999
50/10	0.99975
60/15.6	0.99907
70/21	0.99802
80/26.7	0.99669
90/32.2	0.99510
100/37.8	0.99318
120/48.9	0.98870
140/60	0.98338
160/71.1	0.97729
180/82.2	0.97056
200/93.3	0.96333
212/100	0.95865

C. Normality (N) = The number of gram-equivalents (n) per liter

H_2SO_4 has 2 hydrogen ions to donate, therefore has $n = 2$ gram-equivalents $\cdot$ mole⁻¹

Here, $N = n\,M = 2 \times 17.89$ mol $\cdot$ L⁻¹ = 35.78 Eq $\cdot$ L⁻¹ or 35.78 N

The amount of acid needed to neutralize the alkalinity of a 20,000 m³ lake having a pH of 9.5 can be determined as follows.

The acid strength here is 17.89 moles of H_2SO_4 in a liter of solution. Neutralizing means:

$$\text{Volume (acid)} \times \text{Concentration (H}^+ \text{ ions from dissociation)}$$
$$= \text{Volume (base)} \times \text{Concentration (OH}^- \text{ ions)}$$

In general, for an acid AH_n at concentration c_1 reacting with a base $B(OH)_m$ at concentration c_2, the volumes are related by:

$$n\,v_1\,c_1 = m\,v_2\,c_2$$

where
$n =$ acid
$m =$ base

As pH is a measure of the concentration of H_3O^+ ions in a solution, a pH of 9.5 contains .0000000005 H_3O^+ ions per mole. It currently has $14 - 9.5 = 10^{-4.5}$ OH⁻ ions per mole of H_2O.

Therefore, the same quantity of H_3O^+ ions are required for neutralization.

PROBLEM 20.5.2
Neutralization measurements

Into a small freshwater lake of neutral pH, an overturned truck accidentally spilled 100 kg of H_2SO_4. Immediately after the spill, the volume of the lake was estimated at 50 m³. What is the final pH of the lake? Assume that the lake is unbuffered, with an inconsequential ANC.

SOLUTION 20.5.2

As H_2SO_4 completely dissociates in water, a 1-L sample of the water would suffice:

$$\text{Number of moles} = \left[100\,kg \times 1000 \ g \cdot kg^{-1} / 98 \ g \right] = 1020.4$$

$$\text{Molar concentration} = 1020.4 / 50 \ m^3 \times 1000 \ kg \cdot m^{-3} \times 1000 \ g \cdot kg^{-1} = 2.04$$
$$\times 10^{-5} \ mol \cdot L^{-1}$$

The reaction is:

$$H_2SO_4 \rightarrow 2H^+ + SO_4^{2-}$$

As stated above, as H_2SO_4 is a strong acid, we assume complete dissociation. In this case, with the concentration of sulfuric acid being $2.04 \times 10^{-5} \ mol \cdot L^{-1}$, during its dissociation:

$2(2.04 \times 10^{-5} \ mol \cdot L^{-1})$ M H^+ occurred. The new pH in the lake is:

$$pH = -\log(4.08 \times 10^{-5}) = 0.00000408$$

Count the number of zeros and the number following the last zero to find the pH.
New pH = 5.4

PROBLEM 20.5.3
Neutralization measurement

For the above problem, what quantity of NaOH would return the lake to a neutral pH?

SOLUTION 20.5.3

To neutralize the acidity, we must aim to match $[H^+]$ with $[OH^-]$. NaOH, a strong base, completely dissociates in water, and there are 1020.4 moles of H_2SO_4.
　　As 1 mole of NaOH = 40 g, 1020.4 moles of NaOH = 40816 g.
　　But $H_2SO_4 = 2[H^+]$
　　Hence, the mass of NaOH needed to return the lake to a neutral pH is 2(40.816) = 81.632 kg.

PROBLEM 20.5.4
Neutralization measurements (4)

To neutralize the extreme alkalinity in a freshwater lake produced by an accidental spill, a batch of concentrated commercial sulfuric acid is to be purchased as a 90% (by weight) solution prior to dumping it in the lake. Assume that the 90% on the label is authentic and that the temperature of the solution is 17°C. Knowing that sulfuric acid at 100% has a specific gravity of 1.839:

    a. How can the true concentration of the purchased sulfuric acid be ascertained?
    b. What is the concentration of this solution in units of milligrams per liter, molarity, and normality?
    c. How much of this acid would be needed to neutralize the alkalinity of a 20,000 m³ lake having a pH of 9.5?

SOLUTION 20.5.4

The density of water at 17°C is 998.778 per liter (Table 20.2). Hence, a 90% solution of H_2SO_4 would have a density of:

$$(998.778 \text{ g} \cdot \text{L}^{-1})(0.10) + (1839 \text{ g} \cdot \text{L}^{-1})(0.90) = 1754.98, \text{ or } 1.7 \times 10^6 \text{ mg} \cdot \text{L}^{-1}$$

Molarity = Concentration of H_2SO_4 divided by the molar weight of 98.08 g · L⁻¹:

$$1754.98 \text{ g} \cdot \text{L}^{-1} / 98.08 \text{ g} \cdot \text{mole}^{-1} = 17.89 \text{ M}$$

In acid-base reactions, normality describes the reactivity (number of OH^- or H^+ ions). Normality (N) = The number of gram-equivalents of acid or base (n) per liter
H_2SO_4 has two hydrogen ions to donate and therefore has $n = 2$ gram-equivalents · mol⁻¹
Here, N = nM = 2 × 17.89 mol · L⁻¹ = 35.78 Eq · L⁻¹ or 35.78 N.
Hence, the acid strength is 17.89 M and 35.78 N.

PROBLEM 20.5.5
Neutralization measurements (5)

Warm effluent water from factories burning fossil fuels dissolves less oxygen, thereby stressing fish. A large ANC buffers (protects) the water against sudden influxes of acidic liquids.
    The pH of lake water was determined to be 7.8. Measurement of the bicarbonate species indicated a concentration of 1.5×10^{-2} M. If this system is closed to the atmosphere, what are the concentrations of carbonate, carbonic acid, and total carbonate (C_T)?

SOLUTION 20.5.5

Diprotic acids (i.e., containing two hydrogen atoms) release the first H^+ at a specific pH value called pK_{a1}. Carbonic acid dissociates to form bicarbonate and hydrogen as follows:

$$H_2CO_3 = H^+ + HCO_3^- \qquad K_{a1} = 10^{-6.35} \text{ at } 25°C \qquad (20.5.1)$$

where K_{a1} = the mole fraction of the acid which breaks up in water to release H^+ ions and following is the smaller dissociation of bicarbonate to release the second H^+:

$$HCO_3^- = H^+ + CO_3^{2-} \quad K_{a2} = 10^{-10.33} \text{ at } 25°C \quad (20.5.2)$$

which, as moles, both can be represented by:

$$K_{a1} = \left[HCO_3^-\right]\left[H^+\right]/\left[HCO_3\right] = 10^{-6.35} \quad (20.5.3)$$

And

$$\left[CO_3^-\right]\left[H^+\right]/\left[HCO_3^-\right] = 10^{-10.33} \quad (20.5.4)$$

Thus, K_{a1} is that part of carbonic acid lost to dissociation at pH 7.8, causing:

$$\left[H_2CO_3\right] = \left[HCO_3^-\right]\left[H^+\right]/p\,K_{a1} = \left(1.5 \times 10^{-2}\right)\left(10^{-7.8}\right)/10^{-6.3} = 1.86 \times 10^{-3} \text{ M}$$

To calculate the bicarbonate, we transpose Equation 20.5.4 as follows:

$$\left[CO_3^{2-}\right] = p\,K_{a2}\left[HCO_3^-\right]/\left[H^+\right] = \left(10^{-10.33}\right)\left(1.5 \times 10^{-2}\right)/10^{-7.5} = 2.21 \times 10^{-5} \text{ M}$$

$$C_T = \left[HCO_3^-\right] + [HCO_3] + \left[CO_3^{2-}\right] = 1.86 \times 10^{-3} + 1.5 \times 10^{-2} + 2.21 \times 10^{-5}$$

$$= 1.688 \times 10^{-2} \text{ M} \equiv 1.7 \times 10^{-2} \text{ M}$$

PROBLEM 20.5.6
Neutralization measurements

Two lakes, Lake Arachina, having pH 4, and Lake Fukuda, with pH 6, are to be treated to decrease acidity due to ongoing exposure to acid rain. If the pH of the rain gets lower than it now is, which lake would show a faster increase in acidity?

SOLUTION 20.5.6

Below pH of 4.5, the only carbonate species present in appreciable amounts is carbonic acid (H_2CO_3), and the concentration of OH^- is negligible. Because carbonic acid does not contribute to alkalinity, in this case the reserve alkalinity is negative (due to the H^+). This water would have no ability to neutralize acids, and any small addition in acids would result in a significant reduction in pH.

On the other hand, in the pH range from 7 to 8.3, HCO_3^- and CO_3^{2-} dominate, but HCO_3^- predominates. Though a small number of H^+ occur, the OH^- ions present neutralize them. Therefore, the pH of Lake Arachina would fall at a faster rate, as Lake Fukuda has a greater buffering capacity.

20.6 RAINWATER RUNOFF IN URBAN AREAS

Increased rainwater collection promises a lessening of the impact of global warming on water shortages. However, the high runoff index of urban areas indicates a need for efficient collection. With a high potential for rainwater harvesting, urban areas may contain up to 90% hard, impervious surfaces such as rooftops, sidewalks, streets, construction sites, parking lots, and pavement where water collects and quickly runs off, and research suggests increased rainfall intensities as urban heating becomes greater (Rahimpour et al. 2018).

PROBLEM 20.6.1
Urbanization effects

What effect can urbanization have on rainfall patterns?

SOLUTION 20.6.1

In a detailed study investigating the impacts of urban land use on precipitation, Rahimpour et al. (2018) showed that extra heat causes more rain in urban heat islands (UHI) compared to rain in adjacent rural areas. In the Netherlands, where natural topographic variations are non-significant, they reported changes in rainfall indices in the urban areas, which occurred to a greater degree than those in the rural areas during the recent multi-decadal period. *The monthly maxima of daily precipitation indicate that the greatest increases occurred in August.* Overall, they found larger trends in the extreme precipitation indices in urban areas than in rural areas, with the monthly maxima of daily precipitation being greater in urban areas than in rural areas in the Netherlands. The extreme precipitation differences between urban and rural areas were persistent but varied from region to region. In all the regions, the urban areas received more intense extreme precipitation than did the rural areas and increased after sunrise in the daytime in summer.

PRACTICE PROBLEM 20.6.1
Rainwater tanks

The main contaminants which make rainwater unsafe are microorganisms, which can cause gastrointestinal illnesses, and traces of toxic metals, which can come from atmospheric depositions and roofing materials (check on types of roofs and dangers).

Have you ever observed any water pollution in your area? Depending on the source(s) from the list below, what measures would you suggest for controlling it?
Pollution sources:

Government
Chemical plants and oil refineries
Sanitary landfills
Pesticides
Sewage and septic systems
Radioactive waste
Petroleum waste
Acid mine drainage

PRACTICE PROBLEM 20.6.2
Activity: Match the pollutant (EPA Drinking Water Standards 1989) with the correct process listed below.
Pollution sources:

Mineral processing
Farm animal waste

Feedlots
Fertilizers
Pulp mills
Roadway salt
Cemeteries
Disinfection
Filtration
Cation exchange
Adsorption
Oxidation
Demineralization
Ion exchange
Anion exchange
pH adjustment

20.7   DETERMINING LIME DOSAGE

As stated above, rainwater is mildly acidic due to the following atmospheric reaction:

$$CO_2\,(aq) + H_2O \rightleftharpoons H_2CO_3 \rightleftharpoons HCO_3^- + H^+ \rightleftharpoons CO_3^{2-} + 2H^+ \qquad (20.7.1)$$

On occasion, the appearance of water is cloudy and/or colored. In removing unwanted color and cloudiness (turbidity), a coagulant binds extremely fine particles suspended in raw water into larger particles that can be removed by filtration and settling. To facilitate coagulation / flocculation, aluminum sulfate $(Al_2(SO_4)_3)$, or filter alum, is often added to the water coming into a purification facility. As alum increases the acidity, lime is sometimes added to provide adequate alkalinity (HCO_3^-) in the solids contact clarification process. Additionally, the process removes the aluminum itself. To determine the necessary lime dose, in $mg \cdot L^{-1}$, three steps are required.

First, the total alkalinity required to react with the alum to be added and provide proper precipitation is determined using the following formula:

Total alkalinity required = Alkalinity reacting with alum + Alkalinity in the water

(20.7.1)

PROBLEM 20.7.1
Lime requirement for alum

Raw water requires an alum dose of $47\ mg \cdot L^{-1}$, as determined by jar testing. If a residual $35\ mg \cdot L^{-1}$ alkalinity must be present in the water to ensure complete precipitation of alum added, what is the total alkalinity required in $mg \cdot L^{-1}$?

SOLUTION 20.7.1

*NB: It is known that $1\ mg \cdot L^{-1}$ alum reacts with $0.45\ mg \cdot L^{-1}$ alkalinity.

First, calculate the alkalinity that will react with $47\ mg \cdot L^{-1}$ alkalinity:

Next, calculate the total alkalinity required:

Total alkalinity required, $mg \cdot L^{-1}$ = Alkalinity reacting with alum, $mg \cdot L^{-1}$
+ Alkalinity in the water, $mg \cdot L^{-1}$

Since 1 $mg \cdot L^{-1}$ alum reacts with 0.45 $mg \cdot L^{-1}$ alkalinity
47 $mg \cdot L^{-1}$ reacts with 0.45 × 47 $mg \cdot L^{-1}$ = 21.15 $mg \cdot L^{-1}$
Total alkalinity required, $mg \cdot L^{-1}$ = 21.15 $mg \cdot L^{-1}$ + 35 $mg \cdot L^{-1}$
Total alkalinity required, $mg \cdot L^{-1}$ = 56.15 $mg \cdot L^{-1}$

PRACTICE PROBLEM 20.7.1
Jar tests indicate that 42 $mg \cdot L^{-1}$ alum is optimum for a particular raw water. If a
residual 30 $mg \cdot L^{-1}$ alkalinity must be present to promote complete precipitation of
alum added, what is the total alkalinity required in $mg \cdot L^{-1}$?

Comparing alkalinities
The next step is comparing the required alkalinity with alkalinity already in the raw
water to determine how much $mg \cdot L^{-1}$ alkalinity should be added to the water:

Alkalinity to be added, $mg \cdot L^{-1}$ = Total alkalinity required, $mg \cdot L^{-1}$
− Alkalinity in the water, $mg \cdot L^{-1}$

PROBLEM 20.7.2
Role of bicarbonate

A total of 52 $mg \cdot L^{-1}$ alkalinity is required to react with alum and ensure proper
precipitation. If the raw water has an alkalinity of 33 $mg \cdot L^{-1}$ as bicarbonate, how
much $mg \cdot L^{-1}$ alkalinity should be added to the water?

SOLUTION 20.7.2

Alkalinity to be added, $mg \cdot L^{-1}$ = Total alkalinity required, $mg \cdot L^{-1}$ − Alkalinity in
the water, $mg \cdot L^{-1}$
Alkalinity to be added, $mg \cdot L^{-1}$ = 52 $mg \cdot L^{-1}$ − 33 $mg \cdot L^{-1}$
Alkalinity to be added, $mg \cdot L^{-1}$ = 19 $mg \cdot L^{-1}$
Finally, after determining the amount of alkalinity to be added to the water, we
determine how much lime (the source of alkalinity) must be added.

PROBLEM 20.7.3
Lime requirement to supply alkalinity

It has been calculated that 19 $mg \cdot L^{-1}$ of alkalinity must be added to a raw water.
How much $mg \cdot L^{-1}$ of lime will be required to provide this amount of alkalinity?
*Hint: 1 $mg \cdot L^{-1}$ alum reacts with 0.45 $mg \cdot L^{-1}$ alkalinity.

SOLUTION 20.7.3

$1 \ mg \cdot L^{-1}$ alum reacts with $0.35 \ mg \cdot L^{-1}$ lime.

First, determine the $mg \cdot L^{-1}$ lime required by using a proportion that relates bicarbonate alkalinity to lime. We calculate this through cross-multiplying.

This means you would multiply $(0.45 \times \text{"x"})$ and (0.35×19) and it becomes:

$$0.45 \times \text{"x"} = 0.35 \times 19$$

We need "x" by itself. So we divide both sides by 0.45, canceling it out on the left:

$$x = (0.35 \times 19) / (0.45)$$

$$x = 14.78 \ mg \cdot L^{-1} \ lime$$

Ans. $= 14.78 \ mg \cdot L^{-1}$ lime

20.8   REMEDIATION: RAINWATER REBATES

With a view to conserve potable water, reduce costs, and, by extension, ameliorated global warming, several municipalities, such as Guelph, Canada, and Austin, Texas, offer rebates for installing residential rainwater collection systems. The following solution to Problem 20.8.1 incorporates a system successfully adopted by City of Austin (2016).

PROBLEM 20.8.1
Operating the tank/cistern

How does one (a) install a rain barrel, (b) maintain the system, and (c) keep mosquitos from breeding?

SOLUTION 20.8.1

Installation

Place the barrel so that it can collect as much rain as possible from the roof. If there are gutters, shorten a downspout and reattach the end elbow joint to direct flow into the barrel. If there are no gutters, place the barrel under a corner where two roof sections meet to create a valley. Position the barrel so that it can catch water from light rain as well as from a forceful downpour.

Maintenance

Maintenance is an ongoing duty that includes:

- trimming back trees near your roof;
- cleaning and repairing gutters, piping, filters, and roofs;
- purging and cleaning the first-flush system;
- watching tank levels; and
- maintaining pumps and other equipment.

Mosquitos

Screens can keep adult mosquitoes out of a tank, but larvae can wash in from the gutters. To prevent breeding, install a first-flush system, empty the tank often, or add "mosquito dunks" to the water. Mosquito dunks are a non-toxic bacterial larvicide available at most garden supply stores.

REFERENCES

Ahrens CD and Henson R (2017) Meteorology Today: An Introduction to Weather, Climate and the Environment 12th Edition. Cengage Publishers, Boston, Massachusetts.

Baird Colin and Michael Cann (2012) Environmental Chemistry. 5th Edition, Macmillan, New York. ISBN:9781429277044

Britannica (2020) Climate: Precipitation. http://www.britannica.com/EBchecked/topic/121560/climate/53266/Precipitation

City of Austin (2016) Rainwater harvesting rebate frequently asked questions. https://www.austintexas.gov/sites/default/files/files/Water/Conservation/Rebates_and_Programs/Rainwater_Harvesting_Rebate_FAQ.pdf

EPA (2021) Effects of Ocean and Coastal Acidification on Ecosystems https://19january2021snapshot.epa.gov/ocean-acidification/effects-ocean-and-coastal-acidification-ecosystems_.html

EPA (2022) Soak up the rain: What's the problem? https://www.epa.gov/soakuptherain/soak-rain-whats-problem

Lenntech (2022) Acids & alkalis in freshwater: Effects of changes in pH on freshwater ecosystems. https://www.lenntech.com/aquatic/acids-alkalis.htm#ixzz7KS1uy1Hl

Rahimpour V, Zeng Y, Mannaerts CM (2018) Urban impacts of air temperature and precipitation over the Netherlands. Clim Res 75(2). doi: 10.3354/cr01512.

USGS (2018) Water. http://water.usgs.gov/edu/watercycleatmosphere.html

21 Water
Salinization

SALINIZATION: HIGH POINTS

- 90% produced by evaporation from water bodies.
- 86% of global evaporation and 78% of global precipitation occur over the ocean.
- Using salt to increase road safety in the winter comes with a cost to the health of our freshwater ecosystems upon which humans depend.
- Freshwater salinization triggers a massive loss of zooplankton and an increase in algae – even at the lowest chloride thresholds established in Canada and the United States and throughout Europe.
- De-icing roads with salt adversely affects ecosystems.

21.1 SALINIZATION: SOURCES AND PATTERNS

When a chemical compound consists of an ionic assembly of positively charged cations and negatively charged anions, which results in a compound with no net electric charge, it is referred to as a salt.

Salts occur during a chemical reaction between two entities such as the following:

- An acid and a base, e.g., $HCl + NH_3 \rightarrow NH_4Cl$ (ammonium chloride)
- An acid and a metal, e.g., $H_2SO_4 + Mg \rightarrow MgSO_4 + H_2$ (magnesium chloride plus hydrogen)
- A metal and a non-metal, e.g., $Cu + Cl_2 \rightarrow CuCl_2$ (copper chloride)
- A base and an acid anhydride, e.g., $2\ NaOH + Cl_2O \rightarrow 2\ NaClO + H_2O$ (sodium hypochlorite, commonly known as bleach when in a dilute solution, and comprising sodium cation, Na^+, and a hypochlorite anion, OCl^- or ClO^-, plus water)
- An acid and a base anhydride, e.g., $2\ HNO_3 + Na_2O \rightarrow 2\ NaNO_3 + H_2O$ (sodium nitrate plus water). An anhydride refers to any chemical compound obtained by removing water from another compound. Two examples of inorganic anhydrides are calcium oxide, CaO, derived from calcium hydroxide $Ca(OH)_2$, and sulfur trioxide, SO_3, which is derived from sulfuric acid.
- In the double displacement reaction, where two different salts mix in water, their ions alternately recombine, and a new salt forms as an insoluble precipitate. For example:

$$Pb(NO_3)_2 + Na_2SO_4 \rightarrow PbSO_4 \downarrow + 2NaNO_3$$

Salinization (mainly sodium chloride but including the "lesser" salts) of water sources occurs from irrigation, seawater encroachment due to rising sea levels

DOI: 10.1201/9781003341826-26

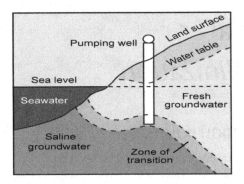

FIGURE 21.1   Salt in freshwater.

(Source: www.epa.gov)

(Figure 21.1), abstraction of groundwater, and salting of icy roads. At a local level, the flushing down of sodium ions used to replace calcium and magnesium ions in home-water softeners increases the salt concentration of wastewater, which, in turn, raises the NaCl concentration of surface streams and groundwaters. Dissolved salts in the water supply also cause soil salinization especially from brackish groundwater through the soil from below. A wide range of management options have been developed over time, yet none may guarantee long-term sustainability.

PROBLEM 21.1.1
Salt in freshwater

A storm sewer carrying snowmelt containing $1.450 \text{ g} \cdot \text{L}^{-1}$ of sodium chloride deposits the salty water in a small stream. The background concentration of sodium chloride in the stream is $15 \text{ mg} \cdot \text{L}^{-1}$. The stream flows at $3.0 \text{ m}^3 \cdot \text{s}^{-1}$ while the incoming water from the storm sewer flows at $2200 \text{ L} \cdot \text{min}^{-1}$. If (a) the incoming water is thoroughly mixed with that of the stream (b) the salt is non-reactive, and (c) the system is at equilibrium (steady state), what is the sodium chloride concentration where the two waters merge?

SOLUTION 21.1.1

$$C_s \text{ in } Q_s \text{ in}$$
$$C_{ss} \text{ in } Q_{ss} \text{ in}$$
$$Q_{mix} = Q_s + Q_{ss}$$
$$[C_{mix} = ?]$$

where
  C_s in = background concentration in stream
  Q_s in = flow rate in stream
  C_{ss} in = concentration in storm sewer
  Q_{ss} in = flow rate in storm sewer
  $Q_{mix} = Q_s + Q_{ss}$
  $C_{mix} = ?$

All in all, the mass flow of salt may be calculated as

$$\text{Mass} / \text{time} = (\text{Concentration})(\text{flow rate}) = (mg \cdot L^{-1})(L \cdot min^{-1}) = mg \cdot min^{-1}$$

Applying the abbreviations in the above equation where the subscripts "sw" refer to sewer and "st" signifies the stream, the mass equilibrium may be depicted as:

$$\text{Accumulation rate of NaCl} = [C_{st}Q_{st} + C_{se}Q_{se}] - \text{Conc}_{mix}Q_{mix}$$

where $Q_{mix} = Q_{st} + Q_{sw}$

As the steady state is assumed, the rate of accumulation $= 0$ and

$$\text{Conc}_{mix}\, Q_{mix} = [\text{Conc}_{st}Q_{st} + \text{Conc}_{se}Q_{se}]$$

Transposing the equation,

$$\text{Conc mix} = [\text{Conc}_{st}Q_{st} + \text{Conc}_{se}Q_{se}] / Q_{st} + Q_{sw}$$

Converting the given units for calculating yields the following:

$$C_{sw} = (1.450\ g \cdot L^{-1} \times 1000\ mg \cdot 1\ g^{-1}) = 1450\ mg \cdot L^{-1}$$
$$Q_{st} = (3.0\ m^3 \cdot s^{-1})(1000\ L \cdot m^3)(60s \cdot min^{-1}) = 180,000\ L \cdot min^{-1}$$
$$\text{Conc mix} = [(15\ mg \cdot L^{-1})(180,000\ L \cdot min^{-1})] + [(1450\ mg \cdot L^{-1})(2200\ L \cdot min^{-1})] /$$
$$180,000\ L \cdot min^{-1} + 2200\ L \cdot min^{-1}$$

Salt concentration where the two waters merge $= 32.32\ mg \cdot L^{-1}$

21.2 ELECTRICAL CONDUCTIVITY (EC25) MEASUREMENTS

Evaporation of water from the surface of a lake concentrates the dissolved solids in the remaining water – thereby increasing the EC. This is a very noticeable effect in reservoirs in the southwestern United States (the major type of lake in arid climates) and has caused high salt concentrations in the Great Salt Lake in Utah, Pyramid Lake, Nevada, and the Caspian Sea. The conductivity of a sample indicates the total number of ions in the sample. Hence, the electrical conductivity (EC) of water varies with density, and thus with temperature, and normalized EC refers to the EC value at "normal" temperature ($25°C$ = EC25). The SI measurement unit for electrical conductivity is the siemens; hence, a microsiemens (μS) is 10^{-6} siemens. As electrical conductance and admittance reciprocate with resistance and impedance, respectively, one siemens is equal to the reciprocal of one ohm (Water on the Web 2016).

PROBLEM 21.2.1
Adjusted conductivity values

The EC value in a stream (Table 21.1) taken at a water temperature of $18°C$ was $57\ \mu S \cdot cm^{-1}$. What is the corrected electrical conductivity value adjusted to $25°C$?

TABLE 21.1

Values for Electrical Conductivity and Total Dissolved Solids

EC25	(µS/cm)	TDS (mg/L)
Divide Lake	10	4.6
Lake Superior	97	63
Lake Tahoe	92	64
Grindstone Lake	95	65
Ice Lake	110	79
Lake Independence	316	213
Lake Mead	850	640
Atlantic Ocean	43,000	35,000
Great Salt Lake	158,000	230,000
Dead Sea	?	~330,000

Source: After values listed in Water on the Web (2016)
https://www.waterontheweb.org/under/waterquality/conductivity.html

SOLUTION 21.2.1

To recalculate non-temperature corrected reading, use:

$$C_{25} = C_m / 1 + 0.0191(t_m - 25)$$

where
  C_{25} = corrected conductivity value adjusted to 25°C
  C_m = actual conductivity measured before correction
  t_m = water temperature at the time of C_m measurement

Hence, $C_{25} = 57 / 1 + 0.0191(18 - 25) = 50.0191 \ \mu S \cdot cm^{-1}$.

PROBLEM 21.2.2
EC25 and total dissolved solids

High levels of salinity deteriorate water quality after rain, thereby producing health concerns. Calculate the concentration of total dissolved salts in the sample.

SOLUTION 21.2.2

TDS concentration of a water sample can be estimated by multiplying its normalized EC (EC25) by a factor of between 0.5 and 1.0 for natural waters, depending upon the type of dissolved solids (Water on the Web 2016).

A widely accepted value to use for a broad approximation is 0.67 (Water on the Web 2016).

$$TDS\,(ppm) \equiv EC@\,25°C \ (\mu S \cdot cm^{-1}) \times 0.67$$

PROBLEM 21.2.3
Ocean productivity

Why are the oceans most productive outside of the tropics?

SOLUTION 21.2.3

Salinity affects dissolved oxygen solubility such that the higher the salinity level, the lower the dissolved oxygen concentration. Oxygen is about 20% less soluble in seawater than in freshwater at the same temperature (Langland & Cronin 2003). This means that, on average, seawater has a lower dissolved oxygen concentration than freshwater sources. Further, the salinity decreases with distance from the Equator. (Notably, the largest aquatic animal species live outside of the tropics.) Conductivity, in particular specific conductance, is one of the most useful and commonly measured water quality parameters (Langland & Cronin 2003).

21.3 REMEDIATION APPROACHES

PROBLEM 21.3.1
Salinity: remediation approaches

What are some correctives for high salt concentration?

SOLUTION 21.3.1

- Though still highly recommended, the leaching of salts further increases salinization of deep groundwaters. Salt-containing drainage waters pose environmental threats through the discharge of toxic trace elements such as selenium Yin et al. (2023).
- Leaching with required drainage management practices is particularly relevant for surface irrigation methods driven by gravity such as through basin, border, and furrow irrigation.
- But, as the salts are leached, other applied substances such as agrochemicals and fertilizers such as nitrates move into the groundwater.
- Either through ditches or perforated drain tubing, groundwater tables are kept sufficiently low to prevent upward salt transport into the rooting zone.
- Typically, drip and sprinkler irrigation are high frequency systems, applying relatively smaller volumes, therefore allowing control of wetted soil volume and root zone salt concentration while minimizing deep percolation below the plant root zone (Taylor & Zilberman 2017).

Besides germination and crop establishment, most crops are more sensitive to salinity during the reproductive phase. Other critical stages vary from crop to crop. Case studies are reported in the literature (Sharma & Minhas 2005).

The effects of the salinity on yields may be estimated as follows (Ayers & Westcot 1985):

$$Y_a / Y_m = 1 - (ECe - ECe_{threshold}) \, b / 100$$

where

Y_a and Y_m = actual and potential crops yields (kg · ha^{-1}), when the crop techniques are appropriate to the local environmental conditions and no water stress affects the crop.

b = crop-specific parameter, which describes the rate of yield decrease per unit of excess salts (% per dS · m^{-1}).

ECe and ECe threshold are, respectively, the actual EC of the soil saturation extract and the crop-specific ECe threshold above which the crop is affected by salinity. The ECe threshold ranges from 1.0 dS · m^{-1} for very sensitive crops, such as carrots and beans, up to more than 7.5 dS · m^{-1} for barley, cotton, and tolerant grasses. The b rate of decrease in yield per unit increase in EC varies from more than 15 (% per dS · m^{-1}) for the sensitive crops down to 5 or less (% per dS · m^{-1}) for tolerant crops.

REFERENCES

Ayers RS and Westcot DW (1985) Water quality for agriculture. FAO Irrigation and Drainage, Paper 29. Food and Agriculture Organization, Rome., https://www.fao.org/3/T0234E/T0234E00.htm

Langland M, Cronin T (eds.) (2003) A Summary Report of Sediment Processes in Chesapeake Bay and Watershed. In Water-Resources Investigations Report 03-4123. U.S. Geological Survey, New Cumberland, PA. http://pa.water.usgs.gov/reports/wrir03-4123.pdf

Sharma BR, Minhas PS (2005) Strategies for managing saline/alkali waters for sustainable agricultural production in South Asia. Agric Water Manage 78(1), 136–151. doi: 10.1016/j.agwat.2005.04.019

Taylor R, Zilberman D (2017) Diffusion of drip irrigation: The case of California. Appl Econ Perspect Policy 39(1), 16–40.

Yin Wang, Chengxiao Hu, Xu Wang, Guangyu Shi, Zheng Lei, et al. (2023) Selenium-induced rhizosphere microorganisms endow salt-sensitive soybeans with salt tolerance, Environmental Research, Vol 236, Part 2, 2023, 116827, ISSN 0013-9351, https://doi.org/10.1016/j.envres.2023.116827.

Water on the Web (2016) Electrical conductivity. Why it is so important. https://www.waterontheweb.org/under/waterquality/conductivity.html

22 Wastewater Treatment

Using practice problems and solutions to problems, this chapter presents some biological and chemical attributes of the water to be cleaned and discusses the processes required.

WASTEWATER: MAIN POINTS

- Domestic, residential, shops/offices/schools, etc. and from toilets, sinks, and bathrooms
- Industrial: industries. Quantity and quality depend on the type of industry
- Storm water: rainwater, may contain pollutants
- Untreated wastewater harmful to health
- Breeding sites for insects, pests, and microorganisms
- Can cause environmental pollution and affect the ecosystem

22.1 WASTEWATER: INCREASINGLY USEFUL

As water supplies become more limited around the world, renovation and reuse of wastewater become more important (Manahan 2005). He reports that wastewater purified by the activated sludge process is good enough to be used for applications such as groundwater recharge and irrigation and that advanced treatment processes can even bring the wastewater up to drinking water standards. The fertile wastewater can support wetlands where algae and plants growing profusely can remove excess nutrients and produce biomass that can be converted to biofuels (Manahan 2005).

Nevertheless, the treatment of wastewater has the potential to deplete resources. Mining, agricultural, chemical, and radioactive wastes all have the potential for contaminating both surface water and groundwater. Sewage sludge spread on land may contaminate water by release of nitrate and heavy metals. Landfills may likewise be sources of contamination. Leachates from unlined pits and lagoons containing hazardous liquids or sludges may pollute drinking water.

22.2 WASTEWATER CHARACTERISTICS

- Physical
- Chemical
- Biological
- Color (depends) mainly on the wastewater constituent
- Odor: not significant if aerobic. Anaerobic wastewater release hydrogen sulfide
- Temperature: higher than water temperature due to the microbiological activities
- Turbidity: solids mainly suspended (SS) from clay, sand, human waste, and plant fibers
- Chemical: Organic carbon based. Usually, combustible lower melting and boiling points

DOI: 10.1201/9781003341826-27

- Less soluble in water
- Have very high molecular weight
- Most organic compounds can serve as a source of food for microorganisms

Sources of pollutants

- Nature: fibers, vegetable oils, animal oils and fats, cellulose, starch, sugar
- Synthesis: a wide variety of compounds and materials prepared by manufacturing processes (e.g., DDT, polyvinyl chloride)
- Fermentation: alcohols, acetone, glycerol, antibiotics, acids

Classification: degradability

- Biodegradable organics
- Non-biodegradable organics
- Food for microorganisms
- Fast and easily oxidized by microorganisms, e.g., starch, fat protein, alcohol, human and animal waste

Non-biodegradable

- Difficult and takes much longer to biodegrade or is toxic to microorganisms
- For example, PVC, pesticide, industrial waste, cellulose, phenol, lignic acid

Effects

- Depletion of the dissolved oxygen in the water
- Destroying aquatic life
- Damaging the ecosystem
- Some organics can cause cancer
- Trihalomethanes (THM – a carcinogenic compound) are produced in water and wastewater treatment plants when natural organic compounds combine with chlorine added for disinfection purposes.

22.3 MEASURING ORGANIC CONTENT

Normally, wastewater has high organic content. The organic content is measured by

a. biochemical oxygen demand (BOD) and
b. chemical oxygen demand (COD) and the value is about 100–400 mg/L.

22.4 BOD DEFINED

BOD is defined as the quantity of oxygen utilized by a mixed population of microorganisms to biologically degrade the organic matter in the wastewater under aerobic conditions:

$$\text{Organic matter} + O_2 \,(\text{microorganisms}) \rightarrow CO_2 + H_2O + \text{New cells}$$

BOD test: 5 days at 20°C or 3 days at 30°C

22.5 INORGANIC COMPOUNDS DEFINED

DEFINITION

When placed in water, inorganic compounds dissociate into electrically charged atoms referred to as ions. All the atoms are linked in ionic bonds.

SOURCE(S)

Heavy metals, nutrients (nitrogen and phosphorus), alkalinity, chlorides, sulfur, and other inorganic pollutants; oxides, carbonates, sulfides, silicates

EFFECT(S)

i. Diseases
   $NO_2^- \rightarrow$ "blue baby syndrome"
ii. Aesthetic
   $Si^{+3+} \rightarrow$ turbidity
iii. Disturb human activities such as the formation of scale in boiler system and excessive usage of soap
   $Ca^{2+}, Mg^{2+} \rightarrow$ hardness

22.6 BIOLOGICAL CHARACTERISTICS

The principal groups of microorganisms found in wastewater are bacteria, fungi, protozoa, microscopic plants and animals, and viruses. *Most microorganisms (bacteria, protozoa) are responsible and beneficial for biological treatment processes of wastewater.*

Pathogenic organisms are usually excreted by humans from the gastrointestinal tract and discharged to wastewater, thereby causing waterborne diseases including cholera, typhoid, paratyphoid fever, and diarrhea. The pathogenic organisms in wastewaters are generally low in density, and hence they are difficult to isolate and identify. Therefore, indicator bacteria such as total coliform (TC) and fecal coliform (FC) are used as indicator organisms.

22.7 WASTEWATER QUALITY STANDARDS

PROBLEM 22.7.1
Test method (BOD_5 @ 20°C)

How is BOD determined?

SOLUTION 22.7.1

1. A water sample containing degradable organic matter is placed in a BOD bottle.
2. If needed, add dilution water (known quantity). Dilution water is prepared by adding phosphate buffer (pH 7.2), magnesium sulfate, calcium chloride, and ferric chloride into distilled water. Aerate the dilution water to saturate it with oxygen before use.

3. Measure DO in the bottle after 15 min (DOi).
4. Closed the bottle and placed it in an incubator for 5 days at temperature 20°C.
5. After 5 days, measure DO in the bottle (DOt).

Calculation of BOD

$$BODt = (DOi - DOt)P$$

where

$BODt$ = biochemical oxygen demand (mg · L^{-1})

DOi = initial DO of diluted wastewater sample at 15 min after preparation (mg · L^{-1})

DOt = final DO of diluted wastewater sample after incubation for 5 days (mg · L^{-1})

P (dilution factor) = volume of sample / (volume of sample + volume of distilled water)

PROBLEM 22.7.2
The need for dilution

Why is dilution needed?

SOLUTION 22.7.2

Dilution is done to get appropriate dissolved oxygen (DO) depletion on the measurement of 5-day BOD. For a valid BOD test, the final DO should not be less than 1 mg · L^{-1}. The BOD test is invalid if the DOt value is near zero. Dilution can decrease the organic strength of the sample. By using the dilution factor, the actual value can be obtained.

Dilution of wastes can be performed by direct pipetting into 300 mL BOD bottle.

22.8   HEAVY METALS IN WASTEWATER

Ongoing obstacles in wastewater treatment include flushing non-degradable objects into the system, a buildup of harsh chemicals having been flushed into the system, dissolved heavy metals, and the very high energy requirements which exacerbate the burning of fossil fuels. The example problems and solutions in this chapter seek, in turn, to reinforce the effects of these obstacles and inasmuch as they accompany remediation.

PROBLEM 22.8.1
Decreasing heavy metals at point of entry

Wastewater from a metal galvanizing factory increased heavy metal concentrations including those of lead, copper, and zinc in a river, thereby killing, and continuing to kill, several species of fish. How can the toxicity of heavy metals be reduced in situ?

SOLUTION 22.8.1

Like Cu^{2+} and Zn^{2+}, Pb^{2+} is less toxic in hard waters than in soft waters because its toxicity is antagonized by ions such as Ca^{2+} and Mg^{2+}. Hence, the strategies include the following:

1. Avoid dumping wastewater into waterways with substrata predominantly consisting of low acid neutralizing capability (ANC) rocks
2. "Seeding" the present waterways with appropriate high-ANC rocks such as limestone or dolomite

PROBLEM 22.8.2
Disinfecting wastewater

The Mt. Diablo wastewater treatment plant must chemically disinfect the treated wastewater before releasing it into the nearby mountain stream. The wastewater contains 5.2×10^5 fecal coliform colony-forming units (CFU) per liter. The upper limit for coliform units that may be discharged is 1800 fecal coliform $CFU \cdot L^{-1}$. The disinfection process will be carried out by a pipe carrying the wastewater. If it is assumed that the pipe behaves as a steady-state plug-flow system, and the fecal coliforms are being destroyed at the rate of 0.25 min $\cdot$ day^{-1}, what length of pipe is required if the linear velocity in the pipe is 0.68 m $\cdot$ s^{-1}?

SOLUTION 22.8.2

The steady-state solution yields:

$$\ln[C_{out} / C_{in}] = -k[L / u]$$

where:
   ln = natural logarithm (i.e., to the base e)
   C = concentration of substance
   −k = disappearance rate constant of a substance as it reacts.
   L/u = liters per second

$$\ln\left[1800 \text{ CFU} \cdot L^{-1} / 5.2 \times 10^5 \text{ CFU} \cdot L^{-1}\right] = 0.25 \text{ min}^{-1}\left[L / \left(0.68 \text{ m} \cdot s^{-1}\right)\left(60 \text{ s} \cdot \text{min}^{-1}\right)\right]$$

$$\ln\left[3.46 \times 10^{-3}\right] = 0.25 \text{ min}^{-1}\left[L / \left(40.8 \text{ m} \cdot \text{min}^{-1}\right)\right]$$

$$-5.66 = 0.25 \text{ min}^{-1}\left[L / \left(40.8 \text{ m} \cdot \text{min}^{-1}\right)\right]$$

$$L = \text{Length of pipe} = 923.7 \text{ m}$$

Almost 1 km of pipe is required to meet the discharge standard. As most wastewater treatment systems are not equipped to facilitate such pipe lengths, an alternate method of treatment may be more feasible in this case.

PROBLEM 22.8.3
Pollutant load calculations

How is the pollutant load of a catchment calculated?

SOLUTION 22.8.3

Stream loading is a function of concentration and flow, and loading can be reported and calculated using several different units. Two frequently used concentration units are $mg \cdot L^{-1}$ and $mg \cdot 100 \ mL^{-1}$; with flow units for both in cubic feet per second; conversion factors respectively are 5.39 and 284.7; and discharge units being $lb \cdot day^{-1}$, and mass $\cdot$ second^{-1} respectively. The smaller concentration unit ($mg \cdot 100 \ mL^{-1}$) measure substances which typically occur at lower concentrations in water.

$$L = F \times C \times D,$$

where:
  F = conversion factor/units
  C = concentration
  D = discharge

PRACTICE PROBLEM 22.8.1

List three major non-point and three major point sources of water pollution. How can each source be measured?

PRACTICE PROBLEM 22.8.2

A water sample had safe levels of EC, TDS, color, taste and odor, pH, and turbidity. What ten attributes could still make that water unsafe for drinking? (Hint: chemical and biological)

PRACTICE PROBLEM 22.8.3

What kind of water pollution would each of the following processes effectively cure – cation exchange, anion exchange, filtration, demineralization, disinfection, oxidation, pH adjustment, adsorption?

PRACTICE PROBLEM 22.8.4

On the list below which depicts water contamination or an unsatisfactory condition, link each contaminant/condition with an appropriate treatment.

1. Leakage from landfill
2. Chemical application to parks and lawns
3. Road salt
4. Irrigation producing salinization
5. Industrial emission
6. Municipal sewage discharge
7. Underground mining causing acidic leachates
8. Water table reduced by heavily pumped well
9. Sewer and pipeline leaks

10. Storm water runoff contaminating reservoir
11. Contaminated underground storage tanks
12. Freshwater aquifer polluted
13. Municipal water supply well with unsafe levels of coliform bacteria
14. Pesticides and fertilizers from agricultural land
15. Private well poisoned by accidental acidic spill
16. Leaks from oil storage tanks
17. Leakage from septic tanks
18. Livestock waste
19. Marine waste
20. River water contaminated with urban waste
21. River water is contaminated with industrial and agricultural pollutants
22. Brine from oil injection well
23. Leaks from the waste lagoon
24. Deep-well hazardous waste disposal

22.9 REMEDIATION OF WASTEWATER

In the activated sludge process used to treat municipal wastewater, a suspension of microorganisms in an aerated tank biodegrades organic wastes represented as $[CH_2O]$, a process which removes oxygen demand from the water so that it will not deplete oxygen when discharged to a stream or body of water. The microorganisms settle in a settling basin and are pumped back to the aeration tank, greatly speeding the biodegradation process, while excess microorganisms, sewage sludge, or biosolids are taken to an anaerobic digester where they produce combustible methane, which can provide fuel sufficient to run the engines that produce all the power needed by the plant (Manahan 2005).

The Orange County California Groundwater Replenishment System, the largest purification and water reuse project of its kind in the world, utilizes a three-step process of microfiltration, reverse osmosis, and ultraviolet light to purify highly treated sewer water to state and federal drinking water standards. The viability of these methods proves that they can be replicated elsewhere, to offset the impact of climate change on worldwide water supplies.

PROBLEM 22.9.1
Lime additions

The lime dose for raw water has been calculated to be 13.0 mg · L^{-1}. If the flow to be treated is 5.0 MGD, how many lb/day of lime will be required?

SOLUTION 22.9.1

$$\text{Lime, lb} \cdot \text{day}^{-1} = \text{Lime, mg} \cdot L^{-1} \times \text{Flow, MGD} \times 8.34 \text{ lb} \cdot \text{gal}^{-1}$$

$$\text{Lime, lb} \cdot \text{day}^{-1} = 13.0 \text{ mg} \cdot L^{-1} \times 5.0 \text{ MGD} \times 8.34 \text{ lb} \cdot \text{gal}^{-1}$$

$$\text{Lime, lb} \cdot \text{day}^{-1} = 542.1 \text{ lb} \cdot \text{day}^{-1}$$

PROBLEM 22.9.2
Lime requirement

The flow to a clarifier is 2,500,000 gpd. If the lime dose required is determined to be 25 mg · L^{-1}, how many lb · day^{-1} of lime will be required?

SOLUTION 22.9.2

First convert the flow from gpd to MGD:

$$2,500,000 \ \text{gal} \cdot \text{day}^{-1} \times (1 \ \text{MG} / 1,000,000 \ \text{gal}) = 2.5 \ \text{MGD}$$

Now determine the lb · day^{-1} of lime required:

$$\text{Lime, lb} \cdot \text{day}^{-1} = \text{Lime, mg} \cdot \text{L}^{-1} \times \text{Flow, MGD} \times 8.34 \ \text{lb} \cdot \text{gal}^{-1}$$
$$\text{Lime, lb} \cdot \text{day}^{-1} = 25 \ \text{mg} \cdot \text{L}^{-1} \times 2.5 \ \text{MGD} \times 8.34 \ \text{lb} \cdot \text{gal}^{-1}$$
$$= 285.0 \times 2.5$$
$$\text{Lime required} = 712.16 \ \text{lb} \cdot \text{day}^{-1}$$

PRACTICE PROBLEM 22.9.1
Converting to the SI system

A total of 344.54 lb · day^{-1} lime will be required to raise the alkalinity of the water passing through a clarification process. How many grams per minute (g · min^{-1}) lime does this represent?

PROBLEM 22.9.3
Orthophosphate in wastewater

A wastewater has a soluble orthophosphate concentration of 5.00 mg · L^{-1} as P. What amount, in theory, of ferric chloride will remove it completely?

SOLUTION 22.9.3

Removing 1 mole of phosphorus requires 1 mole of ferric chloride:

$$FeCl_3 + HPO_4^{2-} \rightleftharpoons FePO_4 + H^+ + 3Cl^-$$

The relevant gram molecular weights are:

$$FeCl_3 = 162.21$$

P = 30.97 g
With a PO$_4$-P of 5 mg · L^{-1}, the amount of ferric chloride, in theory, would be

$$5.00 \times 162.2 / 30.97 = 26.18, \text{or } 26.2 \ \text{mg} \cdot \text{L}^{-1}$$

Note:

Owing to interacting factors such as solubility product limitations, side reactions, etc., the actual amount of chemical to be added can only be ascertained by jar tests on the wastewater. The actual amount required can therefore be between 1.25 and 3 times the theoretically calculated amount (Davis and Masten 2020). Similarly, the theoretical dose of alum would fall short of the actual dose by a factor of 1.25–2.5 times.

22.10   BAMBOO BIOCHAR IN WASTEWATER TREATMENT

Activated carbon is a key component of the filter material used to clean exhaust gases utilized in treating wastewater and drinking water.

PROBLEM 22.10.1
Biochar for water purification

What is the role of biochar in water purification?

SOLUTION 22.10.1

Activated carbon is a popular adsorbent with a huge surface area, a consistent microporous structure, and radiation stability (Mahanim et al. 2011). Biochar can be used for the immobilization of contaminants in water, soils, or sediments (Teixido et al. 2013; Thompson et al. 2016).

REFERENCES

McKenzie Davis and Susan Masten (2020) Principles of Environmental Engineering & Science. 4th edition. ISBN10: 1259893545 | ISBN13: 9781259893544

Hernandez-Mena LE, Pécora AAB, Beraldo AL (2014) Slow pyrolysis of bamboo biomass: Analysis of biochar properties. Chem Eng Trans 37, 115–120. https://doi.org/10.3303/CET1437020

Mahanim SMA, Asma IW, Rafdah J, Puad E, Shaharuddin H (2011) Production of activated carbon from industrial bamboo wastes. J Trop Forest Sci 23, 417–424.

Manahan S (2005) Environmental Chemistry. Taylor and Francis, Boca Raton, Fla.

Teixido C, Hurtado JJ, Pignatello JL, Granados M, Peccia J (2013) Predicting contaminant adsorption in black carbon (biochar)-amended soil for the veterinary antimicrobial sulfamethazine. Environ Sci Technol 47(12), 6197e6205.

Thompson KA, Shimabuku KK, Kearns JP, Knappe DR, Summers RS et al. (2016) Environmental comparison of biochar and activated carbon for tertiary wastewater treatment. Environ Sci Technol 50(20). 11253–11262. doi: 10.1021/acs.est.6b03239.

23 Water Protection

MAIN POINTS – CONDITIONS PRONE TO CONTAMINATION AND/OR RESISTING CONTAMINATION:

- Unconfined aquifers with no cover of dense material.
- Bedrock with large fractures because the fractures provide pathways for contaminants.
- Wells connecting two aquifers have greater chances of cross-contamination between the aquifers.
- Confined, deep aquifers are better protected than surface ones with a dense layer of clay material.
- Water-bearing confined sandstone resists contamination.

23.1 EFFECTIVENESS OF CARBON FILTRATION

According to Franks (2018), who cites "Total Trihalomethanes," a category made up of still uncounted chemicals, assumed to number in the thousands, that are formed when water containing organic matter (i.e., virtually all water) is treated with chlorine, the maximum allowable level for trihalomethanes, which are suspected cancer causers and are present in virtually all chlorinated tap water, is only 1/10 of one part per million. He further observes that for the organics category, the primary treatment in all cases, and the only recommended treatment in most cases, is activated carbon.

The fact that the EPA's Pesticides category lists 14 familiar poisons including Aldicarb, Chlordane, Heptachlor, and Lindane where activated carbon is the only recommended treatment, in addition to a further 12 herbicides listed (2,4-D, Atrazine, etc.) where activated carbon is the only treatment recommended (Franks 2018), proves the effectiveness of that treatment for polluted water, even when it comes from a source water protection area (Figure 23.1a).

PROBLEM 23.1.1
Activated charcoal

Why is activated carbon the only effective treatment for the above-mentioned contaminants?

SOLUTION 23.1.1

Activated charcoal is produced by the thermochemical breakdown of biomass in the absence of oxygen followed by chemical or physical activation (Flomenbaum et al. 2002). Having been processed to have small, low-volume pores that increase the surface area for adsorption and chemical reactions, it is good at trapping other carbon-based impurities and chlorine. Activated charcoal has a large surface area with a

 DOI: 10.1201/9781003341826-28

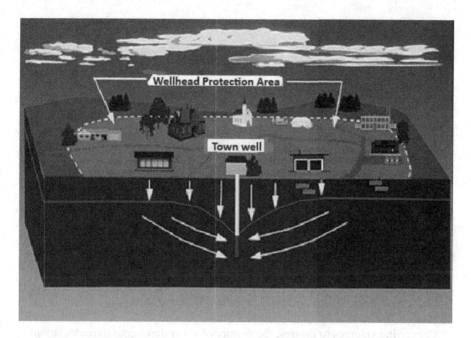

FIGURE 23.1a Parameters of a source water protection area (b) Schematic diagram (shown below) of inflow and outflow of a leaking tank.

Source: www.epa.gov

high sorption capacity (EFSA 2011). Activated charcoal is primarily made up of elemental carbon that lacks functional groups, making van der Waals interactions its primary means of adsorbing compounds. This makes activated charcoal an effective adsorption treatment for large neutral molecules, and each of the above chemicals is composed of neutral molecules. Hence, activated charcoal sorbs large ions whose primary intermolecular forces are also van der Waals interactions (Flomenbaum et al. 2002; Olsen 2010).

23.2 WATER LEAKS

Untreated non-saline groundwater is normally much safer to drink than any untreated surface water since the ground itself provides an effective purifying medium. Such bodies constitute a major source of drinking water supplies. But water leaks = wasted water. Though leaking pollutants can destroy vast amounts of good water, undetected water leaks can be more insidious, causing even more losses over long periods. Incorrectly installed geo-liners are easily ruptured.

PROBLEM 23.2.1
Measuring water leaks

An empty geo-lined reservoir located in the town of Dryden Meer and having a capacity of 3×10^3 m^3 developed a 22-L · min^{-1} leak while being filled. The filling

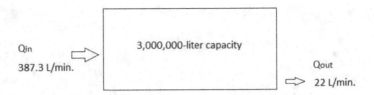

FIGURE 23.1b Schematic diagram of inflow and outflow of a leaking tank.

tap flowed at 387.3 L · min⁻¹. On the assumption that the density of water is 1000 kg · m⁻³, how much time was required to fill the reservoir? How much water was wasted?

SOLUTION 23.2.1

The schematic diagram (Figure 23.1b) is shown here:

$$Qin = 387.3 \ L \cdot min^{-1}$$

$$Volume = (Flow\ rate) \times Time = (Q)t$$

Converting volumes to masses, the density of water dominates the calculation:

$$Mass = (Volume)(Density) = (V)(\rho)$$

$$Since\ 1.0\ m^3 = 1000\ L,\ 3000\ m^3 = 3000 \times 1000\ L = 3000,000\ L$$

$$Accumulation = Mass\ in - Mass\ out$$

$$(V\ ACCUM)(\rho) = (Qin)(\rho)(t) - (Qout)(\rho)(t)$$

$$V\ ACCUM\ (\rho) = (Qin)(\rho)(t) - (Qout)(\rho)(t)$$

$$V\ ACCUM = Qin\ (t) - Qout\ (t)$$

$$V\ ACCUM = 387.3t - 22t$$

$$3000,000\ L = (365.3\ L \cdot min^{-1})t$$

t = 8212.4 min = 136.87 h
The amount of water wasted is:

$$Wasted\ water = \left(22\ L \cdot min^{-1}\right) \times 8212.4\ min = 180,672.8\ L$$

PROBLEM 23.2.2
Wastage from leaks

Water wastage can be reduced if leaks and leak attributes are detected early. Frequent, regular monitoring facilitates early detection. Bellavista reservoir, with a volume of 275 m³, leaks at the rate of 150 L · min⁻¹. The inflow to the lake is 650 L · min⁻¹. How long will the reservoir take to fill?

SOLUTION 23.2.2

Volume accumulated (p) = Qin (p)t − Qout (p)t

If it is assumed that the density of water is 1000 kg · L^{-1},

Volume accumulated = Qin (t) − Qout (t)

Volume accumulated = 650t − 150t

$$275 \times 10^3 \text{ L}^{-1} = (500 \text{ L} \cdot \text{min}^{-1})t$$

t = 275,000 L · min^{-1} = 500 min (as L and L^{-1} cancel out)

PROBLEM 23.2.3
Detention time

A tank has a flow rate of 1000 L · min^{-1} and a volume of 17,000 L. How long is the water being settled for (what is the detention time)?

SOLUTION 23.2.3

Detention time:

$$T_D = V / Q$$

where

T_D = detention time
V = volume of water in tank
Q = average flow rate (volume per unit time)

Detention time = 17 min

23.3 AFFLUENT AND EFFLUENT RIVER FLOWS

PROBLEM 23.3.1
Identifying dry wells in a river basin

Which is the first section of a river basin to have dry wells under sufficiently harsh drought conditions?

SOLUTION 23.3.1

The lower sections ahead of the floodplains dry out first, working gradually back to the source. This is because the water table gradient dips toward lower ground and there is normally more recharge in the upper reaches of a river.

23.4 FLUSHING OF LAKES

Flushing a lake helps to purify it, and there are advantages to flushing it regularly. Hydraulic residence time (HRT) is the time required to refill an empty lake with its natural inflow. A large deep lake with a moderate inflow will have a longer HRT than a small, shallow lake with the same inflow (Lake Dynamics 2017).

PROBLEM 23.4.1
Using the HRT

If a lake becomes polluted with a toxicant, how long does it take to rid itself of the pollutant?

SOLUTION 23.4.1

The HRT (Figure 23.2) is needed to determine annual lake budgets for water, heat, oxygen contaminants, and herbicides. It, therefore, provides the background information for an estimate of the turnover time for water in a lake or "flushing time."

PROBLEM 23.4.2
Determination of HRT

Rainfall plus normal inflow is the only viable way to refill a large lake. How can the natural replenishment time be estimated? How is the HRT of a water body calculated?

SOLUTION 23.4.2

$$\text{Residence time} = \text{Lake volume} / \text{outflow} \qquad (23.4.1)$$

A lake's residence time is calculated by dividing the lake's volume by its average annual water outflow (Figure 23.2). To avoid the undue influence of seasonal variation on the results, lake managers calculate outflow on an annual basis, where volume (V) is usually expressed in acre-feet, and mean outflow is expressed as acre-feet/year (Lake Dynamics 2017). Hence:

$$\text{HRT (years)} = \text{Lake volume (acre-ft)} / \text{mean outflow (acre-ft / year)} \qquad (23.4.2)$$

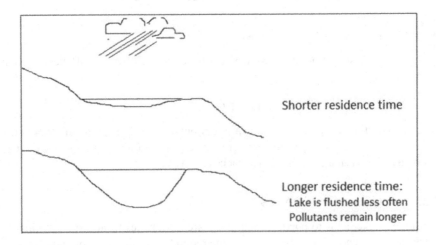

FIGURE 23.2 Hydraulic residence time for a lake.

PROBLEM 23.4.3
Length of HRT

What advantage is a short HRT?

SOLUTION 23.4.3

In lakes with very short residence times (i.e., a lake with a small volume and high inflow and outflow rates), algae may get flushed out of the lake so quickly that they don't accumulate. Intermediate residence times allow algae and aquatic plants enough time to increase biomass based on the nutrients that are present.

23.5   AREAS ADJACENT TO STREAMS

PROBLEM 23.5.1
Riparian areas

What are the benefits of widening riparian areas (vegetated areas adjacent to stream)?

SOLUTION 23.5.1

According to EPA (2022a), wider riparian areas (Figures 23.3 and 23.4) help to:

- Buffer pollutants (intercept and process groundwater)
- Control erosion (of bank and channel)
- Provide habitat (e.g., woody debris, bank vegetation)
- Regulate nutrient inputs (provide organic matter for stream food webs)

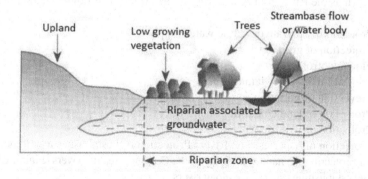

FIGURE 23.3   Major components of a stream or water body riparian area – riparian areas can be symmetrical or asymmetrical in shape. The topography and hydrogeology determine the plant and animal communities associated with the width or meandering area configurations.

Source: www.nrcs.usda.gov

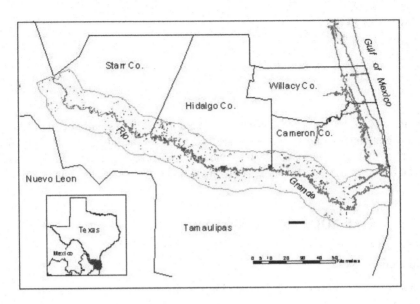

FIGURE 23.4   Lower Rio Grande riparian area.

Source: epa.gov

23.6   PROTECTION OF WATERWAYS

An example of legislation that ensures the continuing integrity of waterways is the
Massachusetts Rivers Protection Act of 1996. The Massachusetts State Legislature
passed the Massachusetts Rivers Protection Act, more formally known as "An Act
Providing Protection for the Rivers of the Commonwealth," thereby providing protection to rivers by regulating activities within a newly established wetland resource
area known as the Riverfront Area. This Act identifies the following eight purposes:

- Protection of private or public water supply
- Protection of groundwater
- Flood control
- Prevention of storm damage
- Prevention of pollution
- Protection of land containing shellfish
- Protection of wildlife habitat
- Protection of fisheries. The Rivers Protection Act establishes a state policy
  for protecting the natural integrity of the Commonwealth's rivers (Figure 23.5)
  and establishing open space along rivers

PROBLEM 23.6.1
Protection of adjacent lands

The most effective control is that of government ownership. Does the Act make any pro-
visions for funding the purchase of open space bordering the state's rivers and streams?

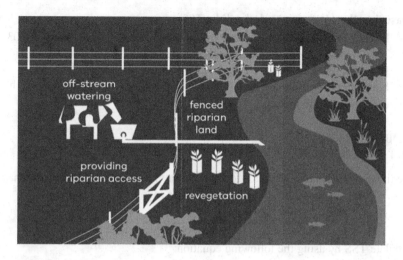

FIGURE 23.5 Physically protected riparian land.

SOLUTION 23.6.1

The Rivers Protection Act has designated $30 million for the acquisition of land bordering rivers and streams (MDEP 2023).

23.7 SEDIMENTS

Sediments are a great water pollutant by volume, can be both suspended in the water column and may be deposited on the bottom of a stream. High levels of suspended sediments produce turbidity.

Causes of sedimentation
- Soil erosion
- Domestic and industrial wastewater discharge
- Urban runoff
- Flooding
- Algal growth due to nutrient enrichment
- Dredging operations
- Channelization
- Removal of riparian vegetation and other stream bank disturbances

Comparing total suspended solids and turbidity
- Both are indicators of the number of solids suspended in the water
- Mineral (e.g., soil particles)
- Organic (e.g., algae, detritus)
- TSS measures the actual weight of material per volume of water ($mg \cdot L^{-1}$)
- Turbidity measures the amount of light scattered
- Therefore, TSS allows the determination of an actual concentration or quantity of material while turbidity doesn't

Measuring the total suspended solids (TSS)

- Filter a known amount of water through a pre-washed, pre-dried (at 103–105°C) pre-weighed (approximately 0.5 mg) filter
- Rinse, dry, and reweigh to calculate TSS in $mg \cdot L^{-1}$ (ppm)
- Save filters for other analyses such as volatile suspended solids (VSS) that estimate organic matter

PROBLEM 23.7.1
Total suspended solids

Five liters of water taken from a stream are filtered through a filter having a mass of 0.8 g. What is the mass of total suspended solids in $mg \cdot L^{-1}$?

SOLUTION 23.7.1

Calculate TSS by using the following equation:

$$TSS \ (mg \cdot L^{-1}) = ([A - B] \times 1000) / C$$

where
  A = final dried weight of the filter (in milligrams)
  B = initial weight of the filter (in milligrams)
  C = volume of water filtered

Therefore, TSS = ([825-800] × 1)/5

$$TSS = 5 \ mg \cdot L^{-1}$$

23.8   THE TESTING OF WELL-WATER

Monitoring wells are designed to provide information on the status of aquifers, not to supply water. They

- may or may not be capable of being pumped and
- may be designed for monitoring of contamination or yield of aquifers or sometimes for other purposes.

PROBLEM 23.8.1
Location of test

Where and how should a monitoring well be located?

SOLUTION 23.8.1

- Locate to detect contamination.
- Typically, at least one up-gradient and three down-gradient, with more for the non-point source of contamination
- Provide access for monitoring personnel

- Monitor the potential source of contamination, not confounding sources
- Access to other properties concerned can be, legally, a problem.

PROBLEM 23.8.2
Monitoring well design

What are the attributes of an effective monitoring well and how is it sampled?

SOLUTION 23.8.2

- Diameter usually narrow 1 – 4 inch. Usually augured but can be driven for a temporary sampling event (*Geoprobe*).
- Depth should be down to aquiclude or lower for dense non-aqueous pollutant liquids (DNAPLs).
- Screened above capillary fringe (to avoid extraneous contamination).
- Can also monitor soil gas but soil moisture requires the use of a suction lysimeter (USEPA 2018) to remove pore water and provide a constant vacuum, bringing gas and water to the surface.

Sampling a monitoring well

- Usually have no pump. Samples collected by a *bailer*.
- Collected sometimes by a surface peristaltic pump system.
- A "popper" or tape is used to determine the depth to the water level (USEPA 2018).

PROBLEM 23.8.3
Monitoring well installation and precautions

What are the hazards involved in establishing a monitoring well?

SOLUTION 23.8.3

- Uncapped holes in the ground are hazardous.
- *Drill cuttings* are a solid hazardous waste.
- Pump and development water is a liquid hazardous waste (USEPA 2018).
- Slug tests are common to avoid generating waste.
- Cap must be locked and clearly identified.
- Grouting is very important.

Monitoring well development

- Monitoring wells are generally not developed (USEPA 2018).
- If they become plugged, compressed air is the usual approach.

Monitoring well operation

- A bailer is used to obtain a sample.
- If water is in the formation but not in the well, it is desired to pump out three *well volumes* with an external pump or temporary submersible pump (USEPA 2018).
- Water samples should be sealed as soon as possible.

PROBLEM 23.8.4
Pumped monitoring wells

What are the ways of testing a monitoring well?

SOLUTION 23.8.4

- Flow through testing of EC, DO, pH, and temperature is possible (USEPA 2018).

Other monitoring methods
- TLC meter is easy and standard.
- *Down-hole video*

Water level monitoring
- Tape
- Popper
- Pressure transducer

PROBLEM 23.8.5
Monitoring well maintenance and abandonment

How are wells maintained and decommissioned?

SOLUTION 23.8.5

- Monitoring wells are only used occasionally (usually every 3 months), but they can get slime like other wells and so may need chlorination (USEPA 2018).

Monitoring well abandonment
- Since monitoring wells are in locations that may be contaminated, abandonment is even more important than for other types of wells.
- Without proper abandonment, holes dug in the earth are dangerous: people, children, animals, or objects (thus also including water) falling into such wells could be fatal.

23.9   WATER WASTAGE FROM LEAKS

Water wastage can be reduced if leaks and leak attributes are detected early. Frequent, regular monitoring facilitates early detection.

PROBLEM 23.9.1
River leakage

Under sufficiently harsh drought conditions, which section of a river basin first reveals dry wells?

SOLUTION 23.9.1

The lower sections ahead of the floodplains dry out first, working gradually back to the source. This is because the water table gradient dips toward lower ground and there is normally more recharge in the upper, higher altitudes ($\geq$precipitation) reaches of a river.

PROBLEM 23.9.2
Results of reservoir leakage

Bellavista reservoir, with a volume of 275 m³, leaks at the rate of 150 L min⁻¹. Inflow to the lake is 650 L min⁻¹. How long will the reservoir take to fill?

SOLUTION 23.9.2

Volume accumulated (p) = Qin (p)t − Qout (p)t

If it is assumed that the density of water is 1000 kg L⁻¹,

$$\text{Volume accumulated} = \text{Qin (t)} - \text{Qout (t)}$$
$$\text{Volume accumulated} = 650t - 150t$$
$$275 \times 10^3 \ L^{-1} = \left(500 \ L \ min^{-1}\right)t$$
$$t = 275{,}000 \ L \ min^{-1} = 500 \ min$$

PROBLEM 23.9.3
A unsuspected leak occurs in a 550-m³ municipal water reservoir at the rate of 45 L · min⁻¹ while the reservoir is being replenished at the rate of 800 L · min⁻¹. How long will it take to fill the reservoir? How much water will be wasted?

SOLUTION 23.9.3

Converting volume (Q) to masses:

Mass = (Volume) (Density)

where
 V = volume, ρ = density, Q = mass
 V = (flow rate)(time) = Q(t)

$$\text{Accumulation} = \text{Mass in} - \text{Mass out}$$
$$\text{Vaccumulated} \times (\rho) = Q\text{in } (\rho)t - Q\text{out } (\rho) \ t$$
$$\text{Vaccumulated} = Q\text{in } (t) - Q\text{out } (t)$$
$$\text{Vaccumulated} = 800 \ (t) - 45 \ (t)$$
$$550{,}000 = 800 \ (t) - 45 \ (t)$$
$$550{,}000 = 55 \ (t)$$
$$t = 10{,}000 \ min$$

Amount of wasted water is determined as:

$$\text{Wasted water} = (45 \text{ L} \cdot \text{min}^{-1})(10{,}000 \text{ min}) = 450{,}000 \text{ L}$$

23.10   REMEDIATION: QANATS

Qanats are systems supplying transported water over long distances in hot dry climates without loss of much of the water to evaporation (Figure 23.6). A qanat consists of one or more drainage galleries of 1–1.4 m high and widths of 0.5–0.6 m.

Constructed as a series of well-like vertical shafts, connected by a gently sloping tunnel that taps into underground water and delivers it to the surface by gravity, qanats have no need for pumping, the vertical shafts along the underground channel being purely for maintenance purposes. The emerging water in arid lands often decreases the salinity of the soils below the emergence of the water.

23.11   REMEDIATION: COLLECTING INTERCEPTED RAINWATER

During a rainfall occurrence, there is substantial water loss from various sources such as interception, evaporation, transpiration, infiltration, and depression storage. As a result, the overland flow and runoff yield against rainfall gets reduced. Such losses are referred to as initial losses. The prediction of initial rainwater loss is imperative for determining runoff and hydrograph derivations. Interception is the amount of rainwater loss due to abstractions from initially dry surfaces of the objects lying on the ground surface, including live vegetation such as herbs, shrubs and trees and any dry surfaces like buildings, etc. From a tree, the interception is mainly from the canopy, which is called canopy interception (Figure 23.7). Intercepted rainwater lost due to evaporation is called interception loss.

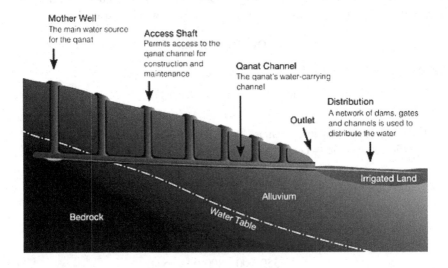

FIGURE 23.6   Qanat technology of prior civilizations.

Source: Bailey (2021)

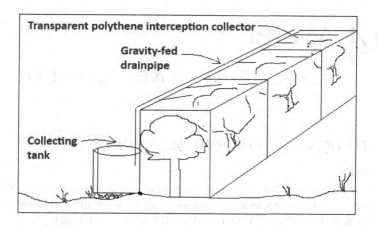

FIGURE 23.7 Water collection from canopy interception. A polyethene sheet placed above trees collects and transfers intercepted rainfall to a storage tank.

PROBLEM 23.11.1
Recovery of interception losses

How could water loss by interception and evaporation be avoided?

SOLUTION 23.11.1

The water intercepted by the tree canopies of the world's coniferous, deciduous, and tropical rainforests comprise substantial quantities. If non-photodegradable polythene sheets were located over those tree canopies (Figure 23.6), all that water could be easily harvested.

REFERENCES

EFSA (2011) Scientific opinion on the substantiation of health claims related to activated charcoal and reduction of excessive intestinal gas accumulation and reduction of bloating (pursuant to Article 13(1) of Regulation (EC)) No 1924/2006. EFSA J 9(4), 2049.

EPA (2022a) Urbanization – Riparian/channel alteration. https://www.epa.gov/caddis-vol2/urbanization-riparianchannel-alteration

Flomenbaum NE, Goldfrank LR, Hoffman RS, Howland MA, Lewin NA, Nelson LS. (2002) Goldfrank's Toxicologic Emergencies. 10th ed. McGraw-Hill, New York, NY.

Franks G (2018) Carbon Filtration: What It Does, What It Doesn't. Ethical H$_2$O https://ethicalh2o.com/carbon-filtration-doesnt-2/

Lake Dynamics (2017) A large deep lake with moderate inflow. https://ilma-lakes.org/sites/default/files/Lake%20Dynamics-Limnology%20101.pdf

MDEP (2023) Rivers Protection Act Questions & Answers. Massachusetts Department of Environmental Protection. https://www.mass.gov/guides/rivers-protection-act-questions-answers

Olsen KR (2010) Activated charcoal for acute poisoning: One toxicologist's journey. J Med Toxicol 6, 190–198.

24 Water
Chlorination Remediation

DISINFECTION WITH CHLORINE: MAIN POINTS

Advantages

- Probably the cheapest and one of the most effective, proven agents of reduction against most bacteria and viruses in water; maintains a chlorine residual $K_a = 3 \times 10^{-8}$ mol $\cdot$ L^{-1} Cl$_2$ (aq) $\rightarrow$ HOCl (aq) $\rightarrow$ ClO (aq)
- HOCl (aq) about 10× more effective than ClO (aq) – the result of the more lipophilic HOCl crossing bacterial membranes more easily
- Residual protection against recontamination
- Ease-of-use and acceptability
- Proven reduction of diarrheal disease incidence
- Scalability and low cost

Disadvantages

- Relatively low protection against protozoa
- Lower disinfection effectiveness in turbid waters
- Potential taste and odor objections
- Must ensure quality control of the solution
- Potential long-term effects of chlorination by-products (Figure 24.1)
- Water with pH >7.5 requires more chlorine – or longer disinfection time – than water with pH <7.5

24.1 ACTION AND DISADVANTAGES OF USING CHLORINE

Chlorine deactivates microorganisms through several mechanisms that can destroy most biological contaminants, including the following:

- Damaging the cell wall.
- Altering the permeability of the cell.
- Altering the cell protoplasm.
- Inhibiting enzyme activity so it's unable to use its food to produce energy.
- Inhibiting cell reproduction.
- Odor from the application of chlorine to water depends on the location of Cl on a phenol molecule.

DOI: 10.1201/9781003341826-29

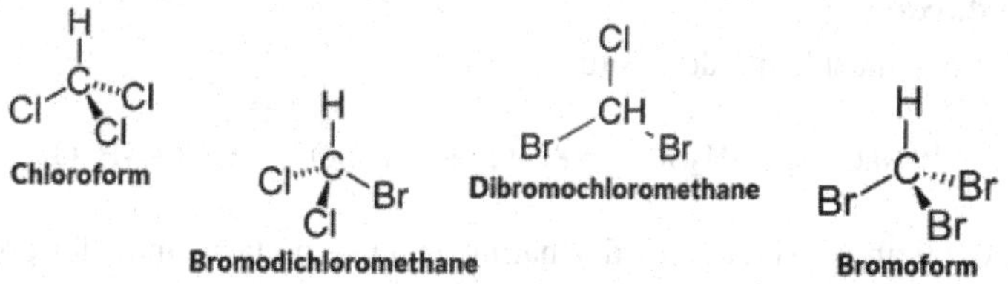

FIGURE 24.1 Disinfection by-products in potable water.

- Taste and odor problems, notably from chlorinated phenols are:

Substituted phenol	Odor threshold, ppb
None	>1000
2–	2
4–	250
2,4–	2
2,6–	3
2,4,6–	>1000

- Chlorination by-products, notably trihalomethanes. Chloroform ($CHCl_3$, Figure 24.1) for example, is often present at 10 ppb or more. The source is natural substances (humic acids (C=O–O–H)):

$$CH_4 + 2Cl_2 \rightarrow + CHCl_3 + HCl + H_2 \qquad (24.1.1)$$

- But $CHCl_3$ is suggested to be weakly carcinogenic according to epidemiology studies (Reuber 1979).
- High repeated doses of chloroform → limited evidence for liver and kidney tumors.
- Chloroform ($CHCl_3$): formerly used as anesthetic and cough suppressant.
- The CDC (2005) has compiled a summary of chloroform carcinogenicity.
- Phosgene hydrolyzes to CO_2 (major route), but also binds to cellular macromolecules and to glutathione (detoxification mechanism).

$$CHCl_3 \rightarrow [HO-CCl_3] \rightarrow O = CCl_2 \,(\text{unstable phosgene}) \qquad (24.1.2)$$

24.2 DECREASING THE THM CONTENT OF CHLORINATED WATER

The trihalomethane (THM) content of chlorinated water could be decreased by using activated carbon either to remove dissolved organic compounds before the water is chlorinated or to remove THMs and other chlorinated organics after the process, although THMs are not efficiently absorbed by the carbon and it is a costly process (Baird and Cann 2013).

24.3 ALTERNATIVES TO CHLORINATION

Ozone, chlorine dioxide, UV irradiation, chloramines, chlorine dioxide

Chlorine dioxide

- Unstable, must be made in situ

$$10NaClO_2 + 5H_2SO_4 \rightarrow 8ClO_2 + 5Na_2SO_4 + 2HCl + 4H_2O \qquad (24.3.1)$$

- An oxidizing agent, not a chlorinating agent → no taste and odor problems

$$ClO_2 + substrate \rightarrow ClO_2 + (substrate) + \qquad (24.3.2)$$

- Can be used as a temporary expedient when taste and odor problems occur – purchase sodium chlorite – storage issues
- No residual effect – rapidly decomposes – must add Cl_2 afterward (when the T&O compounds have been destroyed)
- Some questions about its effectiveness versus Giardia and Cryptosporidium
- Some issues concerning toxicity (Baird 2013)

Ozone (O_3) – a gaseous disinfectant

- Used for greater than 90 years in Europe
- Produced by passing very high electric current into the air (very expensive to produce)
- Very unstable and cannot be stored
- Leaves no measurable residual in water, so some chlorine must be used as a precaution
- Does not produce THM (carcinogen)
- Stronger than chlorine
- Can assist as a coagulant when used with alum (aluminum sulfate – a coagulant), reducing the amount of chemicals needed to adjust pH
- Aids in the filtration process as a coagulant
- Oxidizes and precipitates iron, sulfur, and manganese, so they can be filtered out of the solution
- Ozone treatment creates its own by-product that can be harmful to health if they are not controlled (e.g., formaldehyde and bromate).
- Ozone will oxidize and break down many organic chemicals as well.

Ultraviolet radiation (UV)

- An electromagnetic radiation just beyond the blue end of the light spectrum
- Absorbed by genetic material in microorganisms interfering with their reproduction
- Mercury lamps are best suited for large fractions of UV energy that are needed for disinfection.
- Very unstable and cannot be stored
- Does not involve any chemical handling
- Leaves no residual in water

24.4 AVOIDING THM

THM (TRIHALOMETHANES)

A carcinogenic compound produced by the decay of an organic substance such as vegetation in chlorine (e.g., chloroform).

NB: The lower the pH, the more effective the chlorine disinfection.

Chloramines

Compounds formed from the reaction of ammonia and chlorine (or combined chlorine) (used in swimming pools). They

- are a slower disinfectant but last longer and
- have an objectionable taste and odor.

Free available chlorine

- The radicals HOCL (hypochlorous acid) and OCL (hypochlorite radical). They are a faster type of disinfectant.
- Breakpoint chlorination is when enough chlorine is added to satisfy the chlorine demand and to react with all the dissolved ammonia.

24.5 MONITORING CHLORINE

Chlorine residual

- #1. After filtration, extra chlorine travels with the water through the underground pipes that bring the water to homes. If there are any bacteria in the pipes, they will be killed by the remaining chlorine. The chlorine residual is monitored to make sure it's high enough to kill any bacteria it encounters on the way to homes but low enough to be safe and to reduce odor and taste problems.
- #2 Turbidimeter – this meter displays numbers that identify the turbidity level of the water, thereby measuring the cloudiness of the water. At this stage of the treatment process, the water should be very clear and clean; however,

Though non-clay minerals are chemically rather inert, clays (< 2 μm) are very active in the adsorption of dissolved compounds due to their large specific surface area and their specific chemical layer structure. Clays adsorb organic compounds. Aquatic sediments commonly form the largest reservoir for micropollutants. Micropollutants in aquatic ecosystems have a relatively large tendency to be adsorbed onto suspended matter and sediments.

(USGS 2018)

- Trihalomethanes adhere to organic matter trapped in suspended particles. The elimination of such suspended particles is a major objective in the purification of water using chlorine.

24.6   AVOIDING AND TREATING THM CONTAMINATED WATER

In 1974, trihalomethanes (chloroform – $CHCl_3$, bromodichloromethane – $CHBrCl_2$, dibromochloromethane – $CHBr_2Cl$, and bromoform – CBr_4) were discovered to be formed during the disinfection step of drinking water if free chlorine was the disinfectant (EPA 2022). Together with the perceived hazard to the consumer's health, this led the USEPA to amend the National Interim Primary Drinking Water Regulations to include a maximum contaminant level of 0.10 mg $\cdot$ L^{-1} for total trihalomethanes (EPA 2022).

PROBLEM 24.6.1
Treatment approaches: pros and cons

What are the drawbacks and advantages of various remediation treatments for THMs in potable water?

SOLUTION 24.6.1

The EPA (2022) puts forth the following correctives:

- For trihalomethane removal, aeration – either by diffused air or with towers – and adsorption – either by powdered activated carbon or granular activated carbon – are effective.
- The major disadvantage of this approach is that trihalomethane precursors are not removed by aeration. For trihalomethane precursor control, effective processes are: (1) oxidation by ozone or chlorine dioxide; (2) clarification by coagulation, settling and filtration, precipitative softening, or direct filtration; or (3) adsorption by powdered activated carbon or granular activated carbon.
- In addition, some modest removal or destruction of trihalomethane precursors can be achieved by oxidation with potassium permanganate lowering the pH or moving the point of chlorination to the clarified water.
- Lowering trihalomethane precursor concentrations has the additional advantage of reducing overall disinfectant demand, thereby reducing the possibility of the formation of all disinfection by-products. Chlorine dioxide, ozone, and chloramines do not produce trihalomethanes at significant concentrations when used alone as disinfectants.
- Furthermore, the cost of any of these unit processes is very low. A major disadvantage of using alternate disinfectants for trihalomethane control relates to the lack of any precursor removal.

24.7   CHLORINE RESIDUALS

In adding chlorine, the main goal is to disinfect the water and maintain enough chlorine in the treated water as it travels through the distribution system. As pathogens may regrow where distribution systems are a long distance from the storage tanks or where water movement slows, or stops where water is not used, sufficient remaining chlorine needs to be left in the system. This chlorine residual maintains the integrity of the treatment in the water sent out for consumption. Otherwise, poor water quality results, as well as slime and biofilm growth in the distribution system, which will eventually contaminate the clean, treated water being distributed.

PROBLEM 24.7.1
Residual dosage

What factors determine the chlorine residual for a water distribution system?

SOLUTION 24.7.1

The residual amount of chlorine, determined by testing, is that portion remaining after the demand is satisfied. As chlorine destruction of pathogens takes time, residual becomes lower as time passes after the dosage. After all the demand is satisfied, the free residual remaining needs to be at least 0.2–0.4 ppm (mg · L^{-1}), a concentration that normally provides a high degree of assurance that the disinfection of the water is complete.

PROBLEM 24.7.2
Chloramines as disinfectants

What are the advantages of having chloramines in water?

SOLUTION 24.7.2

Chloramines act as secondary disinfectants, most formed when ammonia is added to chlorine to treat drinking water. By combining free chlorine with nitrogen compounds (to form chloramines), they provide longer lasting disinfection as the water moves through pipes to consumers. Combined residual results.

PROBLEM 24.7.3
Deducing the chlorinator setting

What should the chlorinator setting be as lb · day^{-1} to treat a flow of 1.88 MGD if the chlorine demand is 2.1 mg · L^{-1} and a chlorine residual of 0.6 mg · L^{-1} is desired?

SOLUTION 24.7.3

First, calculate the chlorine dosage in mg · L^{-1}:

Chlorine dose, mg · L^{-1} = Chlorine demand, mg · L^{-1} + Chlorine residual, mg · L^{-1}

Chlorine dose, mg · L^{-1} = 2.1 mg · L^{-1} + 0.6 mg · L^{-1}

Chlorine dose, mg · L^{-1} = 2.7 mg · L^{-1}

Now calculate the chlorine feed rate in $lb \cdot day^{-1}$ (at a dose of 2.7 mg $\cdot$ L^{-1}):

$Chlorine, lb \cdot day^{-1} = Chlorine, mg \cdot L^{-1} \times Flow, MGD \times 8.34 \ lb \cdot gal^{-1}$

$Chlorine, lb \cdot day^{-1} = 2.7 \ mg \cdot L^{-1} \times 1.88 \ MGD \times 8.34 \ lb \cdot gal^{-1}$

$Chlorine, lb \cdot day^{-1} = 42.3 \ lb \cdot day^{-1}$

$Chlorine \ dose, mg \cdot L^{-1} = Chlorine \ demand, \ mg \cdot L^{-1} + Chlorine \ residual, \ mg \cdot L^{-1}$

$Chlorine \ dose, mg \cdot L^{-1} = 1.5 \ mg \cdot L^{-1} + 0.6 \ mg \cdot L^{-1}$

$Chlorine \ dose, mg \cdot L^{-1} = 2.1 \ mg \cdot L^{-1}$

PROBLEM 24.7.4
Chlorine demand

The chlorine dosage for a water distribution system is 2.1 mg $\cdot$ L^{-1}. If the chlorine residual after 30 min of contact time is found to be 0.6 mg $\cdot$ L^{-1}, what is the chlorine demand in mg $\cdot$ L^{-1}?

SOLUTION 24.7.4

$Chlorine \ demand, mg \cdot L^{-1} = Chlorine \ dose, mg \cdot L^{-1} - Chlorine \ residual, mg \cdot L^{-1}$

$Chlorine \ demand, mg \cdot L^{-1} = 2.1 \ mg \cdot L^{-1} - 0.6 \ mg \cdot L^{-1}$

$Chlorine \ demand, mg \cdot L^{-1} = 1.5 \ mg \cdot L^{-1}$

PROBLEM 24.7.5
Chlorinator setting

Chlorinators operate independently after being calibrated to a desired setting. What should the chlorinator setting be (i.e., $lb \cdot day^{-1}$) to treat a flow of 1.65 MGD if the chlorine demand is 1.9 mg $\cdot$ L^{-1} and a chlorine residual of 0.3 mg $\cdot$ L^{-1} is desired?

SOLUTION 24.7.5

First, calculate the chlorine dosage in mg $\cdot$ L^{-1}:

$Chlorine \ dose, mg \cdot L^{-1} = Chlorine \ demand, \ mg \cdot L^{-1} + Chlorine \ residual, \ mg \cdot L^{-1}$

$Chlorine \ dose, mg \cdot L^{-1} = 1.9 \ mg \cdot L^{-1} + 0.3 \ mg \cdot L^{-1}$

$Chlorine \ dose, mg \cdot L^{-1} = 2.0 \ mg \cdot L^{-1}$

Then calculate the chlorine feed rate in $lb \cdot day^{-1}$ (at a dose of 2.0 mg $\cdot$ L^{-1}):

$Chlorine, lb \cdot day^{-1} = Chlorine, mg \cdot L^{-1} \times Flow, MGD \times 8.34 \ lb \cdot gal^{-1}$

$Chlorine, lb \cdot day^{-1} = 2.0 \ mg \cdot L^{-1} \times 1.65 \ MGD \times 8.34 \ lb \cdot gal^{-1}$

$Chlorine, lb \cdot day^{-1} = 27.52 \ lb \cdot day^{-1}$

PROBLEM 24.7.6
Dosage and setting

The chlorinator setting is 38.3 lb · day^{-1} for a flow of 1.71 MGD. What is the chlorine dosage in mg · L^{-1}?

SOLUTION 24.7.6

$$\text{Chlorine,lb} \cdot \text{day}^{-1} = \text{Chlorine,mg} \cdot \text{L}^{-1} \times \text{Flow,MGD} \times 8.34 \text{ lb} \cdot \text{gal}^{-1}$$

To get dosage alone on the left side, transpose the equation.
Or divide both sides by: (flow, MGD × 8.34 lb · gal^{-1}) to cancel it on the right to give:

$$\left(\text{Chlorine lb} \cdot \text{day}^{-1}\right) / \text{Flow,MGD} \times 8.34 \text{ lb} \cdot \text{gal}^{-1} = \text{Chlorine,mg} \cdot \text{L}^{-1}$$

$$\text{Chlorine,mg} \cdot \text{L}^{-1} = \left(38.3 \text{ lb} \cdot \text{day}^{-1}\right) / 1.71 \text{ MGD} \times 8.34 \text{ lb} \cdot \text{gal}^{-1}$$

$$\text{Chlorine,mg} \cdot \text{L}^{-1} = 2.68 \text{ mg} \cdot \text{L}^{-1}$$

24.7.1   BREAKPOINT CHLORINATION

Chlorine reacts immediately when put in water, combining with some chemical species, and oxidizing other chemicals, processes which reduce its disinfecting rate and power. Hence, residual chlorine will be detectable in the water, but in combined form with a weak disinfecting power. Yet, adding more chlorine further decreases the chlorine residual because additional chlorine reacts with the combined chlorine compounds. Further addition (i.e., after the chlorine compounds have been used up) eventually produces a free residual chlorine because free chlorine begins a stage of excess. This point, marking the destruction of most of the combined compounds and the free chlorine starting to form, is the breakpoint, which can be determined only by experimentation.

To determine breakpoint chlorination, we compare the expected increase in residual with the actual increase in residual. The increase in chlorine dose, lb · day^{-1}, indicates the expected increase in residual. If the water is being chlorinated beyond the breakpoint, then any increase in chlorine dose will result in a corresponding increase in chlorine residual. The mg · L^{-1} to lb · day^{-1} equation helps in calculating the expected increase in residual that would result from an increase in the chlorine dose as follows:
Expected increase in residual:

$$\text{Increase in chlorine dose} \left(\text{lb} \cdot \text{day}^{-1}\right)$$
$$= \text{Expected increase, mg} \cdot \text{L}^{-1} \times \text{Flow,MGD} \times 8.34 \text{ lb} \cdot \text{gal}^{-1}$$

Actual increase in residual:

$$\text{Actual increase} \left(\text{mg} \cdot \text{L}^{-1}\right) = \text{New residual} \left(\text{mg} \cdot \text{L}^{-1}\right) - \text{Old residual,mg} \cdot \text{L}^{-1}$$

PROBLEM 24.7.7
Breakpoint chlorination

The chlorinator setting is increased by 1.5 lb · day^{-1}. The chlorine residual before the increased dosage was 0.4 mg · L^{-1}. After the increased chlorine dose, the chlorine residual was 0.5 mg · L^{-1}. The average flow rate being chlorinated is 1.37 MGD. Is the water being chlorinated beyond the breakpoint?

SOLUTION 24.7.7

First, we calculate the expected increase in chlorine residual:

$$\text{Increase in chlorine dose, lb} \cdot \text{day}^{-1}$$
$$= \text{Expected Increase, mg} \cdot \text{L}^{-1} \times \text{Flow, MGD} \times 8.34 \text{ lb} \cdot \text{gal}^{-1}$$
$$1.5 \text{ lb} \cdot \text{day}^{-1} = \text{Expected Increase, mg} \cdot \text{L}^{-1} \times 1.37 \text{ MGD} \times 8.34 \text{ lb} \cdot \text{gal}^{-1}$$

Dividing both sides by (1.37 MGD × 8.34 lb · gal^{-1}) cancels it out on the right, bringing it to the left:

$$(1.5 \text{ lb} \cdot \text{day}^{-1}) / 1.37 \times 8.34 \text{ lb} \cdot \text{gal}^{-1} = \text{Expected increase (mg} \cdot \text{L}^{-1})$$
$$\text{Expected increase (mg} \cdot \text{L}^{-1}) = 0.13 \text{ mg} \cdot \text{L}^{-1}$$

Then, calculate the actual increase in residual:

$$\text{Actual increase, mg} \cdot \text{L}^{-1} = \text{New residual, mg} \cdot \text{L}^{-1} - \text{Old residual, mg} \cdot \text{L}^{-1}$$
$$\text{Actual increase, mg} \cdot \text{L}^{-1} = 0.5 \text{ mg} \cdot \text{L}^{-1} - 0.4 \text{ mg} \cdot \text{L}^{-1}$$
$$\text{Actual increase, mg} \cdot \text{L}^{-1} = 0.1 \text{ mg} \cdot \text{L}^{-1}$$

As we needed an increase of 0.1 mg · L^{-1} (expected) to fulfill the chlorine demand, the answer is no, it is not being chlorinated past the breakpoint because it dropped to 0.1 mg · L^{-1} in chlorine residual. This means there is no free available chlorine residual, and we are now below breakpoint chlorination.

24.8   USING HYPOCHLORITE

24.8.1   CALCULATING DRY HYPOCHLORITE FEED RATE

Calcium hypochlorite is a white solid that contains 65% available chlorine and dissolves easily in water, having the following attributes:

- Very stable and can be stored for an extended period.
- Corrosive with a strong odor.
- Reactions with organic material can generate enough heat to cause a fire or explosion.
- Must be kept away from organic materials such as wood, cloth, and petroleum products.
- Readily absorbs moisture, forming chlorine gas.
- The most used dry hypochlorite, calcium hypochlorite, contains about 65–70% available chlorine, depending on the brand.
- Not 100% pure chlorine, more must be fed into the system to obtain the same amount of chlorine for disinfection.

PROBLEM 24.8.1
Dosage of hypochlorite

A chlorine dose of 7.6 mg · L^{-1} is required to disinfect a flow of 1.6 MGD. If the calcium hypochlorite to be used contains 68% available chlorine, how many pounds per day of hypochlorite will be required for disinfection?

SOLUTION 24.8.1

The following equation allows determination of the pounds per day of hypochlorite required:

$$\text{Feed rate (lb · day}^{-1}) = [\text{Dosage (mg · L}^{-1}) \times \text{Flow (MGD · day}^{-1}) \times 8.34 \text{ gal}] /$$
$$\text{Purity \% (expressed as a decimal)} \tag{24.8.1}$$

$$\text{Feed rate (lb · day}^{-1}) = [7.6 \text{mg · L}^{-1}) \times \text{Flow (1.6 MGD · day}^{-1}) \times 8.34 \text{ gal}] / 0.68\%$$

$$\text{Feed rate (lb · day}^{-1}) = [7.6 \text{ mg · L}^{-1} \times 1.6 \text{ MGD} \times 8.34 \text{ gal · day}^{-1}] / 0.68$$

$$\text{Feed rate (lb · day}^{-1}) = 150.21 \text{ lb · day}^{-1}$$

PROBLEM 24.8.2
Available chlorine in calcium hypochlorite

A tank contains 390,000 gallons of water and is to receive a chlorine dose of 1.6 mg · L^{-1}. How many pounds of calcium hypochlorite (66% available chlorine) will be required?

SOLUTION 24.8.2

$$\text{Feed rate (lb · day}^{-1}) = [\text{Dosage } (\text{mg · L}^{-1}) \times \text{Volume MG} \times 8.4 \text{ gal}] /$$
$$\text{Purity \% (decimal)}$$

$$\text{Feed rate (lb · day}^{-1}) = [1.6 \text{ mg · L}^{-1} \times 0.39 \text{ MG} \times 8.4 \text{ gal}] / 0.66$$

$$\text{Feed rate (lb · day}^{-1}) = 4.96 \text{ lb · day}^{-1}$$

PROBLEM 24.8.3
Gallons per day (gpd) of hypochlorite

A total of 34 lbs of calcium hypochlorite (67% available chlorine) is used in a day. If the flow rate treated is 3,500,000 gpd, what is the chlorine dosage in mg · L^{-1}?

SOLUTION 24.8.3

$$\text{Feed rate (lb · day}^{-1}) = [\text{Dosage (mg · L}^{-1}) \times \text{Volume MG} \times 8.34 \text{ gal}] /$$
$$\text{Purity \% (decimal)}$$

$$34 \text{ lbs · day}^{-1} = [(x \text{ mg · L}^{-1}) (3.5 \text{ MGD}) (8.34 \text{ lb · gal}^{-1})] / 0.67$$

First multiply both sides by 0.67 to cancel it out on the right, leaving:

$$(34 \text{ lbs} \cdot \text{day}^{-1}) (0.67) = x \text{ mg} \cdot \text{L}^{-1} (3.5 \text{ MGD}) (8.34 \text{ lb} \cdot \text{gal}^{-1})$$

Continuing in like manner yields:

$$[(34 \text{ lbs} \cdot \text{day}^{-1}) (0.67)] / (3.5 \text{ MGD}) (8.34 \text{ lb} \cdot \text{gal}^{-1}) = x \text{ mg} \cdot \text{L}^{-1}$$
$$= 0.779$$
$$x \text{ mg} \cdot \text{L}^{-1} = 0.79 \text{ mg} \cdot \text{L}^{-1}$$

24.8.2   CALCULATING HYPOCHLORITE SOLUTION FEED RATE

Bleach, which is liquid hypochlorite (sodium hypochlorite), usually exists as a clear, greenish-yellow liquid in strengths from 5.25% to 16% available chlorine. When used for common household bleaching, it contains 5.25% available chlorine. The following formula helps when calculating the chemical feed pump setting in mL $\cdot$ min^{-1}:

Chemical feed pump setting (mL $\cdot$ min^{-1}):

$$= [(\text{Dosage, mg} \cdot \text{L}^{-1})(\text{Flow, MGD})(3.785 \text{ L} \cdot \text{gal}^{-1})(1,000,000 \text{ gal} \cdot \text{MG}^{-1})] /$$
$$(\text{Feed chemical density}(\text{mg} \cdot \text{mL}^{-1})(1440 \text{ min} \cdot \text{day}^{-1})$$

PROBLEM 24.8.4
Sodium hypochlorite

A hypo-chlorinator dispenses sodium hypochlorite (NaClO) to disinfect the water pumped from a well. For adequate disinfection, a chlorine dose of 1.92 mg $\cdot$ L^{-1} is required throughout the system. What will the feed pump setting need to be in mL $\cdot$ min^{-1} if the flow being treated is 1.18 MGD? [Hint: The typical density of sodium hypochlorite is 138 mg $\cdot$ mL^{-1}.]

SOLUTION 24.8.4

Chemical feed pump setting (mL/min):

$= [(\text{dosage, mg} \cdot \text{L}^{-1}) (\text{flow, MGD}) (3.785 \text{ L} \cdot \text{gal}^{-1}) (10^6 \text{ gal} \cdot \text{MG}^{-1})]/[\text{feed chem. density } (\text{mg} \cdot \text{L}^{-1}) (1440 \text{ min} \cdot \text{day}^{-1})]$

$= [(1.92 \text{ mg} \cdot \text{L}^{-1}) (1.18 \text{ MGD}) (3.785 \text{ L} \cdot \text{gal}^{-1}) (10^6 \text{ gal} \cdot \text{MG}^{-1})]/[\text{feed chem. density } (138 \text{ mg} \cdot \text{mL}^{-1}) (1440 \text{ min} \cdot \text{day}^{-1})]$

$= [(1.92 \text{ mg} \cdot \text{L}^{-1}) (1.18 \text{ MGD}) (3.785 \text{ L} \cdot \text{gal}^{-1}) (10^6 \text{ gal} \cdot \text{MG}^{-1})]/[\text{feed chem. density } (138 \text{ mg} \cdot \text{mL}^{-1}) (1440 \text{ min} \cdot \text{day}^{-1})]$

Chemical feed pump setting (mL $\cdot$ min^{-1}) = 43.15 mL $\cdot$ min^{-1}

PROBLEM 24.8.5
Effect of hypochlorite strength

If the sodium hypochlorite in the above question had a strength of 5.25%, determine the feed setting in mL $\cdot$ min^{-1}.

SOLUTION 24.8.5

Chemical feed pump setting (mL · min^{-1}):

> = [Dosage (mg · L^{-1}) · Flow (m^{-3} · day^{-1})]/(Feed chemical density, g · cm^{-3}) ·
> Active Chemical % (decimal) (1400 min · day^{-1})

We must convert first to the required units in the formula:
Convert flow (MGD) to m^3/day^{-1} with 1 m^3 = 264.2 gal:

$$=\left[\left(1.18\ MG^{-1} \cdot day^{-1}\right) \times \left(10^6\ gal\ /\ 1\ MG\right)\left(1\ m^3\ /\ 264.2\ gal\right)\right] = 4466.3\ m^3 \cdot day^{-1}$$

Density of sodium hypochlorite is 138 mg · mL^{-1}, but the equation requires the density in g/cm^3; it is converted to:

$$(138\ mg \cdot mL^{-1}) \times (1g\ /\ 1000\ mg) \times 1mL\ /\ 1\ cm^3 = 0.138\ g \cdot cm^{-3}$$

Note: There is no difference in volume between mL and cm^3. The primary difference is that milliliters are used for fluid amounts while cubic centimeters are used for solids. Hence, converting between them is 1:1 as above.

Now we can place the values into the formula:
Chemical feed pump setting (mL · min^{-1}):

> = [Dosage (mg · L^{-1}) · Flow (m^3 · day^{-1})] / (Feed chemical density, g · cm^{-3}) ·
> Active Chemical % (decimal) (1400 min · day^{-1})

Chemical feed pump setting (mL · min^{-1}):

= (= 1.92 mg · L^{-1}) (4466.31 m^3 · day^{-1})]/(0.138 g · cm^{-3}) (0.0525) (1440 min · day^{-1})

Chemical Feed Pump Setting = 821.96 (mL · min^{-1})

It can therefore be seen that, compared with the previous problem, the feed pump setting is much higher when the strength of the hypochlorite is only 5.25%.

24.8.3 CALCULATING THE STRENGTH OF SOLUTIONS

The composition of a solution varies within certain limits, such as that between the solid form (calcium hypochlorite) and the liquid form (sodium hypochlorite) of chlorine. Either of two formulas can facilitate the calculation of the percent strength of the hypochlorite:

Hypochlorite strength, %

= Chlorine required, lb / (Hypochlorite solution needed, gal) $\left(8.34\ lb \cdot gal^{-1}\right) \times 100\%$

or

Hypochlorite strength, %

$= \left[(\text{Chlorine required, kg})(100) / (\text{Hypochlorite solution needed, kg})\right] \times 100\%$

*Note: To convert from kg to lbs, use (1 kg / 2.20 lb). For the obverse, use (1 lb / 0.453 kg).

PROBLEM 24.8.6
Unit conversions

If we add 82 ounces of calcium hypochlorite (66% available chlorine) to 32 gallons of water, what is the percent chlorine strength, by weight, of the solution?

SOLUTION 24.8.6

To use the first equation, we convert ounces to pounds of calcium hypochlorite:

$$82 \text{ oz} \times (1 \text{ lb} / 16 \text{ oz}) = 5.12 \text{ lb}$$

The hypochlorite only has 66% available chlorine, i.e.,

$$5.12 \text{ lb} \times 0.66 = 3.38 \text{ lb}$$

For the solution (solute + solvent) needed, add the calcium hypochlorite (3.38 lb) to the amount of water (32 gal) by first converting pounds of hypochlorite to gallons:

$$3.38 \text{ lb} \times (1 \text{ gal} / 8.34 \text{ lb}) = 0.40 \text{ gal}$$

$$32 \text{ gal} + 0.40 \text{ gal} = 32.40 \text{ gal}$$

We can now put in the values:

Hypochlorite strength, %

$$= [\text{Chlorine required (lb)} / \text{Hypochlorite solution needed (gal} \times 8.34 \text{ lb} \cdot \text{gal}^{-1})] \times 100$$

$$\text{Hypochlorite strength, } \% = [3.38 \text{ lb} / 32.40 \text{ gal} \times 8.34 \text{ lb} \cdot \text{gal}^{-1}] \times 100$$

$$\text{Hypochlorite strength, } \% = 1.25\%$$

PROBLEM 24.8.7
Dissociation of hypochlorous acid

Being a weak acid, hypochlorous acid (ClOH, HClO, HOCl, or ClHO, forms when chlorine dissolves in water, partially dissociating and forming hypochlorite, ClO^-. To 18 mg of HOCl in a volumetric flask, water is added to make a solution up to the 1.0 L mark. The final pH was observed to be 7.2. What is the concentration of HOCl and OCl? On the assumption that the temperature was 25°C, what percentage of the HOCl is dissociated?

SOLUTION 24.8.7

The dissociation reaction for HOCl is represented as:

$$\text{HOCL} = \text{H}^+ + \text{OCl}^-$$

The pK_a for HOCl is 7.54, and

$$K_a = 10^{-7.54} = 2.876 \times 10^{-8}$$

Converting the equilibrium expression and substituting the concentration of H+ ions:

$$K_a = [H^+][OCL^-] / [HOCL] = [1.00 \times 10^{-7.2}][OCL^-] / [HOCl] = 2.876 \times 10^{-8}$$

$$\text{Therefore, HOCL} = [1.00 \times 10^{-7.2}][OCL^-] / 2.876 \times 10^{-8}$$

$$= [1 / 10^{7.2}][OCL^-] / 2.876 \times 10^{-8} = 0.000,000,06309 / 0.000,000,0286$$

$$[HOCl] = 2.1936 [OCL^-]$$

Since at equilibrium the combined concentration of HOCl and OCl- must equal that which was added initially,

$$[HOCl] + [OCL^-] = \text{Molar concentration added}$$

Therefore, as HOCl was added, it is necessary to use the molecular weight of HOCl to calculate the molar concentration.

$$\text{Molar concentration} = \left(18 \text{ mg} \cdot L^{-1}\right)\left(10^{-3} \text{ g} \cdot mg^{-1}\right)(1 \text{ mol} / 52.461 \text{ g}) = 3.43 \times 10^{-4} \text{ M}$$

Thus,

$$[HOCl] + [OCL^-] = 3.43 \times 10^{-4} M$$

A substitution from the equation for HOCl concentration yields:

$$2.1936[OCL^-] + [OCl^-] = 3.43 \times 10^{-4} \text{ M}$$

$$\text{Hence, } 2.1936 [OCl^-] + 1[OCl^-](\text{i.e.}, 3.1936 [OCl^-]) = 3.43 \times 10^{-4} \text{ M}$$

$$\text{Therefore, } [OCl^-] = 3.43 \times 10^{-4} \text{ M} / 3.1936$$

$$[OCl^-] = 1.07 \times 10^{-4} \text{ M}$$

To find the concentration of HOCl, either multiply this [OCl-] concentration by 2.1936 or subtract this concentration from 3.43×10^{-4} M.

$$[HOCl] = 3.43 \times 10^{-4} \text{ M} - 1.07 \times 10^{-4} \text{ M} = 2.36 \times 10^{-4} \text{ M}$$

Calculating the percentage of OCl- that is dissociated produces:

$$[OCl^-] / [HOCl] + [OCl^-] = 1.07 \times 10^{-4} \text{ M} / 3.43 \times 10^{-4} \text{ M}$$

$$[OCl^-] = 31.19\%$$

REFERENCES

Baird Colin and Michael Cann (2013) Environmental Chemistry. WH Freeman & Company, New York.

CDC (2005) Chloroform (Trichloromethane) (2005-110) l NIOSH l CDC. www.cdc.gov

EPA (2022) Removing trihalomethanes from drinking water – an overview of treatment techniques. https://cfpub.epa.gov/si/si_public_record_report.cfm?Lab=NRMRL&direntry id=29716#:~:text=For%20trihalomethane%20precursor%20control%2C%20effec tive%20processes%20are%3A%20%281%29,by%20powdered%20activated%20 carbon%20or%20granular%20activated%20carbon.

Reuber MD (1979) Carcinogenicity of chloroform. Environ Health Perspect 31, 171–182. doi: 10.1289/ehp.7931171.

USGS (2018) Effects of turbidity, sediment, and polyacrylamide on native freshwater mussels. https://www.usgs.gov/publications/effects-turbidity-sediment-and-polyacrylamide-native-freshwater-mussels

25 Water
Eutrophication

MAIN POINTS

- Dense algal and plant growth such as surface matting (years or decades)
- Nutrients mainly from animal wastes, fertilizers, and sewage
- Slower type (thousands of years) as lakes grow old and get filled with sediments

25.1 THE PHOSPHORUS PROBLEM

Phosphorus is a major element found in sewage that leads to the eutrophication of large water bodies, thereby enriching the water bodies with nutrients from outside the ecosystem. In animal manures, the ratio of phosphorus to nitrogen (P:N) is typically higher than the required amount of P. Also, the high amount of N in such manures means that delivery of no more than the correct plant needs for N exclusively from animal manures leads to a buildup of excess P over time in agricultural soils. Ferric chloride is used in wastewater treatment plants to precipitate phosphorus. Ferric chloride mainly converts unstable exchangeable phosphorus (Ex-P), ferric phosphate (Fe-P), and organic phosphorus (Or-P) into more stable occluded phosphate (O-P), reducing the possible release of phosphorus from sediments (Li et al. 2020).

PROBLEM 25.1.1
Wastewater and orthophosphate

If wastewater has a soluble orthophosphate concentration of 6 mg $\cdot$ L^{-1} as P, what theoretical dosage of ferric chloride would be required to remove it completely?

SOLUTION 25.1.1

The precipitation reaction which removes a mole of phosphorus requires 1 mole of ferric chloride:

$$FeCl_3 + HPO_4 \rightarrow FePO_4 + 3HCl \qquad (25.1.1)$$
$$56 + 3(35.5) \rightarrow 56 + 31 + 64$$

31 g of phosphorus is precipitated by 162.5 g of ferric chloride.
1 g of P is precipitated by 162.5 g / 31 g of ferric chloride.
Therefore, 6 mg P is removed by 6 (162.5 / 31) mg of ferric chloride.
Answer = 31.45 mg of FeCl$_3$

DOI: 10.1201/9781003341826-30

25.2   COST OF EUTROPHICATION

The process is not only a threat for the viability of many ecosystems, but also constitutes a drain on financial resources. For example,

- Eutrophic water bodies no longer contribute to the production of drinking water.
- Algae overrun touristic zones.
- Expensive measures are required to protect endangered species.
- Freshwater ecosystem losses are estimated at around 2.2 billion dollars per year (1.7 billion EUR) in the freshwater ecosystems of the United States alone.
- The decline of kelp forests along the Norwegian shores due to invasion of filamentous epiphytic algae led to the loss of around 50,000 tons of fish stocks.

25.3   ROLE OF AGRICULTURAL CHEMICALS

Most agricultural chemicals are water-soluble nitrates and phosphates that are applied to fields, lawns, and gardens to stimulate the growth of crops, grass, and flowers, but when not used by the plants, those nutrients can enter streams and lakes during runoff or leaching events. If they reach a body of water, such nutrients continue to promote the growth of plants. The resulting plant detritus is food for microorganisms, and as the population of such organisms grows, the supply of oxygen in the water is depleted.

Oxygen reaches water either through the atmosphere or from the process of photosynthesis carried out by many aquatic green plants during daylight. The other major water pollutant is organic matter such as leaves, grass, trash etc., polluting water because of runoff. Even a moderate amount of organic matter when it decomposes in water can deplete the water of its dissolved oxygen (Davis and Masten 2020).

The bacteria responsible for degrading biodegradable detergent feed on it and grow rapidly. While growing, they may use up all the oxygen dissolved in water, leading to the lack of oxygen which kills all other forms of aquatic life such as fish and plants. Thus, bloom-infested water inhibits the growth of other living organisms in the water body.

PRACTICE PROBLEM 25.3.1
Pollution remediation

For what kind of water pollution would each of the following processes be most effective: cation exchange, anion exchange, filtration, demineralization, disinfection, oxidation, pH adjustment, adsorption?

25.4   CULTURAL EUTROPHICATION

When nutrient enrichment arises from human activities, it is called cultural eutrophication (Baird and Cann 2013).

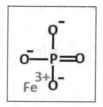

FIGURE 25.1 Iron III phosphate.

PROBLEM 25.4.1
Adsorption of phosphates

P is rapidly removed from solution by algal-bacterial uptake or by adsorption to sediments (Ruzycki 2004). Algal blooms have reduced water quality in a lake located in a watershed dominated by lateritic soils. What is the likely cause, and how could high quality water be restored?

SOLUTION 25.4.1

High levels of iron and aluminum dominate lateritic soils.

Phosphate, or PO_4^{3-}, is strongly absorbed to cationic sites (Al^{3+}, Fe^{3+}, Ca^{2+}). Hence, ferric (+3) produces iron phosphate ($FePO_4$), an insoluble floc (Figure 25.1). But if the ferric iron is reduced to the ferrous state (Fe^{2+}), phosphate is released. Mixing from anoxic bottom waters with high phosphate levels is closely tied to iron redox reactions (Ruzycki 2004).

- When $O_2 > 1 \ mg \cdot L^{-1}$ – insoluble ferric (+3) salts form that precipitate and settle out, adsorbing PO_4^{3-}.
- When $O_2 < 1 \ mg \cdot L^{-1}$ (anoxic) – ferric ion becomes reduced to soluble ferrous ion (Fe^{2+}) – allowing phosphate from sediments to diffuse up into the water.

Wind mixing (storms and fall de-stratification) can re-inject high P water to the surface, causing algal blooms (Ruzycki 2004). But, as ferrous (+2) is soluble, iron II phosphate $[Fe_3(PO_4)_2]$ releases phosphate in water, unless it reacts with sulfide, causing FeS to precipitate (Ruzycki 2004). Therefore, soluble PO_4^{3-} concentration is strongly affected by iron redox (i.e., reducing or oxidizing) reactions, and reducing conditions add electrons (i.e., making it less positive, ferrous = Fe^{2+}), thereby releasing PO_4^{3-} in water. A common source of soluble P is anoxic bottom sediments. Removing such anoxic sediments by dredging could decrease or eliminate algal blooms.

PROBLEM 25.4.2
Determination of phosphate sources

Not all lakes have enormous releases of phosphate from the sediment interstitial (pore) water under anoxia because there may not be much iron in the system. If phosphorus causing algal water degradation did not come from sediments in the water body, how can its most probable source(s) be determined?

SOLUTION 25.4.2

First, sample the possible sources.

Phosphate concentrations are usually very low in the euphotic (epipelagic or sunlight) zone due to rapid assimilation by algae and bacteria, even in relatively productive systems. High levels of phosphate in the upper water column usually indicate an influx of high P water, such as from a wastewater plant or agricultural drainage (Ruzycki 2004). It may also be a short-term result of transient mixing of high-P anoxic bottom waters in productive shallow lakes, as can occur during stormy conditions.

PROBLEM 25.4.3
Role of nitrates

Increased pesticides and fertilizer applied due to poorer soils can increase the toxicity of surface waters and decrease groundwater quality. A farm drain carrying excess nitrates into a stream at the rate of 850 g · L^{-1} has a flow rate of 25 L · min^{-1}. The stream flows at 1.3 m^3 · s^{-1}. Ignoring the time for any decomposition to occur, what is the concentration of nitrates after the discharge point?

SOLUTION 25.4.3

$$Q_{mix} = Q_{drain} + Q_{stream}$$

$$\text{Convert units}: Q_{stream} = \left[1.3 \text{ m}^3 \cdot \text{s}^{-1} \times \left(1000 \text{ L} \cdot \text{m}^{-3} \right) \times 60 \text{ min}^{-1} = 78{,}000 \text{ L} \cdot \text{min}^{-1} \right]$$

$$C_{mix} = 10 \text{ mg} \cdot \text{L}^{-1} \left(78{,}000 \text{ L} \cdot \text{min}^{-1} \right) + 25{,}000 \text{ mg} \cdot \text{L}^{-1} \left(25 \text{ L} \cdot \text{min}^{-1} \right) / Q_{drain} + Q_{stream}$$

$$= 780{,}000 \text{ mg} \cdot \text{min}^{-1} + 625{,}000 \text{ mg} \cdot \text{min}^{-1} / \left(78{,}000 \text{ L} \cdot \text{min}^{-1} + 25 \text{ L} \cdot \text{min}^{-1} \right)$$

$$= 18 \text{ mg} \cdot \text{L}^{-1}$$

PROBLEM 25.4.4
Effects of riffles

What are some useful roles of stream riffles?

SOLUTION 25.4.4

By causing a rapid turnover of exposed surface water, rifles increase oxygen dissolution from the air. A high frequency of riffles as observed upstream or downstream indicates a good habitat for aquatic insects and fish.

PROBLEM 25.4.5
Nitrate dangers

With lowered crop yields occurring due to poorer hillslope soils being brought under cultivation as sea levels rise, more nitrates will be required as fertilizers. Blue baby disease is of worldwide concern, especially in agricultural areas. The cause has typically been attributed to residential drinking water wells having high nitrate-nitrogen concentrations caused by contamination in runoff from fields fertilized with nitrogen fertilizer. How can the danger levels be ascertained before harmful health effects occur?

SOLUTION 25.4.5

Although nitrite is the ion that causes the problem, it is the more oxidized form of nitrogen, nitrate, which is usually the route of the problem because the gut flora of young children, typically <6 months old, have not yet become very acidic (Davis & Masten 2020), and the hydrochloric acid of the stomach normally breaks down nitrates as follows:

$$NaNO_3 + HCl \rightarrow NaCl + HNO_3 \qquad (25.4.1)$$

Without sufficient HCl, conditions are favorable when nitrate is in high concentrations (>10 mg N · L^{-1}) for anoxic, denitrifying bacteria to biologically reduce enough nitrate to nitrite to cause the symptoms of the disease. It is the nitrite (NO^{2-}) ion, by binding to blood hemoglobin competitively with oxygen, which causes the disease. This reduces the oxygen content of the blood and is thereby life-threatening. Though the blood appears brownish, the overall coloring of the child is bluish. The US and World Health Organization drinking water criterion for nitrite is only 1 mg N · L^{-1}, which is 10% of the nitrate standard (EPA 2022). The EPA's maximum contaminant level (MCL) for nitrate set to protect against blue baby syndrome is 10 L^{-1} (EPA 2022).

PROBLEM 25.4.6
Amount of nitrogen in a quantity of nitrate

The WHO criterion for nitrate is 10 mg N · L^{-1}, while that for nitrite is 1 mg N · L^{-1}. What mass of nitrate contains 10 mg of N?

SOLUTION 25.4.6

$$\% \text{ N in molar mass of } NO_3^- = 14 / 62 \times 100\% = 23\% \text{ N}$$

10 mg N represents 23% of NO_3^-.
23% of NO_3^- comprises 10 mg N.
Therefore, 1% of NO_3^- comprises 10/23 mg = 0.43 mg.
Hence, 100% of $NO_3^- = 0.43 \times 100 = 43$ mg of nitrate contains N = 10 mg.
Therefore, the upper threshold of danger levels of nitrate is 43 mg NO_3 · L^{-1}.

PROBLEM 25.4.7
Algal blooms

What is the quantitative impact of P algal blooms?

SOLUTION 25.4.7

Algal composition is approximately:

500 g wet weight = 100 g dry weight : 40 g C : 7 g N : 1 g P

This ratio of C:N:P is called the "Redfield ratio" (Ruzycki 2004) and approximates the composition of algae:
40:7:1 by weight

The atomic ratio is calculated using atomic masses as follows:

$$C = 40/12, N = 7/14, P = 1/30.9$$

$P = 0.032$, being the smallest, becomes the unit which divides into 98.5 (Nitrogen) and 480 (Carbon).

Approximately, this is $= 100:16:1$ by atoms.

Hence, P availability $= 1/100$ algae w/w (by dry weight).

PROBLEM 25.4.8
Oxygen consumption

At 25°C the concentration of oxygen dissolved in water is only about 8 mg $\cdot$ L^{-1}. How many milligrams of oxygen does the breakdown of 50 kg of carbohydrate consume?

SOLUTION 25.4.8

Readily consumed by biodegradation of biomass (abbreviated $\{CH_2O\}$) by oxygen-utilizing bacteria:

$$\{CH_2O\} + O_2 \rightarrow CO_2 + H_2O \qquad (25.4.2)$$

Only about 8 mg of $\{CH_2O\}$ consumes 8 mg of O_2.

30 g of CH_2O consumes 32 g of O_2.

Hence, 1 g of CH_2O consumes 32/30 g of O_2.

Therefore, 50 kg of CH_2O consumes $32/30 \times 50,000$ g of O_2.

So 53,333.33 g $= 53.33$ kg of oxygen consumed

REFERENCES

Baird Colin and Michael Cann (2013) Environmental Chemistry. WH Freeman & Company, New York.

Davis ML and Masten SJ (2020) Principles of Environmental Engineering & Science. 4th edition, McGraw Hill, New York. ISBN10: 1259893545 | ISBN13: 9781259893544

EPA (2022) Estimated nitrate concentrations in groundwater used for drinking. https://www.epa.gov/nutrient-policy-data/estimated-nitrate-concentrations-groundwater-used-drinking#:~:text=While%20nitrate%20does%20occur%20naturally%20in%20groundwater%2C%20concentrations,nitrate%20indicate%20human%20activity%20%28Dubrovsky%20et%20al.%202010%29

Li S, Lin Z, Liu M, Jiang F, Chen J, Yang X, Wang S (2020) Effect of ferric chloride on phosphorus immobilization and speciation in Dianchi Lake sediments. Ecotoxicol Environ Saf 197, 110637. doi: 10.1016/j.ecoenv.2020.110637.

Ruzycki E (2004) Mod10/11-E Stream Surveys - Watershed Assessemnt Module 10/11 Stream Surveys Stream Surveys – February 2004 Part 5 – Watershed Assessment https://view.officeapps.live.com/op/view.aspx?src=http%3A%2F%2Fwww.waterontheweb.org%2Fcurricula%2Fws%2Funit_03%2Fmod_10_11%2Fmod10-11_part5.ppt&wdOrigin=BROWSELINK

Section V

Remediation of Polluted Soils

26 Chemistry of Soil Pollution Remediation

TYPES OF POLLUTANTS

- Heavy metals (such as lead and mercury at excessively high amounts) in the soil are very poisonous to humans.
- PAHs (polycyclic aromatic hydrocarbons) as a class of organic chemicals where only carbon and hydrogen atoms are present.
- Coke (coal) production, automobile emissions, cigarette smoke, and shale oil extraction as sources of PAHs in the soil.
- Industrial waste soil contamination comes from the dumping of industrial waste into soils.
- Pesticides as chemicals (or chemical mixes) that are used to kill or prevent pests from reproducing.

26.1 BIOREMEDIATION OF SOILS: REQUIREMENTS

Soil is subject to water and wind erosion, and much productive soil has been lost or ruined by salinization, heavy metal contamination, toxic spills, and leaching from older landfills. Many important chemical and biochemical reactions occur in soil, and for sustainability, soil conservation is a top priority.

PROBLEM 26.1.1
Gasoline-contaminated soil

What water content is required for bioremediation of a 375 yd³ gasoline-contaminated soil, and what is the required calculation thereof?

SOLUTION 26.1.1

Several studies found that the most effective water content for aerobic soil remediation is approximately 60% of the field capacity (Table 26.1) of each soil (Harris 2016). This is because field capacity, which varies with porosity on a weight over weight (w/w) basis, is always 100% of the soil's water-holding capacity, blocking out all oxygen from aerobic bacteria. It is the amount of soil moisture held in the soil after excess water has drained away and the rate of downward movement has decreased. This usually takes place 2–3 days after rain or irrigation in pervious soils

DOI: 10.1201/9781003341826-32

TABLE 26.1
Optimum Conditions for Soil Bioremediation

Environmental Factor	Optimum Conditions
Available soil moisture	25–85% water holding capacity
Oxygen	>0.2 mg/L DO, >10% air-filled pore space for aerobic degradation
Redox potential	Eh > 50 mV
pH	5.5–8.5
Temperature	15–45°C
Nutrients	C:N:P = 120:10:1 molar ratio

of uniform structure and texture. If we assume, for instance, a certain organic content, porosity, and current saturation, the water content required is:

$$\text{Volume} = 375 \text{ yd}^3$$

$$\text{Organic content} = \text{low}$$

$$\text{Desired water content} = 25\% \text{ to } 85\% \text{ (depending on porosity and organic content)}$$

$$\text{Assume: porosity, n} = 30\% \text{ initial saturation}$$

$$\text{Saturation (S)} = 20\%$$

$$\text{Use} = 60\% \text{ water content}$$

$$\text{Water needed} = 375 \text{ yd}^3 (0.30)(0.6-0.2) = 45 \text{ yd}^3$$

$$= 1215 \text{ ft}^3 = 9090 \text{ gallons or } 34{,}409.39 \text{ L}$$

PROBLEM 26.1.2
Nutrient requirement calculations

What is the nutrient requirement for a bio-pile soil remediation based on a 158 kg spill of gasoline ($\cong C_7H_{16}$), i.e., heptane?

SOLUTION 26.1.2

Nutrient requirement for a bio-pile

Nutrient sources: ammonium sulfate ($(NH_4)_2SO_4$) trisodium phosphate ($Na_3PO_4 \cdot 12H_2O$)

$$\text{Molecular weight (MW) of gasoline } (\cong C_7H_{16}) = 7 \times 12 + 1 \times 16 = 100 \text{ g} \cdot \text{mol}^{-1}$$

$$\text{Moles of gasoline} = 158 \times 10^3 / 100 = 1580 \text{ mole}$$

$$\text{Moles of C} = 7 \times 1580 \text{ mole} = 1.1 \times 10^4 \text{ mole}$$

$$\text{Molar ratio C} : N : P = 120 : 10 : 1$$

$$\text{Moles of N needed} = 10/120 \times 1.1 \times 10^4 = 920 \text{ mole}$$

Moles of $\left(\left(NH_4\right)_2 SO_4\right)$ needed $= 920/2$ (as N is already doubled) $= 460$ mol

$$MW \text{ of } \left(\left(NH_4\right)_2 SO_4\right) = (14+4)\times 2 + 32 + 4\times 16 = 132$$

Mass of $((NH_4)_2 SO_4))$ needed $= 132\ g \cdot mol^{-1} \times 460\ mol = 6.1 \times 10^5\ g = 61\ kg$

By similar calculation: mass of $(Na_3PO_4 \cdot 12H_2O)$ needed $= 35\ kg$

Oxygen requirement for bio-pile

$$C_7H_{16} + 22O_2 \rightarrow 7CO_2 + 8H_2O$$

1 mole (100 g) gasoline requires 22 moles (704 g) O_2

Oxygen content of air $= 21\%$ by volume $= 210,000$ ppmv

$$1\ \text{ppmv} = \frac{\dfrac{1\ atm \times 32\dfrac{g}{mol}}{1000}\ mg}{g} \times 0.0821\ L\frac{atm}{mol\ K} \times 293\ K$$

$$= 0.00133\ mg/L$$

$210,000$ ppmv or 2.1×10^5 ppmv $= 280\ mg/L = 0.28\ g/L$

Oxygen needed for 158 kg spill of gasoline ($\cong C_7H_{16}$)

100 g gasoline needs ~ 700 g oxygen

Therefore, 158 kg gasoline $\times 7 = 1106\ kg\ O_2 = 1.1 \times 10^6\ g\ O_2$

Water in pile $= 375\ yd^3 (0.30)(0.6) = 67.5\ yd^3 = 52\ m^3 = 52,000\ L$

At saturation at 20°C and 1 atm, DO $= 9.2$ mg/L

Mass of oxygen in soil moisture $=$

$$= 52,000\ L \times 9.2\ mg/L \times 0.001\ g/mg$$

$$= 480\ g\ O_2$$

But 480 g O_2 in soil moisture is much less than 1.1×10^6 g O_2 required
At 0.28 g/L air, air requirement is like an analogy:
I have 15000 cents and to know how many dollars, I divide by 100 cents. Likewise:

1.1×10^6 g (of oxygen)/0.28 g (of oxygen)/L (of air) $= 3.95 \times 10^6$ L air

$$= 3,950\ m^3\ air$$

Air void volume in pile $= 375\ yd^3\ (0.30)\ (0.4)$

$$= 45\ yd^3 = 34\ m^3$$

Need to exchange $3950/34 = 116$ void volumes

to fulfill oxygen requirement.

26.2   ROLE OF BIOREACTORS FOR POTENTIAL LIQUID SOIL POLLUTANTS

As sea levels rise, arable land decreases. Moreover, soil quality and fertility decrease with altitude. Polluted water supplies can damage soil. For acid mine drainage, for example, one way to remediate the pollution of such soils entails applying a bioreactor, like that used by Harris and Ragusa (2001), which decreases the concentration of heavy metals in an acid mine drainage (AMD). First-order reactions are those in which degradation or increase of a substance occurs at a non-exponential rate. For example, the half-life of radioactive materials is a first-order reaction, where 10 g of a substance becomes 5 g after a fixed time, then 2.5 g, and so on.

26.3   ROLE OF FIRST-ORDER REACTIONS

A first-order reaction proceeds at a rate proportional to the concentration of one of the reactants such that doubling the concentration doubles the reaction rate. Other attributes are as follows:

- The concentration of other reactants will not have any effect on the rate of reaction.
- Thus, the order of reaction for a first-order reaction is always one.
- The rate varies depending on the concentration changes of only one reactant.
- Temperature and reactant concentration affect the rate, which is affected by time.

Solutions to problems listed below include two approaches to soil remediation, including pump and treat, and in situ. Nonconservative substances are those which undergo significant short-term degradation or change in the environment other than by dilution. In a strictly chemical sense, the three orders of reaction (first, second, and third) could be depicted as:

$$\text{First order (unimolecular) } A \rightarrow B + C$$

$$\text{Second order (bimolecular) } A + B \rightarrow C + D$$

$$\text{Third order (termolecular) } A + B + M \rightarrow AB + M$$

where M could be described as a ubiquitous surrounding medium at great excess such as air or water.

A good example of a true first-order reaction (where there is no intervening foreign substance, as is not normally the case) is radioactive decay, such as $222Rn \rightarrow 218Po + $ alpha-particles. According to Davis and Masten (2020), when a nonconservative substance decays as a first-order reaction with no influent in the batch reactor, the mass balance is described as:

$$dC/dt = -kC \qquad (26.3.1)$$

where
  d = change (sometimes written as Δ for "delta")
  C = concentration
  t = time
  −k = rate constant (a constant rate)

And where integration of that equation yields:

$$Ct/Co = e^{-kt}$$

where
  Co = initial concentration
  e = 2.718... (in contrast to "normal" logarithms having a base = 10)

PROBLEM 26.3.1
The log (base e) of a number

How is the logarithm to base e (natural logarithm) calculated?

SOLUTION 26.3.1

Type the number into a scientific calculator
Press the [ln] button (i.e., the natural logarithm)
See the answer. This is the log (base e) of the number.
Or the log (base e) of exp [whatever] means: $1 \times$ [whatever]
Press the exp button on the calculator
Press the [ln] button

PROBLEM 26.3.2
Mixed batch reactors

A well-mixed batch reactor has no input or output of mass. The amounts of individual components may change due to reaction, but not because of flow into or out of the system. A site contaminated by a mine pollutant is to be excavated and treated in an anaerobic lagoon. However, a microcosmic model of the treatment, in the form of a completely mixed batch reactor, is necessary to determine the time required for the treatment. Assuming a first-order reaction, and given the information below, calculate the rate constant k and determine the length of time required to achieve 95% decrease in the original concentration.
  Given information:

Time (d)	Waste concentration (mg · L^{-1})
1	240
18	125

SOLUTION 26.3.2

Using Equation 26.3.1 above, we can estimate the rate constant k. Hence, for the 1st and the 18th day, the time interval t = 18 − 1 = 17 d.

$$125 \cdot L^{-1}/240 \cdot L^{-1} = \exp\left[-k(17 \text{ d})\right]$$
$$0.5208 = \exp\left[-k\ (17)\right]$$

Finding the logarithm of both sides of the equation obtains

$$0.6534 = -k(17)$$

Solving for k yields

$$k = 0.0384 \text{ d}^{-1}$$

To reach a reduction of 95%, the concentration at time t must be 1 − 0.95 of the original concentration.

$$Ct/Co = 0.05$$

Therefore, the estimated time is:

$$0.05 = \exp\left[-0.0384(t)\right]$$

To remove "exp," we again take the logarithm (base e) of both sides to give: $-2.9957 = -0.0384(t)$; solving for t produces:

$$t = 78 \text{ days}$$

26.4 NATURAL SOIL REMEDIATION: ROLE OF CATION EXCHANGE CAPACITY

Microscopic particles of clay minerals and organic matter tend to be negatively charged, thus attracting positively charged ions (cations) on their surfaces by electrostatic forces, resulting in the cations remaining within the soil root zone and not easily lost through leaching. The ability of a soil to break down and metabolize pollutants varies directly with the soil cation exchange capacity (CEC), as each clay mineral has a range of exchange capacities because of differences in structure and in chemical composition. Being a measure of the total negative charges within the soil that adsorb plant nutrient cations such as calcium (Ca^{2+}), magnesium (Mg^{2+}), and potassium (K^+), CEC, measured in milliequivalents, can determine how long a pollutant is held as a nutrient source in the presence of destructive microbial degraders. Thus, as chemical and biochemical phenomena in soils operate to reduce the harmful nature of pollutants, many soils can assimilate and neutralize pollutants via oxidation-reduction processes, hydrolysis, acid-base reactions, precipitation, sorption, and biochemical degradation (Manahan 2005). Manahan reports that hazardous organic chemicals may be degraded to harmless products on soil and heavy metals may be sorbed by it but voices concern regarding the disposal of chemicals, sludges, and other potentially hazardous materials on soil, particularly where the possibility of water contamination exists.

PROBLEM 26.4.1
Cation exchange capacity calculations

What is the equivalent weight for calcium carbonate or $CaCO_3$?
  [Hint: MEQ = milliequivalent weight]

SOLUTION 26.4.1

$$CaCO_3 - \text{formula wt.} = 40 + 12 + 48 = 100$$
$$\text{Charges involved} = 2$$
$$EQWT = 50$$
$$MEQ = 0.05 \text{ g}$$
$$\text{Or one MEQ of } CaCO_3 = 0.05 \text{ g}$$

PROBLEM 26.4.2
CEC and clay, organic content

Calculation the CEC using the average CEC for clay and OM.

SOLUTION 26.4.2

$$\text{Assume Avg. CEC for } \% \text{ OM} = 200 \text{ meq}/100 \text{ g}$$
$$\text{Assume Avg. CEC for } \% \text{ clay} = 50 \text{ meq}/100 \text{ g}$$
$$CEC = (\% \text{ OM} \times 200) + (\% \text{ Clay} \times 50)$$

From soil data: soil with 2% OM and 10% Clay

$$200 \times 0.02 + 50 \times 0.1 = 4 + 5 = 9 \text{ meq}/100 \text{ g}$$

PROBLEM 26.4.3
Determination of CEC

What are two ways to determine the sum of cations?

SOLUTION 26.4.3

Method: Find the surrogate cations.

1. Divalent cations can replace monovalent cations
2. Therefore, the soil is saturated with NH_4^+. The NH_4^+ is replaced by Ca^{2+} and the NH_4^+ removed is measured.

Estimation based on texture

1. Sand = 0–3 meq/100 g
2. Loamy sand to sandy loam = 3–10
3. Loam = 10–15
4. Clay loam = 15–30
5. Clay = >30 (depends on kind of clay)

Base saturation vs pH

% Base Saturation = meq bases ÷ CEC × 100
% Hydrogen saturation= meq H ÷ CEC × 100
Example: Ap soil horizon
Cations:

- H^+ Ca^{2+} Mg^{2+} K^+ Na^+

8.6   13   3.5   0.6   0.2
CEC = 27 meq/100g (sum of cations)
% Base saturation = 17.3 ÷ 25.9 × 100 = 65% (as H^+ is not a base)
% Hydrogen saturation = 8.6 ÷ 27 × 100 = 35%

As climate change decreases the availability of fertile, low-lying land, conservation of adequate sources of good soil will be critical.

Clay minerals may attain a net negative charge by ion replacement, in which Si(IV) and Al(III) ions are replaced by metal ions of similar size but lesser charge. Additional cations must compensate for this negative charge deficit by associating with the clay layer surfaces. Since these cations need not fit specific sites in the crystalline lattice of the clay, they may be relatively large ions, such as K^+, Na^+, or NH_4^+ (Table 26.2).

TABLE 26.2

Charge and Relative Size of Some Soil Cations (Ionic Diameter in Picometers*)

Monovalent	Divalent	Trivalent
Sodium – 116	Magnesium – 86	Aluminum – 41
Potassium – 152	Calcium – 11	Boron – 41
Ammonium – 167	Iron – 61	Gallium – 62

* 10^{-12} m.

TABLE 26.3
Equivalent and Milliequivalent Weight of Some Common Soil Cations

Element	Na^+	K^+	Ca^{2+}	Mg
Valence	1	1	2	2
Eq. wt.	23/1 = 23	39/1 = 39	40/2 = 20	24/2 = 12
MEQ wt.	0.023	0.039	0.02	0.012

Plant roots absorb these cations in solution when H^+ ions are pushed out in vast numbers by roots to remove and replace cations on the clay particles. The loosened nutrient cations, in solution, are pulled into the permeable roots. Therefore, the conservation of soil nutrients against intense weathering and erosion depends largely on the binding energies and number of adsorption sites of clay particles. Thus, 1 MEQ of CEC has 6.02×10^{23} adsorption sites (Manahan 2005).

PRACTICE PROBLEM 26.4.1
Soil processing of pollutants

What determines the ability of a soil to (a) oxidize (b) reduce (c) sorb heavy metals?

PRACTICE PROBLEM 26.4.2
Milliequivalents

The following worked examples clarify the concept of milliequivalents of nutrient cations in soils and hence the soil fertility status. For each molar mass, the equivalent weight and milliequivalent (MEQ) weight can be deduced in Table 26.3.

The following data represent the levels of concentration in ppb (parts per billion) of sulfur dioxide in the air over a city during the 15 days of a 15-day time span after an explosion in a chemical plant. Determine (a) the mean, (b) the median, (c) the mode, (d) the standard deviation, and (e) whether this set of numbers represents a normal curve distribution: 15, 16, 17, 14, 13, 10, 9, 9, 8, 8, 8, 7, 6, 4, 4.

[HINT: See Statistics in Chapter 30]

26.5 PESTICIDES IN SOIL

Most liquid contaminants will move either at the same rate or slower than water (Davis and Masten 2020). The retardation coefficient expresses how much slower a contaminant moves than does the water itself. Based on the assumption that sorption of organics occurs to the naturally occurring organic carbon in an aquifer, a sorption coefficient, K_d (also called the soil/water distribution coefficient), is defined from the fraction of organic carbon, f_{oc}, in the aquifer and the organic carbon partition coefficient, K_{oc}, of the chemical:

$$= K_{oc} f_{oc}$$

Therefore, the retardation coefficient depends on the hydrophobicity of the pollutant and on the soil properties. The following equation applies to electrically neutral organic chemicals:

$$R = 1 + (\rho b / \eta)\ K_{oc} \cdot f_{oc}$$

where

R = retardation coefficient
ρb = bulk density of the soil
η = the fraction of the soil which is porous
K_{oc} = partition coefficient for the organic
f_{oc} = fraction of inorganic carbon in the soil

Retardation coefficients of typical groundwater contaminants are published by Priddle and Jackson (1991).

PROBLEM 26.5.1
Pesticide movement in groundwater

The increased warming of planet earth may cause the proliferation of pests. A householder in House A disposes a pesticide down their drain, thereby contaminating the attached septic field. House B is located 70 m downgradient of House A. The linear velocity of the water in the unconfined aquifer used for drinking water is 4.3×10^{-5} m · s^{-1}. The unconfined aquifer rests on a confining layer parallel to the water table. On the assumption that the pesticide has a retardation coefficient of 2.8 and does not degrade in the soil, how long will it take for the pesticide to reach the well of house B?

SOLUTION 26.5.1

According to the EPA (2021), the retardation coefficient is defined as:

$$R = v'_{water} / v'_{cont}$$

where

R = the extent to which the chemical is retarded in the water
v'_{water} = the linear velocity of the groundwater
v'_{cont} = the linear velocity of the contaminant

$$v'_{cont} = v'_{water} / R = 4.3 \times 10^{-5}\, m \cdot s^{-1} / 2.8 = 1.53 \times 10^{-5}\, m \cdot s^{-1}$$

Since the distance between the septic field and the well is 70 m, the travel time is:

$$Distance/v' = 70\ m / 1.53 \times 10^{-5}\ m \cdot s^{-1} = (Days / 86,400\ s) = 52.95\ days$$

Therefore, it will take almost 53 days for the pesticide to reach the drinking well. It is noted that this result fails to include the concentration. The concentration depends on several other factors, some of which include the original mass of the pesticide, the quantity of water flowing, and other substances in the water.

PROBLEM 26.5.2
Disposal by sanitary landfill

The clay liner of an 18-ha sanitary landfill consists of a compacted 1.2-m-thick clay liner having a hydraulic conductivity of 5.8×10^{-9} m $\cdot$ s^{-1}. What is the flow rate of leachate through the landfill if the head of water is 0.7 m?

SOLUTION 26.5.2

The Darcy velocity for the leachate through the clay liner is required. Regardless of the cross-sectional areal extent, that velocity is:

$$V = K \, (\Delta H / \Delta L)$$

where
  K = hydraulic conductivity
  ΔH = hydraulic gradient (the head difference between two locations)
  ΔL = horizontal distance between the two points

That is, $(5.8 \times 10^{-9}$ m $\cdot$ s$^{-1})$ $(0.7$ m/1.2 m$) = 3.83 \times 10^{-8}$ m $\cdot$ s^{-1}
To calculate the flow, use the equation:

$$Q = K(i) A$$

where
  K = hydraulic conductivity
  i = hydraulic gradient (the head difference between two locations)
  i = $\Delta H / \Delta L$ (i.e., head/horizontal distance)
  A = cross-sectional area of an aquifer

Or $Q = v/A$:

$$Q = \left(1.96 \times 10^{-8} \text{ m} \cdot \text{s}^{-1}\right)(18 \, \text{ha}) \left(10^4 \text{ m}^2 \cdot \text{ha}^{-1}\right) = 3.52 \times 10^{-3} \text{ m}^3 \cdot \text{s}^{-1}$$

26.6 UNIVERSAL SOIL LOSS EQUATION

This answer assumes that there is no leachate collection system, but in most sanitary landfills there is a leachate collection system, thereby minimizing leakage to groundwater systems.

   To cause soil erosion, energy is necessary to do work, and excessive global warming can provide the latent energy required to produce high-energy cyclones which in turn produce high-intensity rainfall events. To the extent that erosion is accomplished, there are soil attributes that facilitate or hinder this work. The major factors are included in the universal soil loss equation as follows (FAO 2023):

$$A = (R)\ (K)\ (LS)\ (C)\ (P)$$

where
    A = predicted soil loss (in $Mg \cdot ha^{-1}$)
    R = rainfall erosion index
    K = soil erodibility factor
    LS = topographic relief factor as functions of slope length and %
    C = crop management factor
    P = conservation practice factor

According to FAO (2023),

- The R-factor is calculated as a product of the kinetic energy of a rainfall times its maximum 30-min intensity of fall (i.e., a comparison among all consecutive 30-min period). Yet, despite this conclusion by the FAO (2023), Rengasamy and Malcom Sumner (1993) refuted the raindrop kinetic energy hypothesis of aggregate breakdown as an effective factor. Thus, they found that drops of carbon tetrachloride, a non-polar liquid, of similar size to raindrops but much denser fell on soil aggregates with even greater force than the raindrops but failed to dislodge soil aggregates, while raindrops falling from the same height dislodged and damaged the aggregates. This suggests the need to modify the raindrop factor in the universal soil loss equation to include the physical effects on soil aggregates of the high dielectric constant of water compared to other natural liquids.
- The K-factor represents the ease with which a soil can be eroded and is influenced by the soil texture, organic matter content, soil structural strength, and permeability. The soil erodibility factor is low for soils in which the water infiltrates readily, for example, sandy soils.
- The L and S factors express the influence of the landscape on soil erosion.
- The C-factor is the ratio of erosion under a specified cover and management to the amount of erosion under a continuous bare fallow, considering the type and density of vegetative cover on the soil as well as all related management practices, such as time between operations, weed control, tillage, watering, fertilization, crop residues, etc.
- The P-factor is the ratio of the erosion resulting from the described practice to that which would occur with up-and-down slope cultivation, thereby recognizing the influence of conservation practices, such as contour planting, strip cropping, terracing, and combinations.

PROBLEM 26.6.1
Manipulating the universal soil loss equation

Using the universal soil loss equation, determine the annual soil loss for a farm in New York state that has a Dunkirk silt loam with a slope of 4% and an average slope length of 61 m. The land is under continuous soybeans. The farmer uses conventional tillage and contour-farms his field and leaves the residue.

SOLUTION 26.6.1

From any table of annual values of the rainfall erosion index in the United States, we select a commensurate rainfall erosion index, which in this case at this location is 160. From a table showing K-values for soils on erosion research stations, the value for K in this case is 0.28, and the value for topographic factor (LS) for the combination of slope length and steepness in this case is 0.528. The next value is from the table for crop management or C values for different crop sequences, which in this case is 0.49. Finally, the P value is for ploughing system, which in this case a table shows to be 0.50.

The annual soil loss is therefore:

$$A = (R)\,(K)\,(LS)\,(C)\,(P)$$
$$= (160)(0.69)(0.28)(0.49)(0.50) = 7.57 \text{ Mg} \cdot \text{ha}^{-1}$$

26.7 SOIL REMEDIATION USING AN INVASIVE TREE SPECIES

The invasive species *Spathodea campanulata* grows well in subtropical and tropical areas with an even distribution of rainfall but will tolerate a dry season of up to 6 months, growing on a wide variety of sites, from poorly to excessively drained, and can rehabilitate degraded land through its rapid growth (Rojas-Sandoval & Acevedo-Rodríguez 2013).

26.8 SOIL REMEDIATION USING BIOCHAR

Biochar improved crop growth in contaminated soil due to the stable carbon matrix and large surface area containing various functional groups (Li et al., 2018) and adsorbed heavy metals (Danmaliki & Saleh 2017). According to Tanveer et al. 2019)), there are four valuable aspects that could be provided by applying pyrolysis onto bio-wastes: (1) reduce the negative effects of waste (sanitizing pathogens, balancing C:N ratios), (2) energy generation (bio-gas, bio-oil, and sometimes biochar), (3) use of biochar as soil amendment for decreasing fertilizer dosage and improving soil fertility, and (4) reduce the gas emissions since most of the carbon species are relatively 'stable'.

- The slowed decomposition of biochar extends the long-term stability of carbon in soil (Danmaliki & Saleh 2017).
- Decreases soil bulk density (Blanco-Canqui 2017)) and improves the soil water holding characteristics (Tokova et al. 2020).
- Improved N availability for plants by changing the N-bacterial communities and related soil biological functions (Pokharel et al. 2018) in soil, and as a result reduced N losses by leaching (Wang et al. 2015).
- Several studies found that biochar amendment increased the enzymatic activities involved in N and P cycling but decreased the activities of C related enzymes (Bhaduri et al. 2016; Zhang et al. 2017) in soils.

TABLE 26.4

Banana peel waste products of biochar compared with microbial decomposition.

No amendment	Banana peel waste	Biochar
	Volatile carbon = high	Volatile content = low
	Fixed carbon = low	Fixed content = high
Soil	Soil + Peel waste	Soil + Biochar
	Increased microbial biomass, C & N	Decreased microbial biomass, C & N

- Results from other studies were inconsistent (Table 26.4; Paz-Ferreiro et al. 2014), with some workers having attributed that to variable soil types, temperature, and biochar characteristic due to varied pyrolysis temperatures and raw material feedstock (Lan et al. 2017; Pokharel et al. 2018). Such variables affect soil enzymatic activities such as: major biocatalysts involved in all biochemical reactions including microbial life cycle and metabolism, break down and decomposition of soil organic matter and residues, degradation of inimical organic contaminants, and nutrients cycling.

REFERENCES

Bhaduri D, Saha A, Desai D, Meena HN (2016) Restoration of carbon and microbial activity in salt-induced soil by application of peanut shell biochar during short-term incubation study. Chemosphere 148, 86–98. https://doi.org/10.1016/j.chemosphere.2015.12.130.

Blanco-Canqui H (2017) Biochar and Soil Physical Properties. Soil Science Society of America Journal, 81: 687–711. https://doi.org/10.2136/sssaj2017.01.0017

Danmaliki GI, Saleh TA (2017) Effects of bimetallic Ce/Fe nanoparticles on the desulfurization of thiophenes using activated carbon. Chem Eng J 307, 914–927. https://doi.org/10.1016/j.cej.2016.08.143.

Davis ML and Masten SJ (2020) Principles of Environmental Engineering & Science.4th edition, McGraw Hill, New York. ISBN10: 1259893545 | ISBN13: 9781259893544

EPA (2021) EPA On-line Tools for Site Assessment Calculation. https://www3.epa.gov/ceampubl/learn2model/part-two/onsite/ard_onsite.html

FAO (2023) Land and Water: Universal Soil Loss Equation. https://www.fao.org/land-water/land/land-governance/land-resources-planning-toolbox/category/details/en/c/1236441/

Harris MA (2016) Erodibility of Unconsolidated Mine Wastes Under Simulated Rainfall and Hydraulic Forces After Organic Amendments. In: Geobiotechnological Solutions to Anthropogenic Disturbances. Environmental Earth Sciences. Springer, Cham. https://doi.org/10.1007/978-3-319-30465-6_12

Harris M, Ragusa S (2001) Bioremediation of acid mine drainage using decomposable plant material in a constant flow bioreactor. Env Geol 40, 1192–1204 (2001). https://doi.org/10.1007/s002540100298

Lan ZM, Chen CR, Rezaei RM, Yang H, Zhang DK (2017) Stoichiometric ratio of dissolved organic carbon to nitrate regulates nitrous oxide emission from the biochar-amended soils, Sci Total Environ 576, 559–571. https://doi.org/10.1016/j.scitotenv.2016.10.119.

Li R, Wang JJ, Zhang Z, Awasthi MK, Du D et al. (2018) Recovery of phosphate and dissolved organic matter from aqueous solution using a novel CaO-MgO hybrid carbon composite and its feasibility in phosphorus recycling. Sci Total Environ 642, 526–536. https://doi.org/10.1016/j.scitotenv.2018.06.092.

Manahan Stanley (2005) Environmental Chemistry. Taylor and Francis, Boca Raton, Fla.

Paz-Ferreiro, J, Lu H, Fu S, Méndez A, Gascó G (2014) Use of phytoremediation and biochar to remediate heavy metal polluted soils: A review. Solid Earth, 5, 65–75. https://doi.org/10.5194/se-5-65-2014.

Pokharel P, Kwak J-H, Ok YS, Chang SX (2018) Pine sawdust biochar reduces GHG emission by decreasing microbial and enzyme activities in forest and grassland soils in a laboratory experiment, Sci Total Environ 625, 1247–1256. https://doi.org/10.1016/j.scitotenv.2017.12.343.

Rengasamy P & Sumner M (1993) Dr. Pichu Rengasamy from the University of Adelaide and Regis Professor Sumner University of Georgia, at Athens, personal communication.

Rojas-Sandoval J, Acevedo-Rodríguez P (2013) Spathodea campanulata (African tulip tree). CABI Compendium. https://doi.org/10.1079/cabicompendium.51139.

Priddle MW & Jackson RE (1991) Laboratory column measurement of VOC retardation factors and comparison with field values. Groundwater, 29(2), 260–266.

Tanveer Ali Sial, Muhammad Numan Khan, Zhilong Lan, Farhana Kumbhar, Zhao Ying, Jianguo Zhang, Daquan Sun, Xiu Li (2019) Contrasting effects of banana peels waste and its biochar on greenhouse gas emissions and soil biochemical properties, Process Saf Environ Prot, 122, 366–377, ISSN 0957-5820, https://doi.org/10.1016/j.psep.2018.10.030.

Wang X, Song D, Liang G, Zhang Q, Ai C (2015) Maize biochar addition rate influences soil enzyme activity and microbial community composition in a fluvo-aquic soil. Appl Soil Ecol 96. https://doi.org/10.1016/j.apsoil.2015.08.018.

Zhang A, Cheng G, Hussain Q, Zhang M, Feng H (2017) Contrasting effects of straw and straw–derived biochar application on net global warming potential in the loess plateau of China. Field Crops Res, 205, 45–54. https://doi.org/10.1016/j.fcr.2017.02.006.

27 CO$_2$ Sequestration Using Alkali Mine Waste Treatments in Bauxite Residue Disposal Areas (BRDA)

MAIN POINTS

- Carbonation (aging) worldwide naturally sequesters 7 million tons of CO$_2$ annually.
- There are three kinds of aged BRDAs (bauxite residue disposal areas): soft mud (low strength), "hard mud," and "brick-strength."
- Carbonation capability varies directly with the strength and depth of BRDA.

27.1 ROLE OF ATMOSPHERIC CARBONATION

It has been estimated that over 100 million tons of CO$_2$ worldwide (Figure 27.1) have been accidentally sequestered from the air through the natural weathering of historically produced red mud (Chunhua et al. 2013).

It has also been stated that, based on the current production rate of red mud, it is likely that some 7 million tons of CO$_2$ are sequestered annually through atmospheric carbonation (Chunhua et al. 2013). However, calculations based on the limited surface area exposed to atmospheric carbonation in red mud impoundments mean that the latter estimate is too high. This chapter aims to identify the necessary environmental conditions for maximizing the quantity of CO$_2$ sequestered globally by red mud impoundments. For example, approximately 6 million tons of additional CO$_2$ can be captured and stored in the red mud while simultaneously reducing the hazards, if appropriate technologies are in place for incorporating binding cations into red mud. But the reality has never been the case. This is because the dominant mineral phases responsible for carbon sequestration in many red muds are calcium cations, which are often in short supply partly because they are being sought by titanate and silicate ions, often the two major oxyanions competing with carbonate ions for the available soluble calcium. Using a problem and solution approach, this chapter considers Ca^{2+} ions as a viable pathway for maximizing carbon sequestration in BRDAs and simultaneously reducing the causticity of red mud.

DOI: 10.1201/9781003341826-33

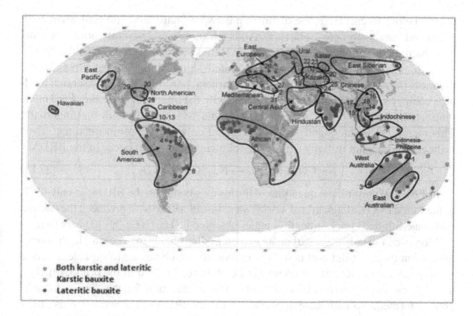

FIGURE 27.1 Global potential quantities of red mud bauxite, lateritic, and karstic (formed from carbonates). Bauxite tailings can significantly reduce the CO$_2$ in the atmosphere by carbonation.

Source: Based on usgs.gov.

ROLE OF DEPTH

Khaitan et al. (2010) showed that carbonation from atmospheric carbon dioxide reduced the pH of red mud from 12.5 to about pH 9.3 in all cells, with the depth of neutralization dependent on the age of the stored residue (up to 1.2 m depth after 35 years). However, Harris (2009) observed a direct relationship between depth below 30 cm and the strength of the residue. This decreasing strength signifies less carbonation.

PROBLEM 27.1.1
Evidence for carbon sequestration in bauxite waste dumps

What occurrences comprise the evidence for carbonation in BRDAs?

SOLUTION 27.1.1

Changes in BRDA strength

Referring to pre-gypsum BRDA, Harris (2009) cited O'Callaghan et al. (1998), who found that ploughing would have made little difference in reducing particle size except to "turn over the lumps of hard mud." However, the strength of this "hard mud" had increased greatly within 7 years of their treatment with gypsum, and in 2005, soil strength tests in Pond 6 indicated a dry, brick-hard surface environment in the above gypsum-treated locations from the surface to a depth of 15 cm, but below 15 cm depth the material was less strong, as gypsum treatment had had little impact below 15 cm. Here, the distinction between "hard mud" (O'Callaghan et al. 1998) and "brick-strength" (Harris 2009) is telling because several "brick-strength" and "hard mud" samples were observed by Harris (2016) who found that the "brick-strength" (gypsum-treated) samples exhibited significantly greater strength and hardness (hardness 2 on the Mohs scale of mineral hardness, based on scratch hardness comparison, i.e., could not be scratched with the fingernail) than the "hard mud" (untreated) samples (Mohs scale = 1, i.e., could be scratched with the fingernail). This increased hardness and strength occurred only after gypsum treatment. Even if the added gypsum initially caused the following reaction to occur in the BRDA:

$$Na_2CO_3 + CaSO_4 \rightarrow Na_2SO_4 + CaCO_3 \qquad (27.1.1)$$

in 1998 and onward, the quantities of intrinsic carbonate in the BRDA, at only 0.3%, as measured by O'Callaghan et al. (1998), were insufficient to have produced the calcium carbonate-lithified "brick-strength" entities observed by Harris (2009) 10 years later.

Moreover, the only other carbonate in their findings likely to occur in a high enough proportion (which it did not) is sodium carbonate, which has a Moh's scale hardness of only 1.3. Hence, as little or no intrinsic carbonate occurred in the original BRDA, atmospheric carbonation seems to be the only explanation for the "brick-strength" entities formed. No carbonate appears in a list of BRDAs in general (Evans 2015). Nevertheless, one contradiction emerged regarding carbonate in BRDAs. For instance, Wong and Ho (1992) note "high concentrations" of dissolved Na^+, OH^-, CO_3^{2-}, and $Al(OH)^{4-}$ in bauxite waste which were to be disposed of in confined impounds. There is, therefore, a lack of certainty as to the proportion of carbonate in BRDAs.

PROBLEM 27.1.2
Mechanism of carbonation. What causes carbonation in a BRDA?

SOLUTION 27.1.2a: Proof from field tests

Gräfe et al. (2011) found that amendments such as applied gypsum can further displace Na^+ from the residue exchange complexes. According to Han et al. (2017), since Ca^{2+} concentration limits CO_2 sequestration in the bauxite residue, extra Ca^{2+} sources were added in a semi-soluble mineral and salt form (flue gas desulfurization gypsum or $CaCl_2$) to verify whether this Ca addition accelerated and enlarged the CO_2 sequestration obtained because of neutralization. Their results of 55 days of batch- and longer term field tests were in good agreement, and the neutralization rate was accelerated through the addition of both Ca^{2+} sources. Without the addition of the extra Ca source, their observations showed that atmospheric CO_2 contributed to the neutralization of pore water alkalinity only, while Ca^{2+} addition induced further neutralization through mineral carbonation of atmospheric CO_2 to $CaCO_3$. The simple addition of environmentally benign Ca to bauxite residue has provided a feasible, cost-effective, and easy to apply BRDA management practice in the field.

SOLUTION 27.1.2b: Proof from changes in pH with depth

Additionally, Khaitan et al. (2010) found that an older BRDA cell had a lower pH than the younger cell at all depths. Since no amendments were applied to either site, and no vegetation was present, they attributed the resulting neutralization and increase in inorganic carbon to slow carbonation from atmospheric CO$_2$. The pH changed with depth in the younger pond, indicating to them that neutralization by atmospheric CO$_2$ is probably limited by the availability of carbon dioxide to the residue.

Proof from several field studies

Carbonation causes neutralization, dewatering, and compaction of the bauxite residue (Evans 2015), the mud farming technique sequestering CO$_2$ from the atmosphere accelerating those processes (Pontikes 2016). This process neutralizes the free OH$^-$ present in the bauxite residue, leading to the formation of carbonates (Han et al. 2017). Depending on the refinery and the advances in residue management steps employed, further pH reductions occur through practices such as atmospheric carbonation (mud farming) (Pontikes 2016):

$$Ca(OH)_2 + CO_2 \rightarrow CaCO_3 + H_2O \qquad (27.1.2)$$

Evidence of the natural weathering (atmospheric carbonation) decreasing the pH was shown by Khaitan et al. (2010), who reported a pH of 10.5 for 14-year-old bauxite residue and 9.5 for 35-year-old bauxite residue, with the decreases attributed to the slow carbonation from atmospheric CO$_2$.

PROBLEM 27.1.2
Rate of carbonation

(a) In BRDAs, what is the rate of carbonation? (b) How can the rate be increased?

SOLUTION 27.1.2

(a) The differences in the extent and depth of neutralization between older and younger cells imply that carbonation of bauxite residue by exposure to atmospheric CO$_2$ is a slow process (Khaitan et al. 2010).

(b) Atmospheric CO$_2$ competes with intrinsic OH$^-$ and CO$_3^{2-}$ (carbonate) for any added Ca^{2+}. Hence, an excess of Ca^{2+} will increase the rate of carbonation. Gypsum addition to soil/red mud mixtures, even at relatively low concentrations (1% w/w), was sufficient to buffer experimental pH to 7.5–8.5, an effect attributed to the reaction of Ca^{2+} supplied by the gypsum with OH ($^-$) and carbonate from the red mud to precipitate calcite (Lehoux et al. 2013).

PROBLEM 27.1.3
Carbonation process

What chemical reactions occur during atmospheric carbonation in a bauxite waste?

SOLUTION 27.1.3

Where the Bayer system, which dumps bauxite into vats of boiling sodium hydroxide (NaOH) under high pressure, is used to extract aluminum, the following reactions occur with gypsum addition to the bauxite waste (Equation 27.1.3) or without gypsum addition (Equation 27.1.4):

$$CaSO_4 + NaOH \rightarrow Ca(OH)_2 + Na_2SO_4 \qquad (27.1.3)$$

$$Ca(OH)_2 + CO_2 \rightarrow CaCO_3 + H_2O \qquad (27.1.4)$$

PROBLEM 27.1.4
CO$_2$ sequestration vertically and horizontally in bauxite waste

How does carbonation vary spatially in an old bauxite waste (BRDA) dump?

SOLUTION 27.1.4

Vertically

At depths below 30 cm, the higher water content observed (Harris 2009) indicates less precipitation of calcium carbonate because downward CO_2 penetration from the atmosphere is blocked by the high strength hardened surface layers. Otherwise, atmospheric exposure would have also carbonated the older buried layers. Hence, vertically, the older, buried layers of BW are the least carbonated due to blockage of CO_2 penetration by the younger overlying deposits.

Horizontally

The positive effect of subaerial exposure on BW carbonation is borne out by experiments of Khaitan et al. (2010), where horizontally older cells that had longer exposure to CO_2 showed lower pH values and hence were more neutralized than younger cells. An older bauxite waste cell had a lower pH than the younger cell at all depths (Khaitan et al. 2010). The pH varied with depth in the younger pond, indicating that neutralization by atmospheric CO_2 is probably limited by the availability of carbon dioxide to the residue. Since they added no amendments to either site and no vegetation had been present, the resulting neutralization and increase in inorganic carbon is attributable to slow carbonation from atmospheric CO_2. Further proof can be seen from Khaitan et al. (2010), who found that carbon dioxide neutralization decreases the tricalcium aluminate content in the residue and increases the calcite content. They identified key controlling reactions for long-term carbonation of bauxite residue in the field including the dissolution of tricalcium aluminate (C3A) as follows:

$$Ca_3Al_2O_6(s) + 12H^+ = 3Ca_2^+ + 2Al_3^+ + 6H_2O \qquad (27.1.5)$$

$$\text{and precipitation of calcite: } Ca_2^+ + CO_3^{2-} = CaCO_3(s) \qquad (27.1.6)$$

where their X-ray diffraction (XRD) results show an increase in the height of calcite peaks and a decrease in C3A for the surface residue compared to the deeper residue.

As atmospheric carbonation increases the rate of neutralization, it is concluded that the longer the subaerial exposure (vertically or horizontally), the greater the carbon sequestration in BW dumps.

PROBLEM 27.1.5
The necessity of subaerial exposure

Why is turning over or removing carbonated BW in the surface zone imperative for efficient CO$_2$ sequestration at greater depths?

SOLUTION 27.1.5

A 30–35-year-old bauxite dump will have a highly lithified subaerial zone several centimeters thick, the bottom part of which had been deposited before the upper part. In addition to carbonation of bauxite residue by exposure to atmospheric CO$_2$ being "a slow process," there is a limit to the depth of atmospheric carbonation. Thus, the increased liquidity and decreased strength of bauxite waste below 15 cm depth observed by Harris (2019) in contrast to the lithified nature of the 0–15 cm depth indicate a drastic reduction of CO$_2$ sequestration below 15 cm. At such a stage, effective carbonation at, and below, that level can only occur *after* the overlying layer is removed, thereby facilitating subsurface contact with the atmosphere.

Therefore, slowing down or cessation of CO$_2$ sequestration at lower depths could have been ongoing for years or decades before Harris (2009) observed it. Hence, the claim of atmospheric carbonation being a slow process (Harris 2009; Khaitan et al. 2010) may have been mistaken for the early, relatively rapid calcification of the 0–15 cm zone restricting airflow and thereby subsequently reducing CO$_2$ access to the subsurface. This condition necessitates the physical turnover of hardened surface residue.

Also, the surface residue had a lower pH of 10.4 compared to the deeper residue which had a higher pH of 12.1. Thus, the data indicate that turning over or removing the consolidated 0–15 cm depth layer would increase carbon dioxide penetration below that depth.

PROBLEM 27.1.6
The potential of bauxite waste sequestration

What is the global potential for CO$_2$ sequestration in bauxite wastes?

SOLUTION: 27.1.6

That depends on the following:

1. Volume = Global acreage × Depth of bauxite residue disposal areas (BRDAs)
2. Mass of C3A (Ca$_2^+$ being the active sequester in the bauxite wastes)
3. Dissolution of C3A as: Ca$_3$Al$_2$O$_6$(s) + 12H$^+$ = 3Ca^{2+} + 2Al^{3+} + 6H$_2$O
4. The physical subaerial exposure of uncarbonated material below lithified surfaces

Thus:

Bauxite residue currently produced = 150 million tons per annum (Pontikes 2016)

Bauxite residue currently reused =<2% generated annually (Ujaczki et al. 2018)

Bauxite residue remaining = ~98%, in BRDAs (Di Carlo et al. 2019)

Depth of the impoundments = 1–16 m (3–52 ft)

BRDA average depth = 7 m

BRDA typical surface area = 44.6–105.3 ha (110 and 260 acres) (Pontikes et al. 2006)

BRDA average volume = 700,000 m^3

The refining used to produce aluminum generates about 2–2.5 tons of solid waste for every 1 ton of aluminum produced (EPA 2023). Therefore, at the annual world production total of 65.3 million tons of aluminum (Harbor Aluminum 2020), red mud waste stands at, at least 130 million tons per annum. Considering that bauxite mining began more than 70 years ago, and that production grew gradually to current levels, the estimate below seems to be a reasonable approximation:

$$0.5 \times 70 \text{ years} \times 130 = 4550 \text{ million tons in total}$$

Of that 4550 million tons, a very small proportion is surface-exposed to carbonation. If half were to be so exposed (= 2275 million tons), with 5% of the mass estimated on average to be of Ca^{2+} ions (Khaitan et al. 2010), and with carbonation occurring mainly by the following reaction:

$$Ca(OH)_2 + CO_2 \rightarrow CaCO_3 + H_2O \qquad (27.1.7)$$

Amount of CO_2 thus removed = CO_2 in 5%

$$= \left[44/100 \left(5\% \times 2.275 \times 10^6 \right) \right]$$

$$= 50,050 \text{ tons of } CO_2 \text{ carbonated in bauxite waste}$$

Compared with the annual tonnage of CO_2 placed in the atmosphere, this is a relatively small mass. Therefore, BRDA surface exposure alone is inadequate to make a big difference in reducing atmospheric CO_2. In this regard, if all the yearly amounts of 130 million tons were carbonated, the tonnage of CO_2 so removed would be:

$$Ans = \left[44/100 \left(1.3 \times 10^7 \times 0.5 \right) \right] = 286,000 \text{ tons of } CO_2 \text{ per year}$$

This is a more than fivefold per annum increase in carbonation, yet still far below the actual amount of 6 million tons estimated as being annually sequestered by atmospheric carbonation according to Si et al. (2013). However, the latter estimate seems to be based on the amount of red mud *produced*, rather than the amount interfacing with the atmosphere. But, as shown above, only a small fraction of that produced becomes carbonated.

PROBLEM 27.1.7
Pros and cons of BRDA sequestration methods

What are the advantages and disadvantages of the existing methods of carbon sequestration in bauxite waste dumps?

SOLUTION 27.1.7

Three methods are atmospheric carbonation, liquid CO_2 carbonation, and reforestation of the BRDAs.

Based on a study by Ramesh et al. (2010), the main constituents of red mud (% w/w) are: Fe_2O_3 (30–60%), Al_2O_3 (10–20%), SiO_2 (3–50%), Na_2O (2–10%), CaO (2–8%), and TiO_2 (trace–10%). Interestingly, carbonate appears neither in this list nor in that of O'Callaghan et al. (1998).

The drawback of neutralization by liquid CO_2 as described by Ramesh et al. (2010) was that initial pH decreases reverted to unacceptable environmental levels as additional alkaline material leached from the mud (Ramesh et al. 2010). One reason is that the alkalinity in BRDAs is largely contained in the solids and slowly released by their dissolution, so a point source addition of the liquid CO_2 allowed time for leaching from this surrounding alkaline material. On the other hand, natural atmospheric carbonation from the air, though slower acting than a liquid CO_2 point source, simultaneously alters all subaerially exposed surfaces, leaving no outlying alkaline zone of BRDA solids, thereby avoiding the reversals associated with liquid point source additions of CO_2.

PROBLEM 27.1.8
Expediting carbonation in bauxite waste

Apart from subaerial exposure of BRDAs, how else can atmospheric carbon sequestration be accelerated in the earth?

SOLUTION 27.1.8

Alkaline solid wastes

Alkaline solid wastes, which include a wide array of rock and mud-like wastes from mining, cement and aluminum production, coal burning, and other large-scale industrial processes, share a similar affinity for carbon dioxide. Having a high pH causes them to react with CO_2, a mild acid, through carbonation, which fixes CO_2 in alkaline materials permanently. Alkaline solid wastes, including red mud, containing Ca and Mg, can be utilized in capturing CO_2 (Rahmanihanzaki & Hemmati 2022).

PROBLEM 27.1.9
The role of atmospheric exposure

What are the two rates of carbonation in BRDAs? How can the "slow carbonation" rate of CO_2 sequestration observed at depth in BRDAs be accelerated?

SOLUTION 27.1.9

Gypsum not only effectively reduces the alkalinity of bauxite residue by continuously releasing Ca^{2+} to react with carbonate and hydroxide (Ren et al. 2022) but facilitates carbonation (mud farming) (Evans 2016) in the BRDAs, which helps in the neutralization, dewatering, and compaction of the bauxite residue (Pontikes 2016). However, this natural neutralization process occurs very slowly, at a maximum rate of up to 1.2 m in 35 years (Khaitan et al. 2010). But Harris (2009) consistently observed much slower lithification with viscous material below depths of just 30 cm in 25-year-old red mud tailings dumps:

$$1200 \text{ cm}/35 \text{ years} = 34.28 \text{ cm/year (Khaitan et al. 2010)}$$
$$20 \text{ cm}/25 \text{ years} = 0.8 \text{ cm/year (Harris 2009)}$$

There is, therefore, "fast carbonation" to lithify the top layer, followed by slow carbonation at depth. However, having measured the depth of carbonation at the Williamsfield, Jamaica, BW in April 2020, the author again measured it 2.75 years later in January 2023 and observed no extension in depth of lithification ("fast carbonation"). The "fast carbonation" process therefore had ceased, at the very least having done so during the preceding 2.75 years. And on balance of probabilities, the cessation had begun before that. Yet, as both processes occur downward from the surface, "slow carbonation" cannot proceed before fast carbonation. Whatever slow carbonation had occurred at depth had ceased or greatly slowed for years after the stoppage of "fast carbonation." Even additives, such as gypsum, as top dressing, applied before lithification occurred in the top layer, made little or no difference to the uncarbonated layer below 20 cm after 10 years of gypsum application (Harris 2009). In other words, the barrier to deeper carbonation formed by lithification at the surface through "fast carbonation" almost totally impedes further carbonation at depth, even despite the addition of additives. The "slow carbonation" rate of CO_2 sequestration observed at depth in BRDAs can therefore be accelerated by using earth moving machines to push off the 0–15 cm surface crust, thereby accelerating atmospheric carbonation of the underlying layer.

PROBLEM 27.1.10
The most effective source of $Ca^{2\pm}$

Why are gypsums the preferred additives for atmospheric carbonation in BRDAs?

SOLUTION 27.1.10

1. At a pH of 10 and higher, lime, $Ca(OH)_2$, in contrast to gypsum, is insoluble (Wong & Ho 1992), and bauxite red mud wastes have an initial pH of at least 11 (Bardossy 1982).

2. Though more quickly acting in soil than gypsum (Shainberg et al. 1989), in hot humid areas, the costly calcium chloride (CaCl$_2$), for example, dissolves so quickly that it is usually leached out of the soil layers in a relatively short period.
3. Because the sparingly soluble gypsum remains in the soil for a longer time, and opportunities are thereby provided for solvation to occur over many more effective exchange steps, it is both efficient and economical (Oster 1982). Further, the Cl$^-$ ion (in CaCl$_2$) can be toxic to plants.
4. Though carbonation using added gypsum produced high-strength masses in a BRDA (Harris 2009), Rahmanihanzaki and Hemmati (2022) noted titanate and silicate ions, the two major oxyanions that appeared to strongly compete with carbonate ions for the available soluble Ca. Hence, a steady, continuous supply of additional soluble Ca and/or Mg could be a viable pathway for maximizing carbon sequestration in red mud and simultaneously reducing the causticity of red mud.

PROBLEM 27.1.11
Why are extra Ca^{2+} ions required for fast sequestration?

Addition of 1% gypsum to a BRDA increased the quantities of Na$^+$, Ca^{2+}, and SO$_4^{2-}$ leached. Therefore, why is deep ripping needed?

SOLUTION 27.1.11

As the high-strength surface consistency of red muds at a thickness of 0–20 cm supports the mass of an average adult individual (Figure 27.2), portable lightweight deep ripping machines could expose the underlying viscous partially carbonated material to similar carbonation.

PROBLEM 27.1.12
Which reactions in BRDAs are evidence of long-term atmospheric carbonation?

SOLUTION 27.1.12

Based on the following key controlling reactions for long-term carbonation of bauxite residue in the field, Khaitan et al. (2010) found that carbon dioxide neutralization decreases the C3A content in the residue and increases the calcite content:

$$Ca_3Al_2O_6(s) + 12H^+ = 3Ca^{2+} + 2Al^{3+} + 6H_2O \qquad (27.1.8)$$

and precipitation of calcite:

$$Ca^{2+} + CO_3^{2-} = CaCO_3(s) \qquad (27.1.9)$$

Their XRD results showed an increase in the height of calcite peaks and a decrease in C3A for the surface residue compared to the deeper residue. Additionally, their

FIGURE 27.2 Carbonated, aged "red muds," lithified on the surface but viscous below a depth of 20 cm. The material, at this stage, is strong enough to support a 170-pound individual. Removing this 20-cm-thick carbonated material would expose the underlying uncarbonated sub-layer to the atmosphere, thereby accelerating the sequestration of CO_2.

surface residue having had a lower pH of 10.4 compared to the deeper residue which had a higher pH of 12.1 indicates a slower carbon dioxide neutralization for the deeper residue than for the surface, which may be due to a long-term slow rate of CO_2 movement to depth and, according to Khaitan et al. (2010), possibly also by the (low) rate of C3A dissolution in the residue.

PROBLEM 27.1.13
Gypsum application procedure for carbonation

How is gypsum physically applied to achieve efficient atmospheric carbonation in bauxite waste dumps?

SOLUTION 27.1.13

Advantages of subsurface application

While gypsum has a solubility in water of about one-fourth of 1% (Lebedev and Kosorukov 2017 and is, therefore, a direct source of soluble calcium cations over a prolonged period, when Abrol and Dahiya (1974) applied gypsum *on* the surface of a soil and leached it, only a small fraction of the soluble carbonates reacted with the released calcium. A major fraction of the soluble carbonates leached did not react with applied gypsum. This suggests a lack of sufficient Ca^{2+} ions emanating from the surface-applied gypsum. Conversely, placing gypsum *into* the ground rather than *on* the ground ensures a greater surface area of contact between Ca^{2+} and the ground.

Deep ripping down to the >15 cm depth improves physical contiguity, and some studies (Webster & Nyborg 1986) have shown gypsum application at high rates can decrease bulk density through increased soil aggregation, and aggregation provides a greater surface area for atmospheric carbonation as compared to no aggregation.

Moisture requirements

Under field conditions, applying the treatment at the onset of the rainy season or irrigating prior to its application would ensure solubilization of gypsum.

Small particle size

The gypsum particles were unsorted when applied by O'Callaghan et al. (1998), such that some particles still exceeded 2 cm in diameter when they were observed in the BRDA 10 years later (Harris 2019). A much smaller particle size, e.g., 2 mm, would have released more Ca^{2+} ions much more rapidly and probably to greater depths than had occurred.

PROBLEM 27.1.14
Maximizing the concentration of $Ca^{2\pm}$ from gypsum

How can the concentration of Ca^{2+} be maximized while reducing the water requirement in a bauxite waste dump?

SOLUTION 27.1.14

Quantities of water required for dissolution are generally calculated based on gypsum solubility in free water. For example, if a sodic soil requires a gypsum application of 12.5 t ha^{-1}, the quantity of water required to dissolve this quantity will be above 50 cm depth. However, this is not likely to be the case in practice, as shown by Gupta and Abrol (1990). Gypsum solubility increases severalfold when it is mixed in a highly sodic soil because of the preference of exchange sites for divalent calcium ions compared to sodium ions (calcium is less strongly pulled to exchange sites dominated by multivalent cations). The higher the sodium saturation, the greater the dissolution of gypsum mixed into soil, and by extension, into BRDAs, and hence the lower the water requirements. Sood et al. (2009) observed that <14 cm water was required to dissolve gypsum applied at 12.4 t ha^{-1} and leach reaction products from the top 15 cm soil in a highly sodic soil having an exchangeable sodium percentage (ESP) of 94.0. This quantity of water is only about one-fourth the quantity of water they calculated from considerations of solubility in water alone, and it explains the rapid vegetation growth in a 2007 study, when rains quickly followed a gypsum application to a BRDA (Harris 2009).

Nevertheless, such favorable conditions for increasing the Ca^{2+} concentration in BRDA did not occur in the O'Callaghan et al. (1998) study which entailed surface application. This probably accounted, at least in part, for the weak carbonation that Harris (2009) observed below the 20 cm level in the same BRDA 10 years after the treatment.

PROBLEM 27.1.15
Potential impact of trees

Which is the more efficient method of carbon sequestration: growing trees on the bauxite waste after treatment or subjecting the bauxite waste to atmospheric carbonation without trees?

SOLUTION 27.1.15

Tree growth or carbonation in a BRDA requires the same gypsum treatment. As shown above, atmospheric carbonation takes several years, wherein the strength of the BRDA changes, such that hard, impermeable masses of mainly calcium carbonate completely restrict plant growth. In the intervening period, with a lowered pH, some plants such as leguminous acacias will grow in the dried-out sections of the BRDA. Tree growth will also accelerate carbonation by loosening the material when roots increase in size, thereby enlarging the surface area of contact of the BRDA with the atmosphere.

On the other hand, concurrent carbonation with plant colonization can occur if volunteer plant species are allowed to proliferate, in addition to such plants being proactively incorporated during the years required for the carbonation process. Such an arrangement improves both physical and chemical aspects of the BRDA while sequestering carbon by two processes: (a) biomass photosynthesis and (b) BRDA carbonation.

REFERENCES

Abrol IP, Dahiya IS (1974) Flow-associated precipitation reactions in saline-sodic soils and their significance. Geoderma 11, 305–312.

Bardossy G (1982) Karst Bauxites. Elsevier Scientific Publishing Co. Amsterdam, New York, New York.

Chunhua Si 1, Yingqun Ma, Chuxia Lin (2013) Red mud as a carbon sink: variability, affecting factors and environmental significance. doi: 10.1016/j.jhazmat.2012.11.024

Di Carlo E, Boullemant A, Courtney R (2019) A field assessment of bauxite residue rehabilitation strategies. Sci Total Environ 663, 915–926.

EPA (2023) TENORM: Bauxite and Alumina Production Wastes. https://www.epa.gov/radiation/tenorm-bauxite-and-alumina-production-wastes#:~:text=The%20refinery%20processes%20used%20to%20produce%20aluminum%20generates,and%20solid%20wastes.%20These%20wastes%20can%20contain%20TENORM.

Gräfe M, Power G, Klauber C (2011) Bauxite residue issues: III. Alkalinity and associated chemistry. Hydrometallurgy 108, 60–79.

Gupta R K, Abrol I P (1990) Reclamation and management of alkali soils. Indian J Agric Sci 60, 1–60. doi: 10.1007/978-1-4612-3322-0_7

Han YS, Ji SW, Lee PK, Oh C (2017) Bauxite residue neutralization with simultaneous mineral carbonation using atmospheric CO_2. J Hazard Mater 326, 87–93.

Harbor Aluminum (2020) Aluminum Production by Country. https://www.harboraluminum.com/en/top-aluminum-producing-countries#:~:text=While%20contribution%20by%20country%20varies%2C%20world%20primary%20aluminum,Europe%2C%20North%20America%2C%20Latin%20America%2C%20Africa%20and%20Asia.

Harris MA (2009) Structural improvement of age-hardened gypsum-treated bauxite red mud waste using readily decomposable phyto-organics. Environ Geol 56, 1517–1522. doi: 10.1007/s00254-008-1249-5

Harris MA (2016) Uniaxial compressive strength of some un-calcined red mud mortars: Geotechnical implications. In *Geobiotechnological Solutions to Anthropogenic Disturbances* (pp.121–139). Springer Cham, Geneva, Switzerland. doi: https://doi.org/10.1007/978-3-319-30465-6

Harris MA (2019) Geotechnical implications of aging on impact resistance of uncalcined red mud mortars in a wet, warm climate. In: Confronting Global Climate Change: Experiments & Applications in the Tropics, 1st Edition, CRC Press, eBook ISBN9780429284847

Khaitan S, Dzombak DA, Lowry GV (2010) Field evaluation of bauxite residue neutralization by carbon dioxide, vegetation, and organic amendments. J Environ Eng, 136, 1045–1053. doi: 10.1061/(ASCE)EE.1943-7870.0000230

Lebedev AL, Kosorukov VL (2017) Gypsum solubility in water at 25°C. Geochem Int 55, 205–210. https://doi.org/10.1134/S0016702917010062

Lehoux AP, Lockwood CL, Mayes WM, Stewart DI, Mortimer RJG et al., (2013) Gypsum addition to soils contaminated by red mud: Implications for aluminium, arsenic, molybdenum, and vanadium solubility. Environ Geochem Health 35(5), 643–656. http://dx.doi.org/10.1007/s10653-013-9547-6

O'Callaghan WB, McDonald SC, Richards DM Reid RE (1998) Development of a topsoil-free vegetative cover on a former red mud disposal site. Alcan Jamaica Rehabilitation Project Paper.

Oster JD (1982) Gypsum use in irrigated agriculture – a review. Fert Res 3, 73–83.

Pontikes Y, Boufounos D, Fafoutis D (2006) Environmental aspects on the use of Bayer's process bauxite residue in the production of ceramics. Adv Sci Technol 45, 2176–2181. doi: 10.4028/www.scientific.net/AST.45.2176

Rahmanihanzaki M, Hemmati A (2022) A review of mineral carbonation by alkaline solid-waste. Int J Greenh Gas Control 121, 103798. https://doi.org/10.1016/j.ijggc.2022.103798

Ren X, Zhang X, Tuo P et al. (2022) Neutralization of bauxite residue with high calcium content in abating pH rebound by using ferrous sulfate. Environ Sci Pollut Res 29, 13167–13176. https://doi.org/10.1007/s11356-021-16622-3

Shainberg I, Sumner ME, Miller WP, Farina MPW, Pavan M A, Fey MV (1989) Use of gypsum on soils: A review. Advan Soil Sci 9, 1–111. doi: 10.1007/978-1-4612-3532-3_1

Sood A, Choudhury BU, Ray SS et al. (2009) Impact of cropping pattern changes on the exploitation of water resources: A remote sensing and GIS approach. J Indian Soc Remote Sens 37, 483–491. https://doi.org/10.1007/s12524-009-0033-7

Ujaczki É, Feigl V, Molnár M, Cusack P, Curtin T et al. (2018) Re-using bauxite residues: Benefits beyond (critical raw) material recovery. J Chem Technol Biotechnol 93(9). https://doi.org/10.1002/jctb.5687

Webster GR, Nyborg M (1986) Effects of tillage and amendments on yields and selected soil properties of two solonetzic soils. Can J Soil Sci 66, 455–470. www.world-aluminium.org

Wong JW, Ho GE (1992) Viable technique for direct re-vegetation of fine bauxite refining residue. In Proceedings of an International Bauxite Tailings Workshop, November 2–6, Perth, Western Australia, pp. 258–268.

28 CO$_2$ Sequestration
Alkali Mine Waste Treatments in Acid Mine Drainage

MAIN POINTS

- Carbonation observed in an acid mine drainage (AMD) treatment pit.
- Calcium hydroxide reacts with heavy metal cations in AMD. Excess Ca^{2+} reacts with atmospheric CO$_2$.

28.1 CARBONATION IN ACID MINE DRAINAGE WASTE HEAPS

Unlike some schemes for drawing excess CO$_2$ from the atmosphere, that of reactive alkaline rocks, and alkali mine wastes can both capture the gas and permanently store it, in a solid mineral (Science.org 2020). Hence, active treatment of acid mine drainage (AMD) involving the addition of alkaline reagents such as NaOH or Ca(OH)$_2$ to increase the pH and precipitate the dissolved metals can initiate carbonation. This occurs because substantial amounts of dissolved ions such as Ca^{2+} and Mg^{2+}, fundamental in removing CO$_2$ from the atmosphere, often persist after the treatments (Hyun-Cheol et al. 2016). For example, the reaction of AMD with limestone causes dissolution and produces HCO$_3^-$ and Ca^{2+}:

$$CaCO_3 + H^+ \rightarrow Ca^{2+} + HCO_3^- \qquad (28.1.1)$$

Atmospheric carbonation can then react with Ca^{2+} to produce thermodynamically stable CaCO$_3$.

With the current number of abandoned mines in the USA, for example, estimated to be more than 557,000, many of which are active sources of AMD (RoyChowdhury et al. 2015), there is scope for globally effective remediation of atmospheric CO$_2$.

Following neutralization of AMD (Harris & Ragusa 2001), the waste, consisting of metal precipitates, gypsum, and cations including Ca^{2+}, was stored in a wide pit in Brukunga, South Australia.

PROBLEM 28.1.1
Carbonation reactions

What are some neutralization reactions, products, and ramifications occurring when Ca(OH)$_2$ is used to treat iron pyrites-caused AMD?

DOI: 10.1201/9781003341826-34

SOLUTION 28.1.1

Generally, the products of the post-neutralization high density settling (HDS) process also contain gypsum ($CaSO_4$) and unreacted lime according to the following reactions:

$$H_2SO_4 + Ca(OH)_2 \rightarrow CaSO_4 + 2H_2O \qquad (28.1.2)$$

$$H_2SO_4 + CaO \rightarrow CaSO_4 + H_2O \qquad (28.1.3)$$

or as ions in aqueous solution:

$$SO_2^{-4} + 2H^+ + Ca^{2+} + O_2^-(aq) \rightarrow Ca^{2+} + SO_2^{-4}(aq) + 2H^+ + O_2^-(aq) \qquad (28.1.4)$$

Or as leftover calcium reacting with CO_2 as follows:

$$2Ca^{2+} + 2CO_2 + 2H_2O \rightarrow 2CaCO_3 + H_2 \qquad (28.1.5)$$

The resulting slurry flowed to a sludge-settling vessel, where a mass of metal precipitates (sludge) which settled was recycled to the AMD treatment tank. After the neutralization cycles, the low-strength, friable waste sludge consisting of gypsum plus metal precipitates and ions in solution including Ca^{2+} was stored in a pit. Less than 5 years later, the dumped material exhibited vastly increased strength (a five-fold increase), akin to that of a continuous pediment of granitic rock, with the active ingredient being calcium carbonate. As carbon or carbonates pre-existed in neither the AMD nor the treatment, it is reasonable to conclude that atmospheric carbon dioxide reacted with Ca^{2+} ions in the waste sludge to produce calcium carbonate. Because it is a strong base, the $Ca(OH)_2$ ionized and dissolved in the AMD. The resulting hydroxide (OH^-) ions combined with metal cations and hydrogen (H^+) ions, taking them out of the solution. So, the calcium (Ca^{2+}) which carbonated to $CaCO_3$ in the waste dump came from acid-neutralizing reactions.

PROBLEM 28.1.2
Proportion of waste carbonated in an AMD sludge waste

How much CO_2 can be fixed by carbonation in the AMD sludge waste dump?

SOLUTION 28.1.2

The estimated capacity of CO_2 sequestration through the carbonation treatment of mine drainage from one mine area was of the considerable level of 0.54 g CO_2/kg mine drainage (Hyun-Cheol et al. 2016). Additionally, for CO_2 immobilization efficiency, it is necessary to estimate the optimum volume of CO_2 injection through thorough preliminary tests.

PROBLEM 28.1.3
Long-term effects and sustainability

How effective in the long term is the atmospheric carbonation of alkali wastes?

SOLUTION 28.1.3

Samples from three mine areas resulted in calcite precipitates of different amounts, and no additional environmental problems occurred through CO_2 injection into the mine drainage. Thermodynamically stable carbonate minerals resulted (Hyun-Cheol et al. 2016). Hyun-Cheol et al. (2016) found that, conclusively, CO_2 sequestration using the neutralization process of mine drainage can be evaluated as a positive technique in terms of sustainable development.

PROBLEM 28.1.4
Acid neutralizing agents

One common process is lime (CaO) precipitation in a high-density sludge (HDS) process. In this application, a slurry of lime is dispersed into a tank containing AMD and recycled sludge to increase water pH to about 9, an alkalinity at which most toxic metals become insoluble and precipitate, aided by the presence of recycled sludge (RoyChowdhury et al. 2015). Which chemical agents are used for neutralizing AMD?

SOLUTION 28.1.4

The range of chemical agents being used during the active treatment of AMD water worldwide include limestone ($CaCO_3$), hydrated lime ($Ca(OH)_2$), caustic soda (NaOH), soda ash (Na_2CO_3), calcium oxide (CaO), anhydrous ammonia (NH_3), magnesium oxide (MgO), and magnesium hydroxide ($Mg(OH)_2$).

PROBLEM 28.1.5
Ramifications of active neutralization

What are the advantages and disadvantages of active AMD neutralization?

SOLUTION 28.1.5

Advantages

- Unlike the passive treatment facilities, it does not require any additional space or construction (RoyChowdhury et al. 2015).
- It is fast (hours, not weeks or years) and effective in removing acidity and metals (Harris & Ragusa 2001).
- Lower cost associated with handling and disposal of sludge in comparison to passive treatment techniques (Coulton et al. 2003).
- Treatments include a wide range of effective chemical reagents (Skousen et al. 2000).

Disadvantages

- A major disadvantage of the active treatment process is that it requires a continuous supply of chemicals and energy to perform efficiently (RoyChowdhury et al. 2015)

- Costly chemicals or costly transport of bulky materials.
- Acquiring sufficient manpower to maintain the system increases the overall cost of this technology significantly (RoyChowdhury et al. 2015).
- Efficiency of these systems is completely dependent on regular maintenance and chemical supply: difficult to control for most remotely located abandoned mine sites (RoyChowdhury et al. 2015).
- Efficiency and cost of the systems vary with the type of neutralizing agent used.
- Limestone is inexpensive but less soluble in water and hence less effective than the other chemical agents.
- The use of excessive ammonia can create problems such as nitrification and denitrification in receiving water bodies (Skousen et al. 2000).

REFERENCES

Coulton R, Bullen C, Hallet C (2003) The design and optimization of active mine water treatment plants. Land Conta Reclam 11, 273–279.

Harris M, Ragusa S (2001) Bioremediation of acid mine drainage using decomposable plant material in a constant flow bioreactor. Env Geol 40, 1192–1204. https://doi.org/10.1007/s002540100298

Hyun-Cheol L, Kyoung-Won M, Eui-Young S (2016) A feasibility study on CO$_2$ sequestration using the neutralization process of acid mine drainage. Geosystem Eng 19(6), 293–301. doi: 10.1080/12269328.2016.1207571

RoyChowdhury A, Sarkar D, Datta, R (2015) Remediation of acid mine drainage-impacted water. Curr Pollution Rep 1, 131–141. https://doi.org/10.1007/s40726-015-0011-3

Science.org (2020) The carbon vault. Industrial waste can combat climate change by turning carbon dioxide into stone. https://www.science.org/doi/10.1126/science.369.6508.1156

Skousen JG, Sexstone A, Ziemkiewicz PF (2000) Acid mine drainage control and treatment. Agronomy 41, 131–168.

29 Pozzolanic Carbon Sequestration in Red-Mud-Treated Mortars

MAIN POINTS

- Without calcining, without additives, red mud waste increases mortar strength.
- Strength increases occurred when red mud added was up to 30%.

29.1 POZZOLANS OR SUPPLEMENTARY CEMENTITIOUS MATERIALS

According to ASTM C125-11 (2011), Shi et al. (1999), and Hanson (2017), pozzolans or supplementary cementitious materials (SCMs) are defined as siliceous and aluminous material which possesses little or no cementitious value, but which will, in finely divided form in the presence of moisture, react chemically with calcium hydroxide at ordinary temperatures to form compounds possessing cementitious properties such as calcium silicate hydrate when Portland cement is hydrated. Calcium silicate, being a pozzolan, is one of a group of compounds that can be produced by reacting calcium oxide (CaO) and silica in various ratios (Merlini et al. 2008). Because their use reduces overall environmental impact and cost when mixed with Portland cement (CEM-I) in blended cement systems, pozzolans are of increasing value and interest.

PROBLEM 29.1.1
Pozzolan mechanism

Why are bauxite residue disposal areas (BRDAs) pozzolanic?

SOLUTION 29.1.1

Calcium hydroxide is liberated during the hydration of Portland cement (Majumdar et al. 1990; Merlini et al. 2008) at ordinary temperatures (Equation 29.1.1) to form compounds possessing cementitious properties (Ephraim et al. 2012; Akeke et al. 2013). Finely powdered red mud from the waste products of aluminum smelting hardens and forms a stable and durable compound (Muller 2005) after mixing with $Ca(OH)_2$ (lime) in the presence of water (Yalçin & Sevinc 2000).

$$2Ca_3SiO_5 + 7H_2O \rightarrow 3CaO \cdot 2SiO_2 \cdot 4H_2O + 3Ca(OH)_2 + 173.6\,kJ \qquad (29.1.1)$$

DOI: 10.1201/9781003341826-35

PROBLEM 29.1.2
Required time for curing

Why did the testing for mortar strength occur 2 years after the mortar was made?

SOLUTION 29.1.2

The contribution of the pozzolanic reaction to cement strength is usually developed at later curing stages, depending on the pozzolanic activity. The required calcium hydroxide is usually supplied during hydration of the cement. In most blended cements, initial lower strengths can be observed compared to the parent Portland cement. However, especially in the case of pozzolans finer than the Portland cement, the decrease in early strength is usually less than what can be expected based on the dilution factor. This can be explained by the filler effect, in which small secondary cementitious material (SCM) grains fill in the space between the cement particles, resulting in a much denser binder (Muller 2005). The acceleration of the Portland cement hydration reactions can also partially accommodate the loss of early strength. In this study, the extended period (2 years) provided opportunities for filling in the fine pores, thereby acquiring the potential strength of the composite mortar.

PROBLEM 29.1.3
Identifying pozzolans (SCM): common attributes

What are the three common characteristics of pozzolans?

SOLUTION 29.1.3

Calcining

The heating of solids to a high temperature can remove volatile substances, oxidize a portion of mass, or render them friable, such as bringing limestone to a temperature high enough to expel the carbon dioxide and producing the lime in a highly friable or easily powdered condition. The pozzolanic activities of siliceous materials – volcanic ash, metakaolin, silica fume, coal fly ash, sewage sludge ash, glass powders, silica fumes, crushed bricks, sewage sludge ash, coal bottom ash, and bauxite waste – occurred at some stage by subjection to prolonged high temperatures which changed their chemical and physical properties (calcining).

Small particle size

The particle size of each above-mentioned material is under 50 μm in diameter.

They are terminal (recalcitrant) siliceous materials

Pozzolans are silicates (including ash) that cannot be further physically transformed before their fusion temperature.

PROBLEM 29.1.4
Pozzolanic role in concrete durability

What are the advantages of adding pozzolanic materials to concrete and mortars?

SOLUTION 29.1.4

Durability

One significant benefit of blended cements is improved durability due to the reaction between the pozzolan and the excess $Ca(OH)_2$, which reduces the porosity of the binder (Donatello et al. 2010). Therefore, one of the main advantages of pozzolan blended cements is increased chemical resistance to the ingress and harmful action of aggressive solutions. Such improved durability of the pozzolan-blended binders in the context of the current exacerbation of global warming by cement manufacturing lengthens the service life of structures and reduces the costly and inconvenient need to replace damaged construction (Shi et al. 1999).

Decrease of atmospheric CO_2

Portland cement manufacture utilizes vast quantities of fossil fuels in kilns. Replacing a proportion of Portland cement in concrete with red mud tailings (RM) may therefore reduce carbon emissions.

PROBLEM 29.1.5
Choice of aged (dried) red mud or liquid red mud

To reduce global warming, which is the better SCM (aged, dry red mud or liquid red mud)?

SOLUTION 29.1.5

X-ray diffraction analysis by Shi et al. (1999) indicates that the main components in red mud are kaolinite ($Al_2O_3 \cdot 2SiO_2 \cdot 2H_2O$), quartz ($SiO_2$), and titanium oxide ($TiO_2$). But the pozzolanic activity of the kaolinite, the major pozzolan, occurs only after the chemically combined water is removed, a process requiring high energy inputs by calcination at 750°C for 5 h, whereupon kaolinite in the mud converts into metakaolinite ($Al_2O_3 \cdot 2SiO_2$), a dehydroxylated form of the clay mineral kaolinite commonly used in the production of ceramics (Pinnock & Gordon 1992). Working with liquid Jamaican red muds (Pinnock & Gordon 1992) also needed similar high-energy calcining to achieve high pozzolanic activity. One of their main reasons for introducing heat was to drive chemically bound water from the then liquid red mud.

But working several years later in the then atmospherically carbonated and naturally dried BRDA, Harris (2016) observed strength increases and found no need for calcining.

PROBLEM 29.1.6
Effectiveness of aged, dry red mud as a SMC

Replacing a proportion of CEM-1 with an SMC puts less CO_2 into the atmosphere. How much stronger is CEM-1 mortar when spiked with dry, aged, untreated red mud?

SOLUTION 29.1.6

2019 Gypsum-treated dry red mud tests

The toughness of a material determines its ability to absorb energy during plastic deformation (Schneider et al. 2014; Ahmad et al. 2017). Impact resistance is the ability of a material to withstand a high force or shock applied to it over a short period of time. Therefore, a high force over a short time has a greater effect than a weaker force over a longer period.

Testing the toughness of Portland cement mortar mixed with bauxite waste by comparing its impact-resistance with that of a pure CEM-1 mortar, Harris (2019) found that the results of the gypsum-treated red mud-spiked CEM-1 mortar exceeded those of the CEM-1 samples.

REFERENCES

Ahmad S, Umar A, Masood A (2017) Properties of normal concrete, self-compacting concrete and glass fibre-reinforced self-compacting concrete: An experimental study. Procedia Eng 173, 807–813. doi: 10.1016/j.proeng.2016.12.106

Akeke GA, Ephraim M, Akobo ZS, Ukpata J (2013) Structural properties of rice husk ash concrete. Int J Eng Appl Sci 3 (3), 8269

ASTM C125-11 (2011) Standard Terminology Relating to Concrete and Concrete Aggregates. https://www.astm.org/c0125-11.html

Donatello S, Tyrer M, Cheeseman CR (2010) Comparison of test methods to assess pozzolanic activity. Cem Concr Compos 32(2), 121–127.

Ephraim M, Akeke G, Ukpata J (2012) Compressive strength of concrete with RHA as partial replacement of ordinary Portland cement. Sch J Eng Res 1(2), 32–36.

Hanson K (2017) SCMs in Concrete: Natural Pozzolans. National Precast Concrete Association 2017 – September-October/SCMs in Concrete: Natural Pozzolans/SCMs in Concrete: Natural Pozzolans, September 22, 2017. https://precast.org/2017/09/scms-concrete-natural-pozzolans/

Harris MA (2016) Dust reduction in bauxite red mud waste using carbonation, gypsum & flocculation. In: Geobiotechnological Solutions to Anthropogenic Disturbances. Environmental Earth Sciences. Springer, Cham. https://doi.org/10.1007/978-3-319-30465-6_2

Harris M (2019) Confronting Global Climate Change Experiments & Applications in the Tropics. CRC Press, Boca Raton.

Majumdar AJ, Singh B, Edmonds RN. Hydration of mixtures of cement, aluminous cement and granulated blast furnace slag. Cem Concr Res 20(2), 197–208.

Merlini M, Artioli G, Cerulli T, Cella F, Bravo A (2008). Tricalcium aluminate hydration in additivated systems. A crystallographic study by SR-XRPD. Cem Concr Res 38(4), 477–486. doi: 10.1016/j.cemconres.2007.11.011

Muller CJ (2005) Pozzolanic activity of natural clay minerals with respect to environmental geotechnics. PhD dissertation Diss. ETH No. 16299, A dissertation submitted to the Swiss Federal Institute of Technology Zurich.

Pinnock W, Gordon JN (1992) Assessment of strength development in Bayer process residues. Cem Concr Compos 18(6), 71–379.

Schneider N, Pang W, Gu M (2014) Application of Bamboo for Flexural and Shear Reinforcement in Concrete Beams. In Structures Congress 2014. http://ascelibrary.org/doi/abs/10.1061/9780784413357.091.

Shi C, Grattan-Bellew PE, Stegemann JA (1999) Conversion of a waste mud into a pozzolanic material. Constr Build Mater 13(5), 279–284.

Section VI

Appendices

30 Appendix A
Sampling and Statistics

30.1 STATISTICAL TESTS

Sampling procedures are pre-eminent prior to applying statistical tests.

For example, when studying polluted soils, samples should be collected in such a manner as to best characterize the extent of contamination of the soil in question (DEC 2022) and, by extension, polluted air and water.

PROBLEM 30.1.1
Parameters of sulfur dioxide

The following data represent the levels of concentration in ppb (parts per billion) of sulfur dioxide in the air over a city during the 15-day timespan after an explosion in a chemical plant. Determine (a) the mean, (b) the median, (c) the mode, (d) the standard deviation, and (e) whether this represents a normal curve distribution: 15, 16, 17, 14, 13, 10, 9, 9, 8, 8, 8, 7, 6, 4, 4.

SOLUTION 30.1.1

a. Mean = 9.8667
b. Median = 9
c. Mode = 8
d. i. Population standard deviation = 4.0475
   ii. Sample standard deviation = 4.1896
e. Yes, it represents a normal curve distribution

Mean, median, mode, population standard deviation (σ), and sample standard deviation (S) from the following grouped data.

X	Frequency
4	2
6	1
7	1
8	3
9	2
10	1
13	1
14	1
15	1
16	1
17	1

DOI: 10.1201/9781003341826-37

x (1)	Frequency (f) (2)	$f \cdot x$ (3) = (2) × (1)	$f \cdot x^2 = (f \cdot x) \times (x)$ (4) = (3) × (1)	cf (5)
4	2	8	32	2
6	1	6	36	3
7	1	7	49	4
8	3	24	192	7
9	2	18	162	9
10	1	10	100	10
13	1	13	169	11
14	1	14	196	12
15	1	15	225	13
16	1	16	256	14
17	1	17	289	15
---	---	---	---	---
	$n = 15$	$\Sigma fx = 148$	$\Sigma f \cdot x^2 = 1706$	--

$$\text{Mean } \bar{x} = \sum = fx / n$$
$$= 148 / 15$$
$$= 9.8667$$

Median:

$$M = \text{Value of } \left(\frac{n+1}{2}\right)^{th} \text{observation}$$

$$= \text{Value of} \left(\frac{16}{2}\right)^{th} \text{observation}$$

$$= \text{Value of } 8th \text{ observation}$$

From the column of cumulative frequency cf, we find that the 8th observation is 9. Hence, the median of the data is 9.

Mode:
The frequency of observation 8 is maximum.

$$\text{Mode} = 8$$

$$\text{Population standard deviation: } \sigma = \sqrt{\frac{\Sigma f \cdot x^2 - \dfrac{(\Sigma f \cdot x)^2}{n}}{n}}$$

$$= \sqrt{\frac{1706 - \dfrac{(148)^2}{15}}{15}}$$

$$= \sqrt{\frac{1706 - 1460.2667}{15}}$$

$$= \sqrt{\frac{245.7333}{15}}$$

$$= \sqrt{16.3822}$$

$$= 4.0475$$

Sample standard deviation: $S = \sqrt{\dfrac{\Sigma f \cdot x^2 - \dfrac{(\Sigma f \cdot x)^2}{n}}{n-1}}$

$$= \sqrt{\frac{1706 - \dfrac{(148)^2}{15}}{14}}$$

$$= \sqrt{\frac{1706 - 1460.2667}{14}}$$

$$= \sqrt{\frac{245.7333}{14}}$$

$$= \sqrt{17.5524}$$

$$= 4.1896$$

PROBLEM 30.1.2
Hypothesis testing

As an applied environmental researcher hired by the Ministry of Forests, you oversee an experiment to determine the effects of the pesticides DDT, lindane, and carbaryl on two tree types. You establish a pilot study involving 200 of each tree type. After 1 year, you observed the following dead trees as shown below. Do a named statistical test to determine whether there is a difference among the three treatments.

SOLUTION 30.1.2

Chi-square tests are appropriate when the outcome is discrete (dichotomous, ordinal, or categorical). The test statistic follows a chi-square probability distribution.

If the null hypothesis (Ho) is true, then both groups of observations must come from the same population (there is no difference, i.e., the treatment failed).

To test the null hypothesis, it is necessary to calculate the probability of having such a large difference between the observed and the expected, if they are both drawn indeed from the same population.

If the null hypothesis is correct, all the frequencies will be alike for cells in the same category. The test method therefore depends on a process which measures the difference between observed (O) and expected (E), and it is $(O - E)^2/E$, a quantity which increases as the difference between O and E grows. Summing this quantity for all cells produces chi-square.

H_o = There is no significant difference among the three treatments

H_a = There is a significant difference among the three treatments

P value (Alpha) = 0.05 (probability that the outcome is due to pure chance)

$$\text{Expected} = (\text{row total} \times \text{column total}) / n$$
$$a_1 = (188 \times 43)/307$$
$$= 26.33$$
$$b_1 = (119 \times 43)/307$$
$$= 16.67$$
$$a_2 = (188 \times 87)/307$$
$$= 53.28$$
$$b_2 = (119 \times 87)/307$$
$$= 33.72$$
$$a_3 = (188 \times 177)/307$$
$$= 108.39$$
$$b_3 = (119 \times 177)/307$$
$$= 68.61$$

In the following table, "expected" values are in blue ink:

Tree Species	DDT	Lindane	Carbaryl [Sevins]	Total
Coniferous	24 (26.33) a_1	48 (53.28) a_2	116 (108.39) a_3	188 T_a
Deciduous	19 (16.67) b_1	39 (33.72) b_2	61 (68.61) b_3	119 T_b
Total	43 T_1	87 T_2	177 T_3	307 (n)

$$\chi 2 = \Sigma((O - E)^2 / E)$$
$$= ((24 - 26.33)^2 / 26.33) + ((19 - 16.67)^2 / 16.67) + ((48 - 53.28)^2 / 53.28) +$$
$$((39 - 33.72)^2 / 33.72) + ((116 - 108.39)^2 / 108.39) + ((61 - 68.61)^2 / 68.61)$$
$$= 0.21 + 0.33 + 0.52 + 0.83 + 0.53 + 0.84$$

$$\chi 2_{calc} = 3.26$$
$$df* = (\text{\# of rows} - 1 \times \text{\# of columns} - 1)$$
$$= (2 - 1) (3 - 1)$$
$$= 2$$

As the degrees of freedom and the chi-square calculated ($\chi 2_{calc}$) are now known, we consult Table 30.1, which shows the critical value of chi-square.

$$\chi 2_{calc} = 3.26$$
$$\chi 2_{crit} = 5.991$$
$$\chi 2_{calc} < \chi 2_{crit}$$

Therefore, H_o is accepted. There is no significant difference between treatments.

---

* The number of independent pieces of information that goes into the estimate of a parameter is called the degrees of freedom (df). In this case, the df number ends when simple subtraction can replace the long calculation for the "expected" numbers.

TABLE 30.1
Chi-Square Listing of Critical Values, P Values, and Degrees of Freedom

df	\multicolumn{12}{c}{p value}
	0.25	0.2	0.15	0.1	0.05	0.025	0.02	0.01	0.005	0.0025	0.001	0.0005
	25%	20%	15%	10%	5%	2.5%	2%	1%	0.05%	0.025%	0.01%	0.005%
1	1.32	1.64	2.07	2.71	3.84	5.02	5.41	6.63	7.88	9.14	10.83	12.12
2	2.77	3.22	3.79	4.61	5.99	7.38	7.82	9.21	10.6	11.98	13.82	15.2
3	4.11	4.64	5.32	6.25	7.81	9.35	9.84	11.34	12.84	14.32	16.27	17.73
4	5.39	5.59	6.74	7.78	9.49	11.14	11.67	13.23	14.86	16.42	18.47	20
5	6.63	7.29	8.12	9.24	11.07	12.83	13.33	15.09	16.75	18.39	20.51	22.11
6	7.84	8.56	9.45	10.64	12.53	14.45	15.03	16.81	13.55	20.25	22.46	24.1
7	9.04	5.8	10.75	12.02	14.07	16.01	16.62	18.48	20.28	22.04	24.32	26.02
8	10.22	11.03	12.03	13.36	15.51	17.53	18.17	20.09	21.95	23.77	26.12	27.87
9	11.39	12.24	13.29	14.68	16.92	19.02	19.63	21.67	23.59	25.46	27.83	29.67
10	12.55	13.44	14.53	15.99	18.31	20.48	21.16	23.21	25.19	27.11	29.59	31.42

PROBLEM 30.1.3
Analysis of variance (ANOVA) activity

Cloud seeding, an attempt to offset one of the potential effects of global warming, caused the rainfall yields (in mm of rainfall) shown in the table located under "Solution 30.1.3" below. Use a named test procedure to calculate the F-statistic. Show each step of your procedure. Is there any significant difference among the four seeding treatments?

Terminology:

- F-cal = Calculated statistic = MSB/ MSW (MS between / MS within)
- F-crit = Critical value (found in a table of critical values)
- SS = Sum of the squared differences between each individual observation and the grand mean.
- MS = Mean square. When a difference is significant, the variation (MS between) between treatments must exceed that of the within (MS within) treatments. Mean squares (or error effects) are calculated by dividing the between- and within-group sum of squares by the appropriate number of degrees of freedom. It requires three calculations:
  MS between
  MS within
  MS total

Therefore, the mean sum of squares (MS) for each term shows the source of variation.

SOLUTION 30.1.3

Treatment 1	Treatment 2	Treatment 3	Treatment 4
50	40	38	47
57	42	39	67
32	33	40	54
57	57	45	67
46	57	46	68
52	49	51	65
54	57	51	65
49	54	50	56
62	53	48	60

H_o = There is no significant difference among the four treatments.
H_a = There is a significant difference among the four treatments.

Summary of Data

	Treatments				
	1	**2**	**3**	**4**	**Total**
N	10	10	10	10	40
ΣX	520	497	462	614	2,093
Mean ($\bar{x}$)	52	49.7	46.2	61.4	52.33
ΣX^2	27,724	25,351	21,628	38,138	112,841
Std. Dev.	8.72	8.50	5.61	6.98	

Result Details

Error	SS	Df	MS	
Between treatments	1,268.72	3	422.91	F = 7.41
Within treatments	2,056. 05	36	57.11	
Total	3,324.77	39	82.25	

$$S^2 = \Sigma(x - \bar{x})^2 / n - 1$$

$$
\begin{array}{llll}
S_1^2 = 684 / 10 - 1 & S_2^2 = 650.1 / 10 - 1 & S_3^2 = 283.6 / 10 - 1 & S_4^2 = 438.4 / 10 - 1 \\
\quad = 684 / 9 & \quad = 650.1 / 9 & \quad = 283.6 / 9 & \quad = 438.4 / 9 \\
\quad = 76 & \quad = 72.23 & \quad = 31.51 & \quad = 48.71
\end{array}
$$

$$SS = \Sigma(n_i - 1) \, s_i^2$$
$$= ((10-1) \times 76 + (10-1) \times 72.23 + (10-1) \times 31.51 + (10-1) \times 48.71)$$
$$= 9 \times 76 + 9 \times 72.23 + 9 \times 31.51 + 9 \times 48.71$$
$$= 684 + 650.07 + 283.59 + 438.39$$
$$= 2056.05$$

$$\text{Total SS} = \Sigma x^2 - (\Sigma x)^2 / n$$
$$= 112,841 - (2,093)^2 / 40$$
$$= 112,841 - 438,0649 / 40$$
$$= 112,841 - 109,516.23$$
$$= 3,324.77$$

$$\text{SS Between} = \text{Total SS} - \text{SS Within}$$
$$= 3324.77 - 2056.05$$
$$= 1268.72$$

DF between $= k - 1 = 4 - 1$	DF within $= N - k = 40 - 4$	DF total $= N - 1 = 40 - 1$
$= 3$	$= 36$	$= 39$

$$\text{MS} = \text{SS} / \text{DF}$$

MS Between $= 1268.72 / 3$	MS Within $= 2056.05 / 36$	MS Total $= 3324.77 / 39$
$= 422.91$	$= 57.11$	$= 85.25$

$$F_{cal} = \text{MSB} / \text{MSW}$$
$$= 422.91 / 57.11$$
$$= 7.41$$
$$F_{cal} = 7.41 > F_{crit} = 2.9223$$

Therefore, H_a is accepted. There is a significant difference among the four treatments.

PROBLEM 30.1.4
Correlations

Correlations range from -1 to 0 to $+1$. Imagine that the experiment above was tested for the correlation r value using the Spearman Rank Correlation Test (SPRT). The SPRT is designated for non-parametric data. Non-parametric data does not assume a normal distribution or anything about the form or parameters of the assumed distribution, such as means and standard deviations. The SPRT is an acceptable method for parametric data when there are fewer than 30 but greater than 9 paired variables. The SPRT formula is:

$$r_s = 1 - \left[6\Sigma d^2 / n(n^2 - 1) \right]$$

where
  n = number of pairs
  d = difference between paired ranks
  Σd^2 = sum of the squared differences between ranks

Acid rain produced from atmospheric sulfuric acid downwind of several coal-burning power plants in an industrialized country increases the weathering rate of limestone rocks. Several rock samples were taken from various streams with different pH values. Workers compared the weathered condition of the rock samples with the pH of the water in the stream from which each rock was taken.

Find the total for Σd^2. What is the r_s of the SPRT test?

SOLUTION 30.1.4

Rock	pH	Weathering range: 1–10 1 = highly weathered 10 = no weathering	d $(x - y)$	d^2 $(x - y)^2$
1.	1	2	−1	1
2.	4	1	3	9
3.	7.5	9.5	−2	4
4.	2	3	−1	1
5.	7.5	8	−0.5	0.25
6	3	4	−1	1
7	5	9.5	−4.5	20.25
8	6	6	0	0
9	9	5	4	16
10	10	7	3	9
	n = 10			$\Sigma d^2 = 61.5$

$$r_s = 1 - \left[6\Sigma d^2 / n \left(n^2 - 1 \right) \right]$$
$$r_s = 1 - \left[6 \times \Sigma d^2 / n^3 - n \right]$$
$$= 1 - \left[6 \times 61.5 / 10^3 - 10 \right]$$
$$= 1 - [369 / 990]$$
$$= 1 - 0.37$$

Correlation r value $= 0.63$

Though this r value is not high, it is a significant positive correlation between the acidity of the water and weathering rates of the rocks.

PROBLEM 30.1.5
Pearson's product moment correlation

The Pearson's product moment correlation is the most often used and most precise coefficient of correlation and is used with parametric data. The basic formula, symbolized by r, is:

$$r = N\Sigma XY - (\Sigma X)(\Sigma Y) / \sqrt{\left[N\Sigma X^2 - (\Sigma X)^2 \, N\Sigma Y^2 - (\Sigma Y)^2 \right]}$$

As an agronomist, you moved your crops to higher ground to avoid the problems of rising sea levels. But the cooler elevations lengthen the number of days required for fruit appearance. In an effort to lessen the time required for fruiting, you experiment with the timing of added fertilizers. What is the resultant correlation between the time of first fruiting of trees and the time of first addition of fertilizers?

SOLUTION 30.1.5

The table below illustrates the data and the necessary calculations.

Age at 1st fruiting X	Age at 1st fertilizing Y	X²	Y²	XY
11	11	121	121	121
15	12	225	144	180
18	12	324	144	216
12	11	144	121	132
11	11	121	121	121
16	13	256	169	208
14	13	196	169	182
13	12	169	144	156
19	11	361	121	209
14	10	196	100	140
16	15	256	225	240
17	14	289	196	238
15	13	225	169	195
16	16	256	256	256
15	13	225	169	195
15	14	225	196	210
$\Sigma X = 237$	$\Sigma Y = 201$	$\Sigma X^2 = 3,589$	$\Sigma Y^2 = 2,565$	$\Sigma XY = 2,999$

$N = 16$

$$r = N\Sigma XY - (\Sigma X)(\Sigma Y) / \sqrt{\left[N\Sigma X^2 - (\Sigma X)^2 \, N\Sigma Y^2 - (\Sigma Y)^2\right]}$$

$$r = 16(2999) - (237)(201) / \sqrt{\left[16(3589) - (237)^2\right]\left[16(2565) - (201)^2\right]}$$

$$r = 47,984 - 47,637 / \sqrt{(57,424 - 56,169)(41,040 - 40,401)}$$

$$r = 347 / \sqrt{1255 \times 639}$$

$$r = 347 / \sqrt{801,945}$$

$$r = 347 / 895.51$$

Ans. $= 0.39$

Therefore, there is a weak but positive correlation.

PROBLEM 30.1.6
The Mann-Whitney U test

In a fish farm adjacent to a coal-based power plant, the effects of thermal pollution on the growth rate of fish were tested by head-to-tail length after 6 months. First, the fish pool was partitioned into two halves by an insulating wall, thereby separating the thermally polluted side from the normal side. Is there a difference between the two sides of the pool in fish length after the 6-month duration of the treatment?

SOLUTION 30.1.6

Use the Mann-Whitney procedure to evaluate the following data.

Normal condition	8	6	7	5	10
Thermal pollution	3	7	6	2	1

H_o = No significant difference in fish length occurs after the duration of the treatment.

H_a = There is a significant difference in fish length after the 6-month duration of the treatment.

$$\text{Level of significance} = \alpha = 0.05$$

$$n_1 = 5, n_2 = 5$$

$$\text{U critical} = 2$$

$$\text{Decision rule} = \text{Reject } H_0 \text{ if } U \leq 2$$

		Total Sample (Ordered Smallest to Largest)		Ranks	
Formula	Placebo	Formula	Placebo	Formula	Placebo
8	3		1		1
6	7		2		2
7	6		3		3
5	2	5		4	
10	1	6	6	5.5	5.5
		7	7	7.5	7.5
		8		9	
		10		10	
				$R_1 = 36$	$R_2 = 19$

$$U_1 = n_1 n_2 + \left(n_1\left(n_1 + 1\right) / 2\right) - R_1$$
$$= \left(5 \times 5\right) + \left(\left(5 \times 6\right) / 2\right) - 36$$
$$= \left(25 + 15\right) - 36$$
$$= 40 - 36$$
$$= 4$$
$$U_2 = n_1 n_2 + \left(n_2\left(n_2 + 1\right) / 2\right) - R_2$$
$$= \left(5 \times 5\right) + \left(\left(5 \times 6\right) / 2\right) - 19$$
$$= \left(25 + 15\right) - 19$$
$$= 40 - 19$$
$$= 21$$

Test statistic $= U = 4$

(The test statistic for the Mann-Whitney U test is denoted U and is the smaller of U_1 and U_2.)

Conclusion: $U_{stat} = 4 > U_{crit} = 2$. Therefore, the H_o is accepted. There is no significant difference in fish length after the 6-month duration of the treatment.

*NB: Had the sample size been larger, this test might have detected a difference.

PROBLEM 30.1.7
When to use the t-test

The two sets of data are for average monthly wave incursions (in meters) on a particular beach for eight months in the year 1995 and 2015 ("incursions" are the distances that the water reaches up on the beach during a wave swash):

$$1995: 1.17, 0.70, 0.82, 0.94, 0.86, 0.91, 0.79, 1.03$$

$$2015: 0.63, 0.49, 0.53, 0.51, 0.57, 0.60, 0.48, 0.64$$

With the aid of a t-test, determine if there is a significant difference between the two sets of data and, by extension, make a statement on the evidence for sea-level rise.

SOLUTION 30.1.7

1995	2015	D	D²
1.17	0.63	0.54	0.291
0.70	0.49	0.21	0.044
0.82	0.53	0.29	0.084
0.94	0.51	0.43	0.184
0.86	0.57	0.29	0.084
0.91	0.60	0.31	0.096
0.79	0.48	0.31	0.096
1.03	0.64	0.39	0.152
		$\Sigma = 2.77$	$\Sigma = 1.031$

- H_o: There is no difference between the two years.
- H_a: There is a difference between the two years.

Using the two-tailed test at 0.05

- $df = n - 1$
- $df = 8 - 1$
- $df = 7$

$$t = \Sigma D / \sqrt{\left[\left\{ n\Sigma D^2 - (\Sigma D)^2 \right\} / n - 1 \right]}$$

$n = 8$
$\Sigma D = 2.77$
$\Sigma D^2 = 1.031$

$$t = 2.77 / \sqrt{\left[\left\{(8 \times 1.031) - (2.77)^2\right\} / 8 - 1\right]}$$

$$t = 2.77 / \sqrt{[(8.248 - 7.672) / 7]}$$

$$t = 2.77 / \sqrt{[0.576 / 7]}$$

$$t = 2.77 / \sqrt{0.082}$$

$$t = 2.77 / 0.286$$

$$t = 9.685$$

A larger t value shows that the difference between group means is greater than the pooled standard error, indicating a more significant difference between the groups.

- Critical value = 2.365
- T_{stat} = 9.685
- T_{crit} = 2.365
- T statistic is greater than t critical value and therefore rejects Ho.
- There is a significant difference between the two sets of data as it relates to the sea-level rise.

REFERENCE

DEC (1992) Petroleum-Contaminated Soil Guidance Policy. New York State. Section VI – Sampling. Department of Environmental Conservation. https://www.dec.ny.gov/regulations/30902.html#Sampling

31 Appendix B
Hydrological Applications

31.1 DRAINAGE BASINS

One of the major impacts of global warming is likely to be on hydrological parameters and water resources, which in turn will have significant effects across many sectors of the economy, society, and environment (Climate-Policy-Watcher 2022). The major units affected are drainage basins, ranging in size from a few square kilometers to subcontinental entities like the Mississippi, Nile, or Loire,basins. On the global scale, climate change is likely to worsen water resource stress in some regions but perhaps ameliorate stress in others (Climate-Policy-Watcher 2022).

PROBLEM 31.1.1
Drainage basins

What delineates a drainage basin?

SOLUTION 31.1.1

According to Climate-Policy-Watcher (2022), a drainage basin, or watershed, is an extent or an area of land where surface water from rain, melting snow, or ice converges to a single point at a lower elevation, usually the exit of the basin, where the waters join another waterbody, such as a river, lake, reservoir, estuary, wetland, sea, or ocean.

Case study:

At Seaman's Valley, Portland Parish, Jamaica WI, water leaving a small stream catchment (Figure 31.1) has minimal velocity because, currently, the available difference in height (which far exceeds 20 ft) from the source to the catchment remains unexploited (Figure 31.2). The inclusion of this hydraulic head in the system would improve the low water pressure currently experienced in the districts served by the catchment. Altimeter measurements reportedly show an 80 ft height increase from the catchment to the water source (personal communication. Nevertheless, in the absence of direct verification, it is conservatively estimated here (Figure 31.1) at 20 ft.

PROBLEM 31.1.2
Calculating the water pressure

How is the present hydraulic gradient, and by extension, water pressure, calculated?

DOI: 10.1201/9781003341826-38

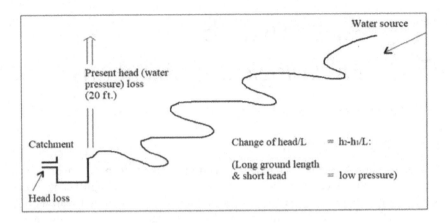

FIGURE 31.1 Water from a stream source runs on the ground surface, picking up sediments and dissolved and undissolved organic matter before reaching the catchment.

SOLUTION 31.1.2

The hydraulic gradient, which depicts the difference, can be calculated as:

$$\Delta h \,/\, L = h_2 - h_1 \,/\, L$$

where

$\Delta h \,/\, L$ = the hydraulic gradient
h_2 = the head at location 2
h_1 = the head at location 1
L = the linear distance between any location 1 and location 2

But the catchment (h), receiving slow-moving water which flows tortuously (Figure 3.2) for 380 ft on the ground from the water source under almost no pressure, therefore has:

$$h_2 - h_1 = 0$$

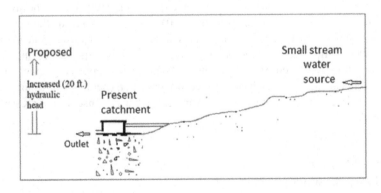

FIGURE 31.2 The catchment at Seaman's Valley.

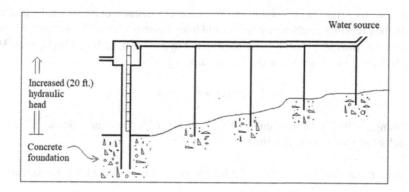

FIGURE 31.3 Proposed changes to a water catchment and distribution facility. Note increased hydraulic head, reduced horizontal water distance on the ground = decreased sediment and organic matter, water off the ground = less dissolved minerals.

Therefore,

$$\Delta h \, / \, L = h_2 - h_1 \, / \, L = h_2 - h_1 \, / \, 380 \text{ ft} = 0$$

As $h = 0$, hydraulic gradient $= 0$ and there is no hydraulic gradient above the catchment tank, water pressure from the source to the catchment is presently nonexistent because *there is no hydraulic head.*

PROBLEM 31.1.3
Determination of actual increase in hydraulic gradient

If the hydraulic head is raised by 20 ft (Figure 31.3), what is the new difference in hydraulic gradient?

SOLUTION 31.1.3

Figure 31.3 depicts the proposed changes to rectify the five problems including $h_2 - h_1 \, / \, L = 20 - 0 \, / \, 380 \text{ ft} = 0.053 \text{ ft} \cdot \text{ft}^{-1}$

PROBLEM 31.1.4
Calculated example

For the above question, the catchment is approximately 1000 ft above mean sea level (MSL). A household exists at 200 ft above MSL and is 500 ft away on a straight-line road from the catchment. What is the proportional difference in the hydraulic gradient (and, by extension, water pressure) coming from the catchment before and after the catchment modifications?

SOLUTION 31.1.4

The present catchment lies at a height of 1000 ft. Therefore, the present hydraulic gradient is as follows:

$$\Delta h \, / \, L = h_2 - h_1 \, / \, L = (1000 - 200) \, / \, 500 = 1.6 \text{ ft} \cdot \text{ft}^{-1}$$

(i.e., water comes from a height of 1.6 ft for every linear foot it runs on the ground)

As the proposed catchment water would hang directly above the old catchment on the ground, thereby potentially increasing the hydraulic head by 20 ft (Figures 31.2 and 31.3), this increases the hydraulic gradient by:

$$(1020 - 200) / 500 = 1.64 \text{ ft} \cdot \text{ft}^{-1}$$

Change in hydraulic gradient = $[(1.64 - 1.6) / 1.6] \times 100 =$ an increase of 2.5%. Before modifications, at a linear distance (L):

$$\Delta h / L = h_2 - h_1 / L = (1000 - 200) / 500 = 1.6 \text{ ft ft}^{-1} \text{ (vertical ft per linear ft)}$$

After the modifications to produce a hydraulic head of 80 ft:

$$\Delta h / L = h_2 - h_1 / L = (1000 + 80) - (200) / 500 = 1.76 \text{ ft} \cdot \text{ft}^{-1} \text{ (vertical ft per linear ft)}$$

Percentage increase in hydraulic gradient from 160 to 176 ft $\cdot$ ft^{-1} = 10%.
It so happens that if water flows freely, and not through rock or soil,

$$v = \Delta h / L$$

where v = velocity.

Hence, both above equations produce the same result.

Therefore, in this case, increased hydraulic gradient = The same proportional increase in water velocity.

Thus, in this case the increase in water pressure = 10%.

Even a water tower half this height would produce a significant increase in water pressure.

PROBLEM 31.1.5

For the above question, what are the effects on chlorinated water?

SOLUTION 31.1.5

Being off the ground in a closed conduit between source and catchment, the water to be chlorinated will be devoid of much organic matter (fewer leaves and less sediment). This reduces the production of tri-halomethanes.

PROBLEM 31.1.6
Unconfined aquifers

The head in an unconfined aquifer has been measured in four locations, as shown in Figure 31.4. What is the direction of flow and the hydraulic gradient?

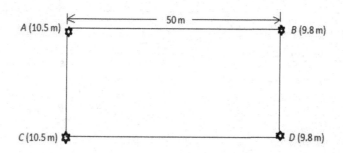

FIGURE 31.4 Four piezometric readings of an unconfined aquifer.

SOLUTION 31.1.6

The diagram (Figure 31.4) depicts a rectangular area of ground with a piezometric reading at each vertex. Water flows in an unconfined aquifer from areas of higher piezometric head to areas of lower piezometric head. Hence, it is from WX to YZ (north to south).

The hydraulic gradient ("rise over run") is determined by the equation:

$$\Delta h / L = h_2 - h_1 / L = 10.5 - 9.8 \text{ m} / 50 \text{ m} = .014 \text{ m} \cdot \text{m}^{-1}$$

PROBLEM 31.1.7
Confining layer

While digging the foundation for a hospital, a construction crew finds water at 6 m below ground surface (bgs). Ninety-two meters away they find water at 5.2 m bgs. The datum is chosen as the confining layer at 22 m bgs. If it is assumed that the confining layer is parallel to the surface, how deep is the piezometric surface at each point? What is the direction of groundwater flow and the hydraulic gradient?

SOLUTION 31.1.7

A drawing of the problem (Figure 31.5) shows depths to the water table of 6 and 5.2 m, respectively, for points P and Q. The datum of 22 m bgs given can be used to calculate the total head of water at each point.

$$\text{Point P: Total head} = 22 - 6 = 16 \text{ m}$$

$$\text{Point Q: Total head} = 22 - 5.2 = 16.8 \text{ m}$$

As the higher piezometric surface is at Q, groundwater flow is from Q to P.

Based on these two piezometric surfaces, the hydraulic gradient can be determined thus:

$$\Delta h / L = h_2 - h_1 / L = 16.8 - 16 \text{ m} / 92 \text{ m} = 0.0087 \text{ m} \cdot \text{m}^{-1}$$

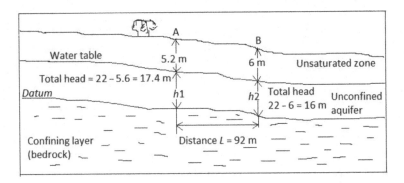

FIGURE 31.5   Confining layer showing depth of water table.

PROBLEM 31.1.8
Darcy velocity

The 19th century French geologist, Darcy, found that the velocity of laminar flowing groundwater in an aquifer is proportional to both its hydraulic conductivity and the hydraulic gradient, providing that the aquifer is fully saturated. If, in the above example, the aquifer is a sand and gravel layer having a cross-sectional area of 875 m through which water flows, what is (a) the Darcy velocity of the groundwater in this zone and (b) the specific discharge?

SOLUTION 31.1.8

In general, a layer of mixed sand and gravel has an approximate hydraulic conductivity of 6×10^{-4} m · m^{-1} (Davis & Masten 2004). As the hydraulic gradient was determined to be 0.0087 m · m^{-1}, the Darcy velocity can be calculated as follows:

$$v = K\,(\Delta h\,/\,L)$$

$$= (6 \times 10^{-4}\ \text{m} \cdot \text{s}^{-1})(0.0087\ \text{m} \cdot \text{m}^{-1})(86{,}400\ \text{s} \cdot \text{day}^{-1}) = 0.449\ \text{m} \cdot \text{day}^{-1}$$

The specific discharge (being through a sand-gravel bed) is the product of v (A):

$$0.449\ \text{m} \cdot \text{day}^{-1} \times 875\ \text{m}^2 = 392.87\ \text{m}^3 \cdot \text{day}^{-1}$$

As for the case of a squeezed water hose, restricting the flow aperture increases the confining pressure, which in turn increases the velocity of a flowing fluid.

Similarly, the velocity of water flowing out of an aquifer varies inversely with the porosity.

PROBLEM 31.1.9
Linear velocity of emerging spring

The daily amount of water available from an aquifer can be calculated if the linear velocity of the water is known. Two piezometric surfaces $h_1 = 150$ cm and $h_2 = 100$ cm exist at different vertical levels above an aquifer of a mixture of sand and gravel

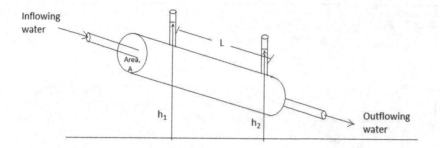

FIGURE 31.6 Distance separating points in an aquifer.

having a porosity of 20%, as depicted in the schematic drawing of Figure 31.6, The distance separating the two points is 550 cm. Drill cores into the ground indicated an aquifer cross-sectional area of 120 m³. What is the linear velocity of the spring which can be seen emerging from the aquifer?

SOLUTION 31.1.9

The hydraulic gradient must be calculated as follows:

$$\Delta h \,/\, L = h_2 - h_1 \,/\, L = 150 - 100 \text{ cm} / 550 \text{ cm} = 0.090 \text{ cm} \cdot \text{cm}^{-1}$$

The hydraulic conductivity, K, of coarse sand is estimated to be 6.0×10^{-4} (Davis & Masten 2004). Hence, the Darcy velocity can be calculated thus:

$$v = K \, \Delta h \,/\, L = (6.0 \times 10^{-4} \text{ m} \cdot \text{s}^{-1})(0.0009 \text{ m} \cdot \text{m}^{-1}) = 5 \times 10^{-5} \text{ m} \cdot \text{s}^{-1}$$

Based on a porosity of 0.20 for gravelly sand (Nimmo 2004), the linear velocity would be

$$v'_{water} = v \,/\, \eta \text{ (where } \eta = \text{porosity of the geologic material)} = 5 \times 10^{-5} \text{ m} \cdot \text{s}^{-1} / 0.20$$

Ans. $= 2.5 \times 10^{-4} \text{ m} \cdot \text{s}^{-1}$

PROBLEM 31.1.10
Stream discharge measurement

According to Brady & Ruzycki (2009), the volume of water passing through the cross-sectional area of a stream channel (Figure 31.7) per unit time can be determined by the following equation:

$$Q = V(\text{m/s}) \times A$$

where
  Q = volume
  V = average velocity of the water
  A = cross-sectional area of stream at that point (m²)

How is the average stream discharge measured?

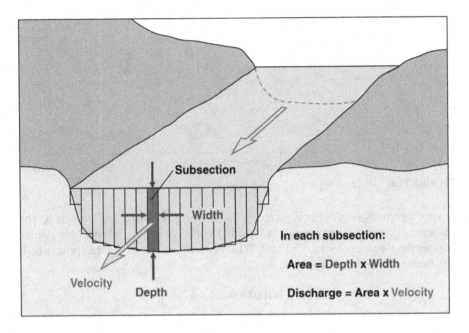

FIGURE 31.7   Stream discharge measurement setup:

Source: After USGS

Source: www.usgs.gov

SOLUTION 31.1.10

According to Richards et al. (2004), stream discharge can be measured as follows:

- Stretch tape across river at right angle to direction of flow.
- Choose measurement intervals.
- 10 intervals are the minimum.
- Width of the subsections can be variable across the cross section.
- No more than 10% of expected discharge should be allotted per interval.
- No interval should be more than 3 m wide.
- At each point – measure velocity with meter at either 0.6 d (d = depth) or 0.2 and 0.8 d.
- Note: As point velocity is constantly changing (pulsation), average it over time.
- Operator position should NOT affect or obstruct flow pattern near flow meter – operator should be downstream of meter, at arm's length.

PROBLEM 31.1.11
Stage-discharge relationship

How can you know how much water is being transported by a stream at any stage of flow?

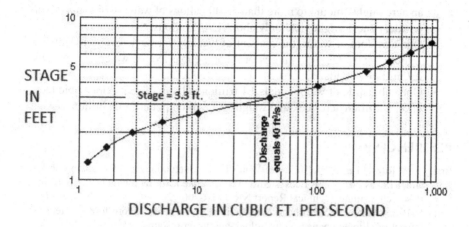

DISCHARGE IN CUBIC FT. PER SECOND

FIGURE 31.8   Stream discharge in cubic feet per second.

Source: https://www.usgs.gov/media/images/usgs-stage-discharge-relation-example

SOLUTION 31.1.11

First, compute the discharge at a range of stages. This can be done by plotting depth vs flow, followed by fitting a curve (Figure 31.8).

- Measure velocity at various depths and positions across the stream to estimate the true "average flow."
- If done over a range of flow conditions, from base flow to high flow while simultaneously measuring the height of the stream, one can generate a graph relating flow to stage height. This curve is then used to convert measured stage height into flow values.

PROBLEM 31.1.12
Parameters of Precipitation

How are the attributes of rainfall quantitatively determined?

SOLUTION 31.1.12

According to USGS (2019):

- One inch of rain falling on 1 acre of ground is equal to about 27,154 gallons and weighs about 113 tons.
- An inch of snow falling evenly on 1 acre of ground is equivalent to about 2,715 gallons of water. This figure, however, based upon the "rule-of-thumb" that 10 inches of snow is equal to 1 inch of water, can vary considerably, depending on whether the snow is heavy and wet, or powdery and dry. Heavy, wet snow has a very high water content—4 or 5 inches of this kind of snow contains about 1 inch of water. Thus, an inch of very wet snow over

an acre might amount to more than 5,400 gallons of water, while an inch of powdery snow might yield only about 1,300 gallons.

- One acre-foot of water (the amount of water covering 1 acre to a depth of 1 foot) equals 326,000 gallons or 43,560 cubic feet of water, and weighs 2.7 million pounds.
- One cubic mile of water equals 1.1 trillion gallons, 147.2 billion cubic feet, or 3.38 million acre-feet, and weighs 9.2 trillion pounds (4.6 billion tons).

REFERENCES

Brady V & Ruzycki E 2009. Effects of Water Level Fluctuations and Regulation on Upper Great Lakes Nearshore Ecosystems: An Annotated Bibliography. Natural Resources Research Institute Technical Report NRRI/TR-2009/20.

Climate-Policy-Watcher (2022) https://www.climate-policy-watcher.org/forest-meteorology/impact-of-climate-change-on-hydrology-and-water-resources.html

Davis M, Masten S (2004) Principles of Environmental Engineering & Science. 3rd ed. McGraw-Hill, New York, p. 214.

Nimmo JR (2004) Porosity and Pore Size Distribution. In Hillel, D., ed. Encyclopedia of Soils in the Environment. Volume 3. Elsevier, London, pp. 295–303.

USGS (2019) Rain and Precipitation. https://www.usgs.gov/special-topics/water-science-school/science/rain-and-precipitation

32 Appendix C
Renewable Energy

Renewable energy includes energy from a source that is not depleted when used, including wind, solar, hydro, tidal, and geothermal. It stands in contrast to fossil fuels, which are being used far more quickly than they are being replenished.

32.1 WIND ENERGY FORMULA

Wind is air that is moving. As the air has mass, it produces kinetic energy as it moves. Based on the First Law of Thermodynamics, the energy in the wind must come from somewhere else, which in this case is solar energy.

Wind power calculations depend on the following equation:

$$P = 1/2\rho Av^3 \tag{32.1.1}$$

where

ρ (rho, a Greek letter) = density (kg $\cdot$ m^{-3})

A = swept area (m^2)

v = wind speed (m $\cdot$ s^{-1})

P = power (W)

PROBLEM 32.1.1
Wind power

If the wind speed is 25 m $\cdot$ s^{-1} & the blade length is 40 m, calculate the wind power.

SOLUTION 32.1.1

Given:

$$\text{Wind speed } v = 25 \text{ m} \cdot \text{s}^{-1}$$
$$\text{Blade length } l = 40 \text{ m}$$
$$\text{Air density } \rho = 1.23 \text{ kg} \cdot \text{m}^{-3}$$
$$\text{The area of a circle is given by } A = \pi r^2$$
$$A = \pi \times 1600 = 5034 \text{ m}^2$$

The wind power formula is given as

$$P = 1/2\rho AV^3$$
$$P = 1/2 \times (1.23) \times (5034) \times 25^3$$
$$P = 48373 \text{ W}$$

DOI: 10.1201/9781003341826-39

PRACTICE PROBLEM 32.1.1
Effect of changing wind speed on power

In the above problem, how much power is generated when the wind speed increases from 25 to 50 m · s⁻¹?

SOLUTION 32.1.1

As can be seen, when the velocity doubles, the power increases by a factor of 8 and when the velocity triples, it increases by a factor of 27. This is because the velocity is cubed: $2^3 = 8$ and $3^3 = 27$.

PROBLEM 32.1.2
Wind power from speed of blade

A wind turbine has a blade length of 25 m and runs at a speed of 12 m · s⁻¹. Determine the amount of wind power available.

SOLUTION 32.1.2

Given:

$$\text{Wind speed } v = 12 \text{ m} \cdot \text{s}^{-1}$$
$$\text{Blade length } l = 25 \text{ m}$$
$$\text{Air density } \rho = 1.23 \text{ kg} \cdot \text{m}^{-3}$$
$$\text{Area, } A = \pi r^2$$
$$= \pi \times 625$$
$$= 1966 \text{ m}^2$$

The wind power formula is given as

$$P = 1/2 \rho A V^3$$
$$= 0.5 \times 1.23 \times 1966 \times 1728$$
$$P = 2,089,308 \text{ W}$$

PROBLEM 32.1.3
Effect of multiple blades

Calculate the wind power. Given:

$$\text{Blade length, } l = 24 \text{ m}$$
$$\text{Number of blades} = 3$$
$$\text{Average island wind speed, } v = 12 \text{ m} \cdot \text{s}^{-1}$$
$$\text{Air density, } \rho = 1.23 \text{ kg} \cdot \text{m}^{-3}$$

SOLUTION 32.1.3

$$\text{Area}, A = \pi r^2$$
$$= \pi \times 576$$
$$= 1812 \text{ m}^2$$

The wind energy formula is given by

$$P = 1/2\rho A V^3$$
$$= 1/2 \times (1.23) \times (1812) \times 12^3$$
$$P = 1,925,648.6 \text{ W}$$

This should be times the number of blades.

PROBLEM 32.1.4
Effect of other conditions on power output

What is a realistic power output (in megawatts) deliverable under the following conditions from a wind turbine?

$$\text{Blade length} = 22 \text{ m}$$
$$\text{Number of blades} = 3$$
$$\text{Average wind speed}, v = 10 \text{ m} \cdot \text{s}^{-1}$$
$$\text{Air density}, \rho = 1.23 \text{ kg} \cdot \text{m}^{-3}$$
$$Ct = 40\% \, (\text{turbine efficiency rating})$$
$$Ca = 65\% \, (\text{alternator / generator efficiency rating})$$

SOLUTION 32.1.4

$$P = 1/2 \times \rho \times A \times v^3 \times Ct \times Ca$$
$$P = 1/2 \times 1.23 \text{ kg} \cdot \text{m}^{-3} \times \left(\pi \times 22^2 \right) \times \left(10 \text{m} \cdot \text{s}^{-1} \right)^3 \times 0.4 \times 0.65$$
$$P = 0.24 \text{ MW}$$

PROBLEM 32.1.5
Effect of other conditions on power output

What is a realistic power output (in megawatts) deliverable under the following conditions from a wind turbine?

$$\text{Blade length}, l = 25 \text{ m}$$
$$\text{Number of blades} = 3$$
$$\text{Average island wind speed}, v = 9 \text{ m} \cdot \text{s}^{-1}$$
$$\text{Air density}, \rho = 1.20 \text{ kg} \cdot \text{m}^{-3}$$
$$Ct = 38\% \, (\text{turbine efficiency rating})$$
$$Ca = 60\% \, (\text{alternator / generator efficiency rating})$$

$$P = 1/2 \times \rho \times A \times v^3 \times Ct \times Ca$$
$$P = 1/2 \times 1.20 \ kg \cdot m^{-3} \times \left(\pi \times 25^2\right) \times \left(9 \ m \cdot s^{-1}\right)^3 \times 0.38 \times 0.65$$
$$P = 212,217.1 \ W = 0.21 \ MW$$

32.2 WIND VECTOR RESULTANTS THROUGH RESTRICTED SEA STRAITS

Sea straits provide problems and opportunities for efficient capture of downwind power. This exercise aims to determine the use of preceding sea straits to assess the most effective locations for the installation of coastal wind power generators in Balkanized but windy coastal areas such as the Scandinavian, Alaskan, north-east Canadian, Far South American, The Bahamas, and the north coast of Cuba.

32.2.1 WIND STRAITS EXTRAPOLATIONS

The wind power depends on the source width and the spatial attributes of the intervening area traversed. The distance between two degrees of longitude is widest at the Equator with 69.172 miles (111.321 km). The distance gradually shrinks to zero as they meet at the poles (Figure 32.1). Because lines of longitude taper poleward, the horizontal distance between two lines of longitude at a particular latitude is:

$$\text{Distance} = \text{Cos (latitude)} \times 111.325 \ km = X \times 111.325 \qquad (32.2.1)$$

where

Cos = adjacent/hypotenuse

111.325 = distance between two lines of longitude (meridians) at the Equator

PROBLEM 32.2.1
Distance between meridians

Let latitude 25°N be the winter northern boundary of the Northeast Trade (NET) winds. Then, what is the distance between any two lines of longitude at 25°N?

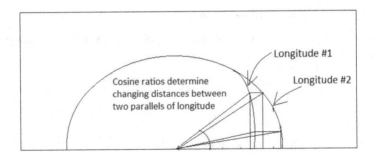

FIGURE 32.1 Distance between two lines of longitude = Cosine of the angle × distance at the Equator (111.325 km).

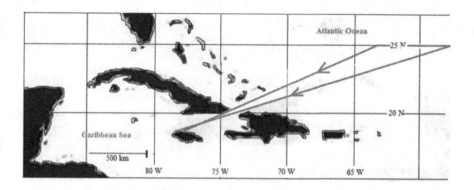

FIGURE 32.2   Limits of unrestricted north-east trade winds affecting the north-western coast of Jamaica: "Zone 1". Wind source: width = 505 km, source area = 166, 633 km² (ratio = 1). Conclusion: The size of the wind source (the triangle) varies directly with wind speed through straits (see also Figures 32.3–4).

SOLUTION 32.2.1

Distance = Cosine (25°) × 111.325 km = 0.9063 × 111.325 = 100.893 km

PROBLEM 32.2.2
Effects of sea straits of wind strength

What changes in wind strength and area of contact occur downwind of a sea strait?

SOLUTION 32.2.2

This problem can be depicted by aspects of the NET winds blowing through the Windward Passage between the islands of Cuba and Hispaniola toward the north coast of Jamaica (Figures 32.2–32.4). Though diurnal sea breezes (sea-land breezes)

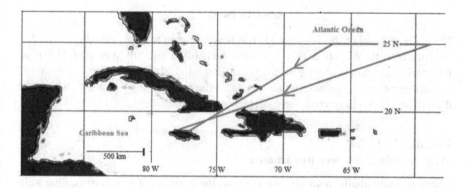

FIGURE 32.3   As F = m/a, the greater number of air particles in the larger triangle here must produce a stronger wind than that of Figure 32.2 (smaller triangle). "Zone 2". Wind source: width = 734 km, source area triangle = 255,504 km² (ratio = 1.4).

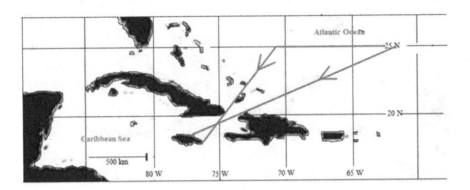

FIGURE 32.4 This is the largest mass of air. It explains the greatest wind force being on the north-east coast of the island. "Zone 3". Wind source: width = 751 km, source area = 277,722 km² (ratio = 1.6), the largest of the three wind sources.

blow perpendicularly toward every corresponding coastal section, the trade wind directions are pre-set from the northeast and are directionally consistent. Therefore, three wind vectors exist, that is, (a) trade wind, (b) diurnal, and (c) a combination of trade wind and diurnal, called the resultant (more will be said later about the resultants). Thus, the north-eastern districts offer the greatest potential (Figure 32.2) for wind energy in Jamaica. But it is facilitated by the width of the northern straits known as the Windward Passage. This is proven by the incremental reduction in wind strength with increasing distance from the northeast (Figures 32.2–32.4).

PROBLEM 32.2.3
Wind vectors and wind power

How can prevailing and diurnal winds combine to increase or decrease wind strength and direction on a coast and inland?

SOLUTION 32.2.3

Normally, diurnal onshore (sea-land) winds arrive at an angle of 90° (Figure 32.5).

Therefore, the direction as it moves inland is often difficult to predict. However, the more these winds align with the prevailing wind, the greater the speeds. Wind vectors combine to increase wind speeds in the north-east of Jamaica, where the diurnal and prevailing winds have vectors with similar velocities (Figure 32.5).

PROBLEM 32.2.4
Wind farming and sleep disturbances

There are individuals in society who oppose the building of wind farms. One well-known complaint against the operation of wind turbines is the alleged sound they produce, which, it is claimed, prevents people from sleeping. What is the evidence for or against?

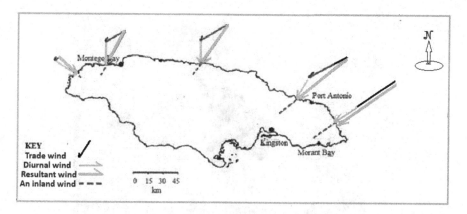

FIGURE 32.5   Potential inland penetration of resultant winds (broken lines) on the north coast of Jamaica. Three winds produce three vectors: prevailing, diurnal, and resultant. Sea-land breezes always blow perpendicularly to coasts. The most parallel, and hence most auspicious, alignment of incoming winds occurs in the northeast, thereby producing the largest resultant vectors. Wind power decreases from the northeast to the northwest coast of Jamaica. Hence, the northeast has the greatest wind power potential.

Source: Adapted from Harris MA (2019) Confronting Global Climate Change: Experiments and Applications in the Tropics. CRC Press, London, New York, Boca Raton.

SOLUTION 32.2.4

As reported in The Sydney Morning Herald (2019)), a South Australia study using recordings of wind turbines to study the sleep patterns of 68 people divided them into four groups of 18 under differing conditions in a laboratory setting. The study found no significant effects on any of the sleep measures even in the group who had earlier reported disruption by wind farm noise. Although they were not testing the worst-case wind turbine noise levels which can occur, the results show that at the very least, noises from average wind turbines do not cause significant sleep loss.

PROBLEM 32.2.5
Remediating wind farming through sea straits

Which large regions exhibit the potential for maximizing wind power downwind of sea straits?

SOLUTION 32.2.5

The answer lies in the following:

- The strength, direction, and vertical depth of the prevailing winds.
- The level of obstructions as determined by the height of the landforms comprising the straits, since prevailing winds in the planetary wind system are

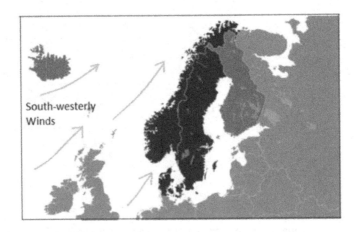

FIGURE 32.6 Wind farming straits of Northwestern Europe: note the plethora of straits downwind of the south westerlies.

Source: Wikimedia Commons.

    effective up to an altitude of 2000 m. For example, the landforms creating the Windward Passage between Jamaica and Hispaniola exceed 2000 m.
- Thus, the ability of straits to affect the downwind wind parameters vary inversely with the height of the landforms creating the straits.

Therefore, world regions which exhibit a potential for maximizing wind power through straits are shown in Figures 32.6–32.9.

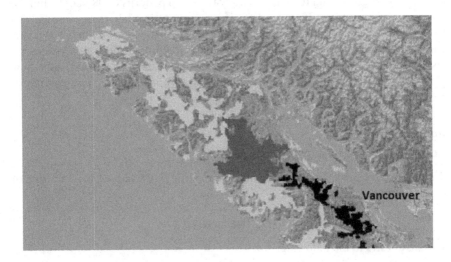

FIGURE 32.7 Potential wind farming straits of the Pacific Northwest. Note the dozens of straits which can modify the downwind characteristics of the prevailing south-westerlies.

Source: www.sciencebase.gov.

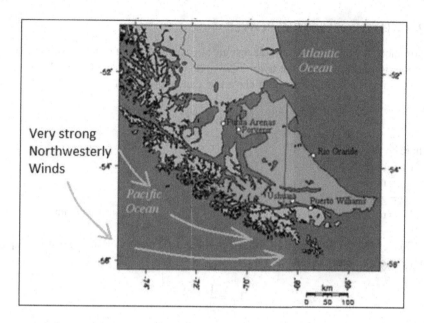

FIGURE 32.8 Wind farming straits of Southern South America.

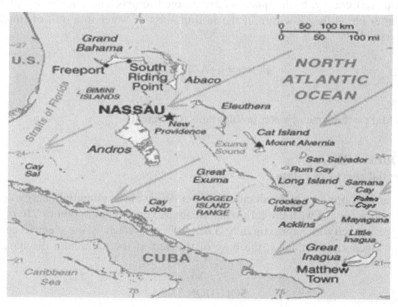

FIGURE 32.9 Not all straits produce downwind wind changes. The Bahamas seemingly present potential wind farm locations in the northern Caribbean based on auspicious positioning with respect to upwind sea straits. However, despite there being dozens of straits among the approximately 800 islands, they are all below 500 m in altitude, thereby presenting little meaningful interaction with the trade winds.

Source: www.en.wikipedia.org.

32.3   HYDROELECTRIC POWER

Hydroelectric power is a renewable resource because the spent water recycles through evaporation back through the hydrologic cycle to the source as rainfall. The energy of position, or potential energy of elevation, can be presented as:

$$Ep = mg(\Delta Z)$$

where
    Ep = potential energy
    m = mass
    g = gravitational acceleration = 9.81 m · s^{-2}
    ΔZ = difference in elevation between the water surface and the turbine

Converting the energy of falling water, the kinetic energy of falling water is equal to the energy of motion as

$$Ek = \frac{1}{2} mv^2$$

where the mass and velocity of the water are the determinants. Being equal to potential energy, both equations are merged to give an equation for power, the latter being the flow rate of the falling mass of water in a unit of time, as follows:

$$Power = g\ (\Delta Z)\ (d\ M/d\ t)$$

where
    d M/d t = flow rate of water in kilograms per second

PROBLEM 32.3.1
Potential energy of a dam

A river is to produce hydroelectrical power. What is the potential energy of the dam and reservoir with the following attributes: maximum height of 216 m and storage capacity of approximately 3.6×10^{10} m^3? What is the electric capacity of a generating plant at the base of the dam if the maximum discharge is 834 m^3 · s^{-1}?

SOLUTION 32.3.1

As the height of dams vary with the seasonal weather changes within the river basin, an average height of 216 / 2 (as roughly half-way between the highest and lowest flows) will be used for the calculation. Units are:

$$\left(kg \cdot m/s^2\right)(m) = N \cdot m = J$$

If the density of water is taken as 1000.0 kg · m^{-1}, the potential energy equation is:

$$Ep = \left(3.6 \times 10^{10} \text{ m}^3\right) \left(1000 \text{ kg} \cdot \text{m}^{-1}\right) \left(9.8 \text{ m} \cdot \text{s}^{-2}\right) (108 \text{ m}) = 3.8 \times 10^{16} \text{ N} \cdot \text{m}$$
$$= 3.8 \times 10^{16} \text{ J or 38 PJ} * (\text{i.e., peta joules})$$

At a flow rate of 834 m^3 · s^{-1}, the electrical capacity is:

$$\text{Power} = \left(9.8 \text{ m} \cdot \text{s}^{-2}\right)(216 \text{ m})\left(1000 \text{ kg} \cdot \text{m}^{-1}\right)\left(834 \text{ m}^3 \cdot \text{s}^{-1}\right) = 1.76 \times 10^9 \text{ W} = 1760 \text{ MW}$$

32.4   DISCARDED TIRE PYROLYSIS

Pyrolysis applies heat energy to break down chemical compounds without completely destroying the components by oxidation, thereby separating, and releasing, intact gases, liquids, and solid components.

PROBLEM 32.4.1
Energizing discarded tires

What are the environmental benefits of tire pyrolysis?

SOLUTION 32.4.1

Pyrolysis releases organic components of waste tires such as gases, condensable oil (Quek & Balasubramanian 2013), and solid char (Pakdel et al. 2001; Choi et al. 2014). The conversion back to fuel in transforming waste tires to alternative fuels reduces the burning of fossil fuels by avoiding the equivalent energy required to mine and process the equivalent in raw petroleum. Therefore, waste tire pyrolysis has been demonstrated to be a feasible way to indirectly reduce greenhouse gas emissions (Dębek & Walendziewski 2015). Additionally, the steel reinforcement of the tires is recoverable from the residual char for recycling back into the iron and steel industry (González et al. 2006).

PROBLEM 32.4.2
Disadvantages of tire pyrolysis

What are the harmful effects of tire pyrolysis?

---

* According to Blakers et al. (2020), the energy storage capability of a pumped hydro energy storage system is the product of the mass of water stored in the upper reservoir (in kilograms), the usable fraction of that water, the gravitational constant, the head (in meters), and the system efficiency. Renewable energy such as wind or solar pumps water upwards through a pipe to a reservoir. When demand increases, or wind or solar drops, water runs downhill from the upper reservoir through a turbine, creating electricity. About 80% of the electricity used to pump the water uphill is recovered, and 20% is lost.

SOLUTION 32.4.2

According to Czajczyńska et al. (2017), tire rubber resists abrasion, water, heat, electricity, many chemicals, bacteria, and mechanical damage. Long life and safety, regardless of weather conditions, make safe disposal very difficult. They also say that tire dumps are also a high threat to the environment and human health because of the risk of fire and because they are habitats for mosquitos and rodents, which are strongly associated with many diseases. These problems must be overcome before large-scale repurposing of discarded tires becomes a reality.

REFERENCES

Choi G-G, Jung S-H, Oh S-J, Kim J-S (2014) Total utilization of waste tire rubber through pyrolysis to obtain oils and CO2 activation of pyrolysis char. Fuel Process Technol 123, 57–64. https://doi.org/10.1016/j.fuproc.2014.02.007.

Czajczyńska D, Krzyżyńska R, Jouhara H, Spencer N (2017) Use of pyrolytic gas from waste tire as a fuel: A review, Energy 134, 1121–1131. https://doi.org/10.1016/j.energy.2017.05.042.

Davis M & Masten S (2020) Principles of Environmental Engineering & Science. 4th edition. McGraw Hill, New York. MISBN10: 1259893545 | ISBN13: 9781259893544

Dębek C, Walendziewski J (2015) Hydro refining of oil from pyrolysis of whole tyres for passenger cars and vans. Fuel 159, 659–665. https://doi.org/10.1016/j.fuel.2015.07.024.

González JF, Encinar JM, González-García CM, Sabio E, Ramiro A (2006) Preparation of activated carbons from used tyres by gasification with steam and carbon dioxide. Appl Surf Sci 252(17), 5999–6004. https://doi.org/10.1016/j.apsusc.2005.11.029.

Pakdel H, Pantea DM, Roy C (2001) Production of dl-limonene by vacuum pyrolysis of used tires. J Anal Appl Pyrolysis 57(1), 91–107. https://doi.org/10.1016/S0165-2370(00)00136-4.

Quek A, Balasubramanian R (2013) Liquefaction of waste tires by pyrolysis for oil and chemicals—A review. J Anal Appl Pyrolysis 101, 1–16. https://doi.org/10.1016/j.jaap.2013.02.016.

The Sydney Morning Herald (2019) Can wind turbines disturb sleep? Research finds pulsing audible in homes up to 3.5 km away. June 18, 2019. https://www.smh.com.au/environment/sustainability/can-wind-turbines-disturb-sleep-research-finds-pulsing-audible-in-homes-up-to-3-5km-away-20190617-p51yik.html

Index

Page numbers in **bold** refer to tables and those in *italic* refer to figures.

Printed in the United States
by Baker & Taylor Publisher Services